EICOSANOIDS, APOLIPOPROTEINS, LIPOPROTEIN PARTICLES, AND ATHEROSCLEROSIS

ADVANCES IN EXPERIMENTAL MEDICINE AND BIOLOGY

Recent Volumes in this Series

Volume 236
NEURORECEPTORS AND SIGNAL TRANSDUCTION
Edited by Shozo Kito, Tomio Segawa, Kinya Kuriyama, Masaya Tohyama, and Richard W. Olsen

Volume 237
HISTOPHYSIOLOGY OF THE IMMUNE SYSTEM: The Life History, Organization, and Interactions of Its Cell Populations
Edited by Sigbjørn Fossum and Bent Rolstad

Volume 238
BIOTECHNOLOGICAL APPLICATIONS OF LIPID MICROSTRUCTURES
Edited by Bruce Paul Gaber, Joel M. Schnur, and Dennis Chapman

Volume 239
HOST DEFENSES AND IMMUNOMODULATION TO INTRACELLULAR PATHOGENS
Edited by Toby K. Eisenstein, Ward E. Bullock, and Nabil Hanna

Volume 240
PROTEASES: Potential Role in Health and Disease II
Edited by Walter H. Hörl and August Heidland

Volume 241
MOLECULAR BIOLOGY OF HEMOPOIESIS
Edited by Mehdi Tavassoli, Esmail D. Zanjani, Joao L. Ascensao, Nader G. Abraham, and Alan S. Levine

Volume 242
VASCULAR ENDOTHELIUM IN HEALTH AND DISEASE
Edited by Shu Chien

Volume 243
EICOSANOIDS, APOLIPOPROTEINS, LIPOPROTEIN PARTICLES, AND ATHEROSCLEROSIS
Edited by Claude L. Malmendier and Petar Alaupovic

A Continuation Order Plan is available for this series. A continuation order will bring delivery of each new volume immediately upon publication. Volumes are billed only upon actual shipment. For further information please contact the publisher.

EICOSANOIDS, APOLIPOPROTEINS, LIPOPROTEIN PARTICLES, AND ATHEROSCLEROSIS

Edited by

Claude L. Malmendier

Research Foundation on Atherosclerosis
and The Free University of Brussels
Brussels, Belgium

and

Petar Alaupovic

Lipoprotein and Atherosclerosis Research Program
Oklahoma Medical Research Foundation
Oklahoma City, Oklahoma

PLENUM PRESS • NEW YORK AND LONDON

Library of Congress Cataloging in Publication Data

International Colloquium on Eicosanoids, Apolipoproteins, Lipoprotein Particles, and Atherosclerosis (4th: 1988: Brussels, Belgium)
Eicosanoids, apolipoproteins, lipoprotein particles, and atherosclerosis / edited by Claude L. Malmendier and Petar Alaupovic.
(Advances in experimental medicine and biology; v. 243)
p. cm.
"Proceedings of the Fourth International Colloquium on Eicosanoids, Apolipoproteins, Lipoprotein Particles, and Atherosclerosis, held March 17-19, 1988, in Brussels, Belgium"—T.p. verso.
Includes bibliographies and index.

1. Atherosclerosis—Pathogenesis—Congresses. 2. Eicosanoic acid—Derivatives—Congresses. 3. Apolipoproteins—Congresses. 4. Blood lipoproteins—Congresses. I. Malmendier, Claude L. II. Alaupovic, Petar. III. Title. IV. Series.
[DNLM: 1. Apolipoproteins—congresses. 2. Arteriosclerosis—congresses. 3. Eicosanoic Acids—congresses. 4. Lipoproteins—congresses. WG 550 I6035e 1988]
RX692.I4675 1988
616.1'36071—dc19
DNLM/DLC 88-28929
for Library of Congress CIP

ISBN-13: 978-1-4612-8055-2 e-ISBN-13: 978-1-4613-0733-4
DOI: 10.1007/978-1-4613-0733-4

Softcover reprint of the hardcover 1st edition 1988

Proceedings of the Fourth International Colloquium on Eicosanoids, Apolipoproteins, Lipoprotein Particles, and Atherosclerosis, held March 17-19, 1988, in Brussels, Belgium

PREFACE

Plasma lipoproteins constitute a unique macromolecular system of lipid-protein complexes responsible for the transport of lipids from their sites of origin to their sites of utilization either as metabolic fuel or as structural components of cell membranes. Although studies on the role of lipoproteins in the mechanism of lipid transport are meritorious in their own right, the ever-increasing interest in chemical and functional properties of this remarkable class of conjugated proteins stems from the impressive evidence of their direct involvement in the genesis and development of atherosclerotic lesions. The initial emphasis on neutral lipids and phospholipids as the most characteristic constituents of operationally-defined lipoprotein classes has shifted in recent years to their protein moieties or apolipoproteins. The discovery of a number of apolipoproteins and characterization of familial hypolipoproteinemias as apolipoprotein deficiency disorders indicated that apolipoproteins play an essential role in maintaining the structural stability and integrity of lipoprotein particles. In addition to their role in the formation of lipoproteins, apolipoproteins were shown to perform a variety of functions in metabolic conversion of lipoproteins and their interactions with cellular surfaces. Results from several laboratories have demonstrated that the chemical and metabolic heterogeneity of operationally-defined lipoprotein classes is due to the presence of several discrete lipoprotein particles with similar physical properties but different and characteristic apolipoprotein composition. Thus, the apolipoproteins have emerged not only as essential structural and functional constituents of lipoproteins but also as unique chemical markers for identifying and classifying lipoprotein particles. The crowning achievement of this "apolipoprotein phase" of lipoprotein research has been the elucidation of the primary structure of almost all recognized apolipoproteins including apolipoprotein B.

These developments, providing new information on the role of apolipoproteins in normal and impaired lipid transport processes and atherogenesis were presented and discussed expertly by the participants of the Fourth International Colloquium on Eicosanoids, Apolipoproteins, Lipoprotein Particles and Atherosclerosis held at the University of Brussels, Belgium, under the sponsorship of the "Fondation de Recherche sur l'Athérosclérose".

The opening topic of this volume was devoted to recent advances in studies on the interactions between lipoproteins and eicosanoids and their significance in atherogenic processes. The next major topics included genetic disorders of lipid transport and their effect on coronary artery disease, new findings on primary, secondary and tertiary structures of apolipoproteins and lipoproteins, and the structure, genetics and potential function of serum amyloid A protein in lipid transport. Another timely topic covered the structure and role of lipid transfer proteins in modulating lipid composition and metabolic fate of lipoprotein particles. The topic on apolipoprotein defined lipoprotein particles presented preliminary evidence for the chemical and functional specificities of discrete ApoA- and ApoB-containing lipoproteins. The closing topic was devoted to the chemical and metabolic abnormalities of lipoproteins in hypertriglyceridemic states.

We hope this volume will be of interest to all investigators in the area of lipoprotein and atherosclerosis research as a useful review of recent accomplishments and a stimulating guide for future studies.

P. Alaupovic

C.L. Malmendier

CONTENTS

EICOSANOIDS

GENETICS OF APOLIPOPROTEINS

STRUCTURE OF APOLIPOPROTEINS AND LIPOPROTEINS

ACUTE PHASE APOLIPOPROTEINS

LIPID TRANSFER PROTEINS

LIPOPROTEIN PARTICLES

HYPERTRIGLYCERIDEMIA

PROSTACYCLIN, EDRF AND ATHEROSCLEROSIS

Salvador Moncada

The Wellcome Research Laboratories, Langley Court
Beckenham, Kent, BR3 3BS, U.K.

Introduction

The main metabolic product of the fatty acid arachidonic acid (AA) in vascular tissue is prostacyclin, a potent vasodilator and inhibitor of platelet aggregation (1). In contrast, in platelets AA is mainly converted to thromboxane A_2, which is a potent vasoconstrictor and inducer of platelet aggregation. An imbalance in the formation or the interaction between these two AA metabolites has been implicated in the underlying process of various thrombotic conditions as well as of atherosclerosis (2).

Another vasodilator mechanism, independent of prostacyclin and mediated by vascular endothelial cells, was described in 1980. This vasodilatation has been shown to be mediated by the release of a humoral agent which became known as endothelium-derived relaxing factor (EDRF) (3). We have identified the chemical nature of EDRF as nitric oxide (NO) (4). Prostacyclin and NO both inhibit platelet aggregation, moreover NO is a potent inhibitor of platelet adhesion. It is likely that, in addition to their role as regulators of vascular smooth muscle tone, prostacyclin and NO also synergise to regulate platelet-vessel wall interactions.

Vascular production of prostacyclin

Prostacyclin is generated by blood vessel microsomes or fresh vascular tissue from AA itself or from the unstable intermediate in AA metabolism, the prostaglandin (PG) endoperoxide, PGH_2. In common with other prostaglandins, prostacyclin release can be stimulated by mechanical, immunological or chemical means, as well as by trauma. Prostacyclin is a vasodilator and the most potent naturally-occurring inhibitor of platelet aggregation yet discovered. It is chemically unstable, with a half-life of 2-3 min, breaking down to 6-keto-$PGF_{1\alpha}$ (for review see 2).

The inhibition of platelet aggregation by prostacyclin is correlated with an activation of the adenylate cyclase system, leading to a substantial rise in platelet intracellular cyclic AMP levels. Other properties of prostacyclin include cytoprotection, enhanced fibrinolysis and stimulation of cholesterol metabolism (for review see 5). Unlike most properties of prostacyclin, cytoprotection is independent of an effect on cyclic AMP.

The ability of medium sized vessels to synthesise prostacyclin is greatest at the intimal surface and decreases progressively towards the adventitia (6). Production of prostacyclin by vascular cells in culture shows that endothelial cells are the most active producers of prostacyclin, although smooth muscle cells also convert AA to prostacyclin (7). Stripping of the endothelium from rabbit aorta in vivo removes virtually all ability of the luminal surface to produce prostacyclin from exogenously-added AA (8). Recent studies have shown that although the subendothelium has a significant capacity to produce prostacyclin, this capacity is only expressed briefly (9).

Prostacyclin and atherosclerosis

There is evidence to suggest that a number of diseases, such as atherosclerosis, are associated with an imbalance in the prostacyclin/TXA_2 system. A decrease in prostacyclin formation by atherosclerotic vascular tissue has been demonstrated both in experimental animals and in man. It has recently been shown that aortae from rabbits made atherosclerotic by cholesterol feeding show a transient increase in prostacyclin production followed by a long-lasting reduction (10). Moreover, smooth muscle cells obtained from atherosclerotic lesions and cultured in vitro consistently produce less prostacyclin than normal vascular smooth muscle cells. These findings probably reflect the increased generation by atherosclerotic vessels of lipoxygenase products, which are selective inhibitors of prostacyclin formation (11).

Atherosclerosis is not invariably associated with reduced generation of prostacyclin as there have been reports of increased release of prostacyclin, or a metabolite, both in experimental models of atherosclerosis (12) and in patients with severe diffuse atherosclerosis (13). This may be the result of the release from activated platelets of factors, such as 5-hydroxytryptamine and platelet-activating factor, with potential prostacyclin-releasing properties.

Prostacyclin has recently been shown to have a regulatory role in aortic cholesterol metabolism. As an atherosclerotic lesion progresses, cholesterol and cholesteryl esters are deposited in the extracellular matrix of the aortic smooth muscle cells as well as in lysosomal and cytosolic compartments of these cells. In cultured smooth muscle cells from the thoracic aorta of rabbits, prostacyclin at low concentrations increases the activity of the enzymes that metabolise cholesteryl esters, while PGE_2 inhibits cholesteryl ester synthetic activity (14). In human atherosclerotic cells in culture, cholesteryl ester metabolism is enhanced by two stable prostacyclin analogues so that the triglyceride and cholesteryl ester levels in the cells are substantially reduced (15).

Prostacyclin can also inhibit mobilisation of fibrinogen-binding sites on human platelets in vitro, which thus may limit the extent of fibrinogen-platelet interactions. In addition, prostacyclin enhances fibrinolytic activity in the canine lung, in human subjects and in human skin fibroblasts where it induces the protease plasminogen activator, a process which could contribute to its long-term clinical actions in chronic obstructive diseases of the circulation (for references see 16).

A new, potentially anti-atherosclerotic effect of prostacyclin has been described by Willis and co-workers (17) who showed that prostacyclin inhibits the release of mitogenic activity ('growth factors') from stimulated human platelets. Such growth factors are thought to mediate the progression of atherosclerosis by promoting smooth muscle cell and fibroblast proliferation (18).

Cigarette smoking is a well-recognised risk factor in atherosclerosis and is known to increase platelet reactivity and to reduce both basal and stimulated levels of urinary 6-keto-$PGF_{1\alpha}$ in man. Prostacyclin formation by kidney microsomes and rabbit isolated hearts is reduced by nicotine. Umbilical arteries from mothers who smoke produce significantly less prostacyclin than those from non-smoking control subjects (for references see 16).

Prostaglandin production by vascular tissue and platelets is also altered in a number of diseases associated with increased risk of arterial thrombosis, for example diabetes and thrombotic thrombocytopenic purpura. In general it seems that in diseases where there is a tendency for thrombosis to develop, TXA_2 production is increased, or prostacyclin production reduced, or both, whilst the opposite is found in some diseases associated with an increased bleeding tendency (for review see 19).

The use of prostacyclin, as a stable freeze-dried preparation (Epoprostenol), has been studied in several clinical conditions (for reviews see 19,20). In addition to its use in extracorporeal circulation systems such as cardiopulmonary bypass operations, renal dialysis and charcoal haemoperfusion, prostacyclin has been shown to be effective in the treatment of atherosclerotic peripheral vascular disease and Raynaud's syndrome. It is also being tested for the treatment of other thrombotic conditions such as stroke and myocardial infarction. Prostacyclin has been used in patients with pulmonary hypertension and has been found to have pulmonary haemodynamic effects similar to those of hydralazine and nifedipine. What is becoming clear is that the efficacy of prostacyclin, particularly its long-lasting effect in certain clinical conditions like peripheral vascular disease and Raynaud's syndrome, cannot be explained solely in terms of its short-lasting vasodilator and anti-aggregatory actions. Consequently, interest is being focused on the other activities of prostacyclin such as cytoprotection, fibrinolysis and stimulation of cholesterol metabolism.

Biological properties of endothelium-derived relaxing factor

The role of the endothelium in the relaxation of vascular tissues was discovered accidentally when it was observed that strips or rings of contracted rabbit aorta (RbA) exhibited great variation in their ability to relax to acetylcholine (ACh). Although a potent vasodilator *in vivo*, ACh is frequently inactive or occasionally produces a small contraction of vascular strips or rings *in vitro*. The explanation of this phenomenon was provided by Furchgott and Zawadzki (3) when they showed that ACh-induced vascular relaxation was dependent on the presence of the intact endothelium. Endothelial cells stimulated with ACh release a factor (later called endothelium-derived relaxing factor, EDRF) which diffuses to the underlying smooth muscle to cause relaxation.

It is now known that, in addition to ACh, many other vasodilators are also endothelium-dependent. These include adenine nucleotides, thrombin, substance P, the calcium ionophore A23187 (calcimycin), vasopressin, vasoactive intestinal polypeptide (VIP), pancreozymin, calcitonin gene-related peptide, bradykinin, bee venom (melittin) and saturated and unsaturated fatty acids (21). Other substances, however, including the nitrovasodilators, atrial natriuretic factor, bovine retractor-penis inhibitory factor and prostacyclin, cause vascular relaxation by endothelium-independent mechanisms.

The humoral nature of EDRF was first demonstrated with a variety of donor/detector systems (for reviews see references 21 and 22). Such bioassay methods revealed that EDRF was a labile substance with a half-life of between 3 and 50 seconds. It also became apparent that a variety of seemingly unrelated compounds could inhibit endothelium-dependent relaxation in intact vascular preparations.

Later it was shown that superoxide anions (O_2^-) contribute to the instability of EDRF (23,24). Superoxide dismutase (SOD), by inactivating O_2^-, potentiates the activity of EDRF. It was also demonstrated that compounds such as phenidone, BW755C, dithiothreitol, hydroquinone and other redox compounds inhibit the action of EDRF via the generation of O_2^- (25). Haemoglobin, however, is another inhibitor of the action of EDRF and of endothelium-dependent relaxation which acts by binding to the EDRF molecule rather than via a mechanism involving the generation of O_2^- (26).

EDRF is also an inhibitor of platelet aggregation. We have observed EDRF-induced inhibition of platelet aggregation in human platelet-rich plasma and washed platelets, induced by collagen, U46619, ADP and thrombin, and have shown that EDRF is equi-active against all these aggregating agents (27,28). EDRF also induces disaggregation of platelets aggregated with collagen and with U46619.

The inhibitory action of EDRF on platelet aggregation can be clearly differentiated from that of prostacyclin. Unlike prostacyclin, the effect of EDRF on platelets is potentiated by SOD and M&B 22948, a selective inhibitor of the cyclic GMP phosphodiesterase, and inhibited by Fe^{2+}, haemoglobin and hydroquinone. These effects are consistent with those observed on vascular strips. Moreover, the inhibitory effect on platelet aggregation is accompanied by an increase in platelet cyclic GMP. The duration of the effect of EDRF on platelets is short, with a half-life of 2 minutes, while that of prostacyclin has a half-life of 4 minutes.

A number of authors have demonstrated a rise in smooth muscle cell cyclic GMP levels associated with endothelium-dependent relaxation or with EDRF-induced vascular relaxation. EDRF, in common with the nitrovasodilators, has been shown to activate soluble guanylate cyclase from smooth muscle cells (29). Endothelium-dependent relaxation and the action of EDRF on platelets are prevented by agents that interfere with the activation of this enzyme, such as haemoglobin and methylene blue (30). These data support the concept that stimulation of soluble guanylate cyclase underlies the vascular relaxant and anti-aggregatory actions of EDRF.

Chemical identification of EDRF

Widespread speculation about the chemical nature of EDRF developed soon after its discovery. In 1986, however, Furchgott (31) and Ignarro et al. (32) independently suggested that EDRF may be NO or a closely-related species. We decided to investigate whether EDRF was indeed NO by comparing first, the pharmacological profile of EDRF and authentic NO on vascular strips and on platelets, and second by measuring directly the release of NO from porcine aortic endothelial cells in culture (4).

Both EDRF and NO caused a relaxation of the vascular strips which declined at the same rate during passage down the cascade. Furthermore, the rate of decay was slower, but similar for both compounds, during transit in polypropylene tubes. Both compounds also inhibited platelet aggregation (27), induced the disaggregation of aggregated platelets (28) and inhibited platelet adhesion (33). Moreover, their biological half-life as inhibitors of platelet aggregation was similar (27).

The actions of EDRF and NO on vascular strips and on platelets were similarly potentiated by SOD and cytochrome c and inhibited by Fe^{2+} and some redox compounds (27,34). Furthermore, the potency of redox compounds as inhibitors of EDRF- and NO-induced vascular relaxation was attenuated by SOD to a similar extent. Haemoglobin also inhibited the effect of EDRF and NO through a mechanism not involving O_2^-. Finally, direct measurements of cyclic GMP, or studies with selective inhibitors of its specific phosphodiesterase, have demonstrated that both compounds act on vascular smooth muscle and platelets via the stimulation of soluble guanylate cyclase and elevation of cyclic GMP.

Nitric oxide may be measured directly as the chemiluminescent product of its reaction with ozone. Using this method we have shown that the concentrations of bradykinin which induce the release of EDRF from endothelial cells in culture also cause a concentration-dependent release of NO (4). Moreover, we have established that the amounts of NO released by the cells are sufficient to account both for relaxations of the vascular strips and for the anti-aggregating and anti-adhesive activity of EDRF (4,28,33). Recently, we have also observed that the vascular relaxing activity released from fresh, perfused arteries of the rabbit, cat and dog by a number of agents, including ACh, substance P and bradykinin, is accounted for by the amounts of NO released (unpublished observations).

All this pharmacological and biochemical evidence clearly demonstrates that EDRF is NO and that it fulfils all the criteria necessary to be classified as a biological mediator (35).

Interactions between NO and prostacyclin

Prostacyclin and NO potentiate each other as inhibitors of platelet aggregation and inducers of platelet disaggregation (28). The supernatants of endothelial cells stimulated with low concentrations of bradykinin contain amounts of NO and prostacyclin too low to explain the anti-aggregating activity observed when these supernatants are added to platelets. This activity is, therefore, the result of a synergistic interaction between NO and prostacyclin. As a result, we have suggested that the very low concentrations of prostacyclin found in plasma may have a physiological effect in regulating platelet aggregability if acting on a background of NO release (36).

Interestingly, NO differs from prostacyclin in that it is also an effective inhibitor of platelet adhesion. The fact that we did not observe a synergistic interaction between these two compounds on platelet adhesion (33) suggests that the physiological process of platelet adhesion and repair of the vessel wall may proceed under circumstances in which both substances, acting in concert, are exerting a powerful anti-thrombotic action.

Prostacyclin and NO are both powerful vasodilators. The interaction between NO and prostacyclin as vasodilators remains to be studied. Preliminary evidence from our laboratory has not shown synergy between NO and prostacyclin in the rabbit mesenteric artery strip (unpublished observations).

The subcellular mechanisms underlying the actions of NO and prostacyclin are the cyclic GMP and the cyclic AMP systems, respectively. It is interesting that in some situations, such as inhibition of platelet aggregation and induction of disaggregation, there is a synergy between the two while in others, such as inhibition of platelet adhesion, there may even be some antagonism (33).

Physiological and pathological implications of EDRF

Although the release of EDRF or NO has not been demonstrated *in vivo*, endothelium-dependent vasodilatation has been demonstrated *in vivo* in a number of species. The changes in diameter which follow changes in blood flow in a number of perfused artery preparations (37-39) are endothelium-dependent and it has been suggested that endothelium-dependent relaxation coordinates increased flow responses through vascular beds (40).

Endothelium-dependent relaxation, or the release of NO, may be impaired in atherosclerosis, as a decrease in the ability of the vascular endothelium to release EDRF (41) or a decrease in the endothelium-dependent relaxation (42,43) has been demonstrated in vascular tissue obtained from rabbits with dietary-induced atherosclerosis. Furthermore, low density lipoproteins, which are associated with the development of atherosclerosis, have been shown to inhibit endothelium-dependent relaxation (44).

Studies in strips of human coronary artery with severe atherosclerosis have shown reduced endothelium-dependent relaxation in response to substance P, bradykinin and A23187 (45). In patients with moderate to severe atherosclerosis, infusions of ACh caused dose-dependent coronary vasoconstriction whereas dose-dependent vasodilatation was observed in control subjects (46). In contrast, glyceryl trinitrate caused dilatation in both normal and atherosclerotic subjects.

Reduced endothelium-dependent relaxation in animals with spontaneous or experimentally-induced hypertension has been reported by some workers (47-51). Restoration of blood pressure to normal can reverse this impairment in the endothelium-dependent response (52,53).

Reduced generation of NO by endothelial cells could play a role in the genesis of coronary vasospasm since endothelial damage renders vessels susceptible to local spasm in experimental animals (54). Inhibition of NO by haemoglobin could also play a role in the vasospasm that follows subarachnoid haemorrhage, which has long been suspected to be mediated by some product of lysed red blood cells (55).

If there is indeed a decrease in the release of NO in the vascular endothelium during hypertension or atherosclerosis, it will be very important to investigate the mechanism by which this takes place. Lipid peroxides and O_2^-, which inhibit the synthesis of prostacyclin (11) and play a role in the destruction of NO (23,24), have long been implicated in the genesis of different types of cardiovascular disease, including atherosclerosis. It would not be surprising if the biochemical basis of their action is related to these effects. Interestingly, O_2^- ions are responsible for the conversion of the ACh-induced cerebral arteriolar dilation to vasoconstriction during an infusion of norepinephrine (56).

Conclusions

Nitric oxide is an important mediator in the vessel wall and may be considered to be the endogenous nitrovasodilator.

The understanding of the release of NO by the vascular endothelium, its interactions with prostacyclin and their role as regulators of vascular tone and of platelet-vessel wall interactions is increasing our knowledge of the physiology and pathophysiology of the cardiovascular system. Manipulation of the soluble guanylate cyclase and/or the specific cyclic GMP phosphodiesterase may offer possibilities for therapeutic intervention in different areas of disease ranging from hypertension to atherosclerosis.

REFERENCES

1. Moncada, S., Gryglewski, R., Bunting, S. and Vane, J.R., An enzyme isolated from arteries transforms prostaglandin endoperoxides to an unstable substance that inhibits platelet aggregation. Nature, 263:663 (1976).

2. Moncada, S., Biological importance of prostacyclin. Brit. J. Pharmacol. 76:3 (1982).

3. Furchgott, R.F. and Zawadzki, J.V., The obligatory role of endothelial cells in the relaxation of arterial smooth muscle by acetylcholine, Nature, 288:373 (1980).

4. Palmer, R.M.J., Ferrige, A.G. and Moncada, S., Nitric oxide release accounts for the biological activity of endothelium-derived relaxing factor. Nature, 327:524 (1987).

5. Moncada, S., Prostacyclin - Discovery and biological importance, in: Prostaglandins: Research and Clinical Update, G.L. Longenecker, S.W. Schaffer eds., Alpha Editions, Minneapolis, pp. 1-39, (1985).

6. Moncada, S., Herman, A.G., Higgs, E.A. and Vane, J.R., Differential formation of prostacyclin (PGX or PGI_2) by layers of the arterial wall. An explanation for the antithrombotic properties of vascular endothelium, Thromb. Res. 11:323 (1977).

7. Weksler, B.B., Marcus, A.J. and Jaffe, E.A., Synthesis of prostaglandin I_2 (prostacyclin) by cultured human and bovine endothelial cells. Proc. Natl. Acad. Sci. USA, 74:3922 (1977).

8. Eldor, A., Falcone, D.J., Hajjar, D.P., Minick, C.R. and Weksler, B.B., Recovery of prostacyclin production by de-endothelialized rabbit aorta. Critical role of neointimal smooth muscle cells. J. Clin. Invest. 67:735 (1981).

9. Boeynaems, J.M., Galand, N. and Ketelbant, P., Prostacyclin production by the de-endothelialized rabbit aorta. J. Clin. Invest. 76:7 (1985).

10. Beetens, J.R., Coene, M-C., Verheyen, A., Zonnekeyn, L. and Herman, A.G., Biphasic response of intimal prostacyclin production during the development of experimental atherosclerosis. Prostaglandins, 32:319 (1986).

11. Moncada, S., Gryglewski, R.J., Bunting, S. and Vane, J.R., A lipid peroxide inhibits the enzyme in blood vessel microsomes that generates from prostaglandin endoperoxides the substance (prostaglandin X) which prevents platelet aggregation. Prostaglandins, 12:715 (1976).

12. Tremoli, E., Socini, A., Petroni, A. and Galli, C., Increased platelet aggregability is associated with increased prostacyclin

production by vessel walls in hypercholesterolemic rabbits. Prostaglandins, 24:397 (1982).

13. Fitzgerald, G.A., Smith, B., Pedersen, A.K. and Brash, A.R., Increased prostacyclin biosynthesis in patients with severe atherosclerosis and platelet activation. New Engl. J. Med., 310:1065 (1984).

14. Hajjar, D.P. and Weksler, B.B., Metabolic activity of cholesteryl esters in aortic smooth muscle cells is altered by prostaglandins I_2 and E_2. J. Lipid Res., 24:1176 (1983).

15. Orekhov, A.N., Tertov, V.V., Masurov, A.V., Andreeva, E.R., Repin, V.S. and Smirnov, V.N., "Regression" of atherosclerosis in cell culture: effects of stable prostacyclin analogues, Drug Dev. Res., 9:189 (1986).

16. Moncada, S. and Higgs, E.A., Arachidonate metabolism in blood cells and the vessel wall. Clin. Haematol., 15:273 (1986).

17. Willis, A.L., Smith, D.L., Vigo, C. and Kluge, A.F., Effects of prostacyclin and orally active stable mimetic agent RS-93427-007 on basic mechanisms of atherogenesis. Lancet, ii:682 (1986).

18. Ross, R., The pathogenesis of atherosclerosis - an update. New Engl. J. Med., 314:488 (1986).

19. Moncada, S. and Higgs, E.A., Prostaglandins in the pathogenesis and prevention of vascular disease. Blood Rev., 1:141 (1987).

20. Moncada, S., Clinical use of prostacyclin. in: Advanced Medicine Vol. 22, D.R. Triger, ed., Baillere Tindall, London, pp. 323-332, (1986).

21. Furchgott R.F., The role of endothelium in the responses of vascular smooth muscle to drugs. Ann. Rev. Pharmacol. Toxicol., 24:175 (1984).

22. Moncada, S., Palmer, R.M.J. and Higgs, E.A., Generation of prostacyclin and endothelium-derived relaxing factor from endothelial cells. in: Biology and pathology of platelet-vessel wall interactions, G. Jolles, J.Y. Legrand, A. Nurden, eds., Academic Press, London, pp. 289-304, (1986).

23. Gryglewski, R.J., Palmer, R.M.J. and Moncada, S. Superoxide anion is involved in the breakdown of endothelium-derived vascular relaxing factor. Nature, 320:454 (1986).

24. Rubanyi, G.M. and Vanhoutte, P.M., Superoxide anions and hyperoxia inactivate endothelium-derived relaxing factor. Am. J. Physiol., 250:H-222 (1986).

25. Moncada, S., Palmer, R.M.J. and Gryglewski, R.J., Mechanism of action of some inhibitors of endothelium-derived relaxing factor. Proc. Natl. Acad. Sci. USA, 83:9164 (1986).

26. Martin, W., Smith, J.A. and White, D.G., The mechanisms by which haemoglobin inhibits the relaxation of rabbit aorta induced by nitrovasodilators, nitric oxide or bovine retractor penis inhibitory factor. Br. J. Pharmacol., 89:562 (1986).

27. Radomski, M.W., Palmer, R.M.J. and Moncada, S., Comparative pharmacology of endothelium-derived relaxing factor, nitric oxide and prostacyclin in platelets. Br. J. Pharmacol., 92:181 (1987).

28. Radomski, M.W., Palmer, R.M.J. and Moncada, S., The anti-aggregating properties of vascular endothelium: interactions between prostacyclin and nitric oxide. Br. J. Pharmacol., 92:639 (1987).

29. Rapoport, R.M. and Murad, F., Agonist induced endothelium-dependent relaxation in rat thoracic aorta may be mediated through cyclic GMP. Circ. Res. 52:352 (1983).

30. Martin, W., Villani, G.M., Jothianandan, D. and Furchgott, R.F., Selective blockade of endothelium-dependent and glyceryl trinitrate-induced relaxation by haemoglobin and by methylene blue in the rabbit aorta. J. Pharmacol. Exp. Ther. 232:708 (1985).

31. Furchgott, R.F., Studies on relaxation of rabbit aorta by sodium nitrite: the basis for the proposal that the acid activatable inhibitory factor from bovine retractor penis is inorganic nitrite and the endothelium-derived relaxing factor is nitric oxide, in: Mechanisms of Vasodilatation, Vol. IV, P.M. Vanhoutte, ed., Raven Press, New York (1988), in press.

32. Ignarro, L.F., Byrns, R.E. and Wood, K.S., Biochemical and pharmacological properties of EDRF and its similarity to nitric oxide radical, in: Mechanisms of Vasodilatation, Vol. IV, P.M. Vanhoutte, ed., Raven Press, New York (1988), in press.

33. Radomski, M.W., Palmer, R.M.J. and Moncada, S., The role of nitric oxide and cGMP in platelet adhesion to vascular endothelium. Biochem. Biophys. Res. Commun., 148:1482 (1987).

34. Hutchinson, P.J.A., Palmer, R.M.J. and Moncada, S., Comparative pharmacology of EDRF and nitric oxide on vascular strips. Eur. J. Pharmacol., 141:445, (1987).

35. Dale, H.H., Progress in autopharmacology. A survey of present knowledge of the chemical regulation of certain functions by natural constituents of the tissues. Bull. Johns Hopk. Hosp. 53:297, (1933).

36. Moncada, S., Palmer, R.M.J. and Higgs, E.A., Prostacyclin and endothelium-derived relaxing factor: biological interactions and significance. in: Thrombosis and Haemostasis, M. Verstraete, J. Vermylen, H.R. Lijnen and J. Arnout, eds., Leuven University Press, Leuven, pp. 587-618, (1987).

37. Rubanyi, G.M., Romero, J.C. and Vanhoutte, P.M., Flow-induced release of endothelium-derived relaxing factor. Am. J. Physiol. 250:H1145, (1986).

38. Holtz, J., Forstermann, U., Pohl, U., Giesler, M. and Bassenge, E., Flow-dependent, endothelium-mediated dilation of epicardial coronary arteries in conscious dogs: effects of cyclooxygenase inhibition. J. Cardiovasc. Pharmacol., 6:1161, (1984).

39. Owen, M.P. and Bevan, J.A., Acetylcholine induced endothelial-dependent vasodilatation increases as artery diameter decreases in the rabbit ear. Experientia, 41:1057, (1985).

40. Griffith, T.M., Edwards, D.H., Davies, R.L.I., Harrison, T.J. and Evans, K.T., EDRF coordinates the behaviour of vascular resistance vessels. Nature, 329:442, (1987).

41. Sreeharan, N., Jayakody, R.L., Senaratne, M.P.J., Thomson, A.B.R. and Kappagoda, C.T., Endothelium-dependent relaxation and experimental atherosclerosis in the rabbit aorta. Can. J. Physiol. Pharmacol., 64:1451, (1986).

42. Verbeuren, T.J., Jordaens, F.H., Zonnekeyn, L.L., Van Hove, C.E., Coene M-C. and Herman, A.G., Effect of hypercholesterolemia on vascular reactivity in the rabbit: 1: Endothelium-dependent and endothelium-independent contractions and relaxations in isolated arteries of control and hypercholesterolemic rabbits. Circ. Res., 58:552, (1986).

43. Henry, P.D., Bossaller, C. and Yammamoto, H., Impaired endothelium-dependent relaxation and cyclic guanosine 5'-monophosphate formation in atherosclerotic human coronary artery and rabbit aorta. Thromb. Res. (Suppl. VII):6, (1987).

44. Andrews, H.E., Bruckdorfer, K.R., Dunn, R.C. and Jacobs, M., Low-density lipoproteins inhibit endothelium-dependent relaxation in rabbit aorta. Nature, 327:237, (1987).

45. Forstermann, U., Properties and mechanisms of production and action of endothelium-derived relaxing factor. J. Cardiovasc. Pharmacol., 8 (Suppl.10):S45, (1986).

46. Ludmer, P.L., Selwyn, A.P., Shook, T.L., Wayne, R.R., Mudge, G.H., Alexander, R.W. and Ganz, P., Paradoxical vasoconstriction induced by acetylcholine in atherosclerotic coronary arteries. New Engl. J. Med., 315:1046, (1986).

47. Luscher, T.F. and Vanhoutte, P.M., Endothelium-dependent responses to platelets and serotonin in spontaneously hypertensive rats. Hypertension, 8:II-55, (1986).

48. De Mey, J.G. and Gray, S.D., Endothelium-dependent reactivity in resistance vessels. Prog. Appl. Microcirc., 8:181, (1985).

49. Winquist, R.J., Bunting, P.B., Baskin, E.P. and Wallace, A.A., Decreased endothelium-dependent relaxation in New Zealand genetic hypertensive rats. J. Hypertens., 2:541, (1984).

50. Lockette, W., Otsuka, Y. and Carretero, O., The loss of endothelium-dependent vascular relaxation in hypertension. Hypertension, 8:II-61, (1986).

51. Van de Voorde, J. and Leusen, I., Endothelium-dependent and independent relaxation of aortic rings from hypertensive rats. Am. J. Physiol., 250:H711, (1986).

52. Luscher, T.F., Raij, L. and Vanhoutte, P.M., Effect of hypertension and its reversal on endothelium-dependent relaxations in the rat aorta. J. Hypertension, 5(Suppl. 5):S153, (1987).

53. Otsuka, Y., DiPiero, A., Hirt, E., Brennaman, B. and Lockette, W., Vascular relaxation and cGMP in hypertension. Am. J. Physiol. 254:H163, (1988).

54. Brum, J.M., Sufan, Q., Lane, G. and Bove, A.A., Increased vasoconstrictor activity of proximal coronary arteries with endothelial damage in intact dogs. Circulation, 70:1066 (1984).

55. Fisher, C.M., Kistler, J.M. and Davis, J.M., The correlation of cerebral vasospasm and the amount of subarachnoid blood detected by computerised cranial tomography after aneurysm rupture. in: Cerebral Artery Spasm, R.H. Wilkins, ed., Baltimore, London, pp. 397-408, (1980).

56. Wei, E.P., Kontos, H.A., Christman, C.W., Dewitt, D.S. and Povlishock, J.T., Superoxide generation and reversal of acetylcholine-induced cerebral arteriolar dilation after acute hypertension. Circ. Res., 57:781, (1985).

CONTROL OF PROSTACYCLIN PRODUCTION BY VASCULAR CELLS : ROLE OF ADENINE NUCLEOTIDES AND SEROTONIN

J.M. Boeynaems, D. Demolle, S. Pirotton, E. Raspe, M. Lecomte, A. Hepburn, A. Van Coevorden and C. Erneux

Institute of Interdisciplinary Research, School of Medicine Université Libre de Bruxelles, Campus Erasme, Route de Lennik 808, B-1070 Brussels, Belgium

The in vivo production of prostacyclin (PGI_2) in man is normally very low. The use of specific analytical methods has shown that the concentration of PGI_2 in peripheral blood is less than 3 pg/ml (10 pM), a concentration too low to inhibit platelets (Blair et al., 1982). The measurement of the main urinary metabolite of PGI_2, prostaglandin 2,3-dinor-6-keto $F_{1\alpha}$, has allowed to estimate that the rate of PGI_2 secretion into the circulation of normal man is $\pm$ 0.1 ng/kg x min, whereas studies with exogenous PGI_2 suggest that infusion rates of 2-4 ng/kg x min are required to achieve the threshold for inhibition of platelet function (FitzGerald et al., 1981). Several diseases involving the intravascular activation of platelets are associated with an increased biosynthesis of PGI_2 which might represent a compensatory mechanism : severe atherosclerosis of the lower limbs (FitzGerald et al., 1984), systemic sclerosis complicated by Raynaud's phenomenon (Reilly et al., 1986), unstable angina, during the episodes of chest pain and acute myocardial infarction (Fitzgerald et al., 1986).

In correlation with these in vivo observations, several in vitro studies have shown that the vascular production of PGI_2 is increased by factors generated or released during the haemostatic response. Endothelial cells convert platelet prostaglandin endoperoxides into PGI_2 (Marcus et al., 1980; Schafer et al., 1984). Thrombin increases the release of PGI_2 from human umbilical vein endothelial cells (Weksler et al., 1978). Serotonin and platelet-derived growth factor produce a synergistic stimulation of PGI_2 synthesis in cultured aortic smooth muscle cells (Coughlin et al., 1981). ATP and ADP enhance the release of PGI_2 from the arterial endothelium.

This last action seems to be a general phenomenon, since it was observed in the rat and rabbit aorta and pulmonary artery (Boeynaems and Galand, 1983), as well as in cultured endothelial cells from porcine (Pearson et al., 1983) and bovine (Van Coevorden and Boeynaems, 1984) aorta. ATP and ADP were active at uMolar concentrations (fig. 1) : since the serum concentration of (ADP + ATP) reaches 20 uM following platelet activation by thrombin and even higher concentrations can be obtained locally when aggregated platelets release the content of their dense granules, it is likely that ATP and ADP might induce a release of PGI_2 in vivo, when platelets aggregate on the site of an endothelial injury

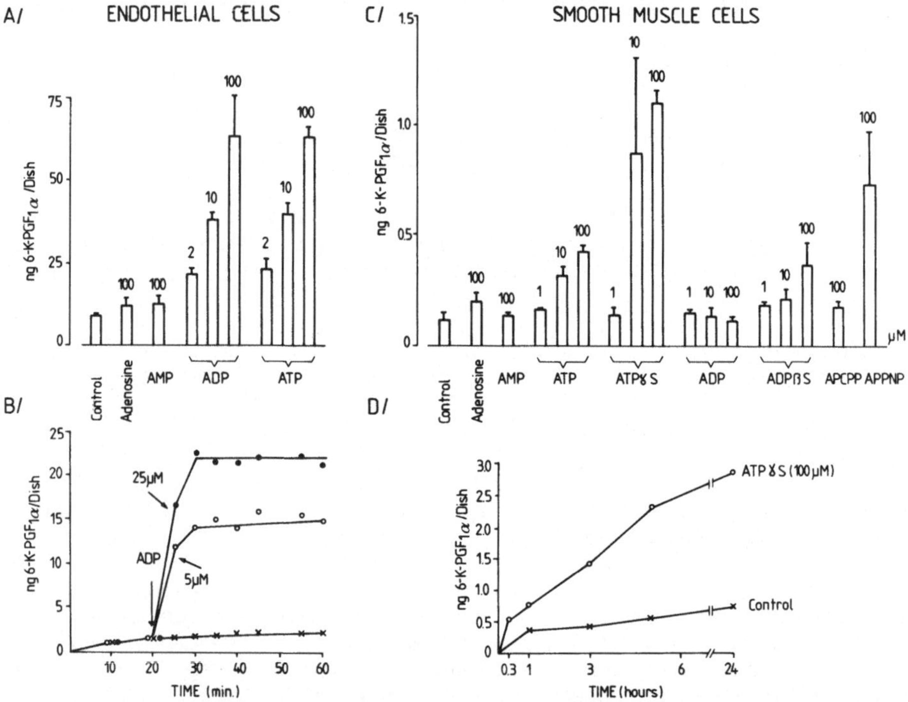

Figure 1. Purinergic stimulation of PGI_2 release from bovine aortic cells: comparison between endothelial and smooth muscle cells. The release of PGI_2 in the incubation medium was evaluated by the radioimmunoassay of PGI_2 stable degradation product, prostaglandin 6-keto-$F_{1\alpha}$ (6-K-$PGF_{1\alpha}$). APCPP : adenosine 5'-(α , β -methylene) triphosphate. APPNP : adenosine 5'-(β , γ -imido) triphosphate. ATPγS : adenosine 5'-O-(3-thiotriphosphate). ADPβS: adenosine 5'-O-(2-thiodiphosphate).

(fig. 2). Other factors released from platelets (serotonin, platelet-derived growth factor) or generated during blood clotting (thrombin) had no effect on the production of PGI_2 by aortic endothelial cells. On the contrary, thrombin increased PGI_2 synthesis in endothelial cells from human umbilical vein, whereas ADP was completely inactive. These discrepancies underscore the heterogeneity of endothelial cells collected from different vascular beds.

The stimulatory action of ADP and ATP on PGI_2 release from aortic endothelial cells is mediated by P_2-purinoceptors of the P_{2Y} subtype (Burnstock and Kennedy, 1986; Houston et al., 1987; Needham et al., 1987). It was recently demonstrated that these P_{2Y} receptors are coupled to a phospholipase C which hydrolyses phosphatidylinositol bisphosphate (PIP_2) into inositol (1,4,5)-trisphosphate (IP_3) and diacylglycerol (DAG) (fig. 3). Indeed, ADP and ATP induced a rapid and transient increase of the IP_3 level (Pirotton et al., 1987a) and of the cytosolic concentration of free Ca^{++} (Luckhoff and Busse, 1986; Hallam and Pearson, 1986; Pirotton et al., 1987a) in aortic endothelial cells. Similar observations have been made with microvascular endothelial cells (Forsberg et al., 1987) and hepatocytes (Charest et al., 1985). The coupling of P_{2Y} receptors to phospholipase C in endothelial cells seems to involve a GTP-binding regulatory protein, sensitive to inhibition by Pertussis toxin (Pirotton et al., 1987b). After the rapid and transient stimulation of PIP_2 cleavage, adenine nucleotides induce a more sustained hydrolysis of phosphatidylcholine (PC) in aortic endothelial cells (Pirotton and Boeynaems, unpublished data). This hydrolysis seems to involve the action of phospholipases C and D, since both choline and phosphorylcholine were released (fig. 3). This pathway might thus generate 2 signals in response to adenine nucleotides : DAG, an activator of protein kinase C, and phosphatidic acid, which can modulate Ca^{++} translocation. In hepatocyte membranes, ATP directly enhances the hydrolysis of PC by phospholipases C and D (Irvine and Exton, 1987; Bocckino et al., 1987). The effects of second messengers appear to be mediated by the activation of specific protein kinases which phosphorylate a variety of substrate proteins. Using two-dimensional gel electrophoresis, we recently characterized the pattern of protein phosphorylation in bovine aortic endothelial cells and its modulation by ATP (Demolle, Lecomte and Boeynaems, submitted for publication). ATP increased the phosphorylation of at least 15 proteins: each of these phosphorylations was fully reproduced by ionophore A23187, while phorbol 12-myristate, 13-acetate (PMA) had only a slight and delayed effect. Two main substrates - 100 kD and 28 kD - were phosphorylated with distinct time courses, rapid and transient for the 100 kD protein, delayed and sustained for the 28 kD protein. The 100 kD protein seems to be identical to an ubiquitous substrate of Ca^{++}-calmodulin kinase III, recently identified as Elongation Factor-2 (Nairn and Palfrey, 1987). The 28 kD protein appears similar to a protein whose phosphorylation is induced by thrombin in platelets and by various mitogens in fibroblasts (Chambard and Pouyssegur, 1983). Tumor Necrosis factor-α also induced the phosphorylation of a 28 kD protein in aortic endothelial cells (Hepburn et al., 1988). No specific substrate of protein kinase C could be detected.

The increase of IP_3 level and cytosolic Ca^{++} appears to be a general signalling pathway for agonists which stimulate the release of PGI_2 from endothelial cells, such as bradykinin (Derian and Moskowitz, 1986; Lambert et al., 1986; Morgan Boyd et al., 1987; Colden-Stanfield et al., 1987),

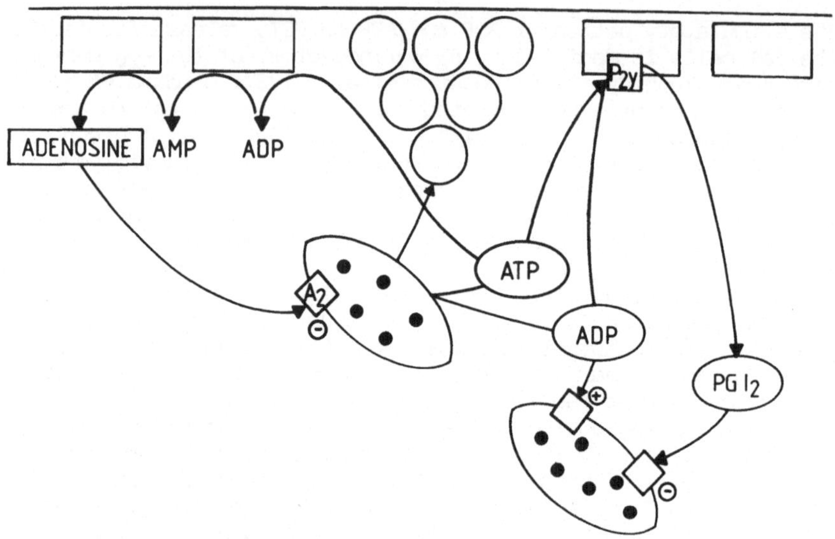

Figure 2. Role of endothelial P_2 purinergic receptors in the interaction between platelets and a lesion of the arterial wall.

endothelial cell

non-aggregated platelet

aggregated platelet

dense granule

During their aggregation on the site of an endothelial lesion, platelets release the content of their dense granules, including ADP and ATP. This triggers the operation of several regulatory loops :

- ADP recruits additional platelets;
- ADP is sequentially degraded by endothelial ectonucleotidases into adenosine, which inhibits platelets via A2 receptors;
- ADP and ATP, co-released from platelets, activate P_{2y} receptors on the surface of endothelial cells, resulting in the release of prostacyclin.

histamine (Lo and Fan, 1987) and thrombin (Jaffe et al., 1987). The time courses of the increase in cytosolic Ca^{++} and of the PGI_2 release are closely related. A plausible mechanism for the stimulation of PGI_2 production would thus involve the activation of a Ca^{++}-dependent phospholipase A_2, providing free arachidonate to prostaglandin H synthase. However, we cannot exclude the existence of a direct coupling of P_{2y} and other receptors to a phospholipase A_2 via a G protein, as demonstrated for α_1-receptors in thyroid cells (Burch et al., 1986) and for bradykinin receptors in fibroblasts (Burch and Axelrod, 1987).

The subendothelium, presumably the smooth muscle cells, constitutes an important source of vascular PGI_2. Deendothelialization of the rabbit aorta in vitro triggers a rapid and sustained release of free arachidonate, but the resulting increase in PGI_2 production is transient, probably as a consequence of prostaglandin H synthase self-inactivation (Boeynaems et al., 1985). Arterial smooth muscle cells can assume two distinct phenotypes : contractile or synthetic (Chamley-Campbell and Campbell, 1981). Most cells in the media of adult arteries are in the contractile state, whereas the smooth muscle cells in atherosclerotic lesions have lost their contractility and acquired the capacity to proliferate and to synthesize components of the extracellular matrix. The mechanisms which control PGI_2 release are very different in preparations of aortic smooth muscle representative of each phenotype : explants of bovine aortic media cultured for a short period ("contractile phenotype") and smooth muscle cells obtained by outgrowth from these explants ("synthetic phenotype"). Serotonin stimulates PGI_2 production in both models, but via distinct receptors : 5-HT_2 receptors in the synthetic cells (Coughlin et al., 1984) and "5-HT_1 like" receptors in the contractile cells (Demolle and Boeynaems, 1986; Hirafuji et al., 1987). PMA and ionophore A23187 mimic the action of serotonin in synthetic, but not in contractile cells (Demolle and Boeynaems, 1988). In the dog saphenous vein, contraction to serotonin is mediated by 5-HT_1 receptors, whereas the release of PGI_2 involves 5-HT_2 receptors (Kokkas and Boeynaems, 1988). In deendothelialized strips of rabbit aorta, epinephrine stimulates the release of PGI_2 via α_1-receptors, but this stimulation does not involve either an increase of cytosolic Ca^{++} or the activation of protein kinase C, which both play a role in the contractile response (Boeynaems et al., 1987). In bovine aortic smooth muscle cells, ATP and mostly its stable analog ATPγ S stimulate the production of PGI_2 (fig. 1; Demolle and Boeynaems, submitted for publication). The actions of ATP on endothelial and smooth muscle cells from the bovine aorta are very different : in smooth muscle cells, ADP or ADPβ S have no effect and the stimulation by ATPγ S is sustained over several hours (fig. 1). This stimulation seems to be mediated by receptors distinct from both the P_{2y}-purinoceptors involved in the release of PGI_2 from endothelial cells and from the P_{2x}-purinoceptors responsible for the contraction of vascular smooth muscle. These various examples emphasize that, in the vascular smooth muscle, distinct mechanisms control contraction and release of PGI_2.

Acknowledgements

This work was performed under contract of the Ministère de la Politique Scientifique ("Action Concertée") and was supported by grants from the Fonds de la Recherche Scientifique Médicale and the Fondation Médicale Tournay-Solvay.

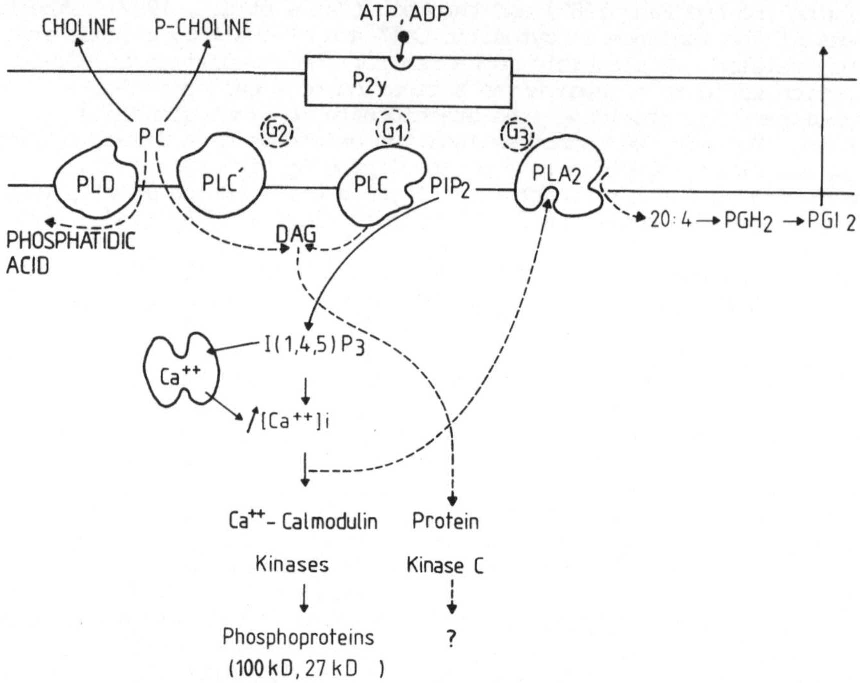

Figure 3. Schematic representation of the biochemical events involved in the action of adenine nucleotides on aortic endothelial cells. PLA_2, PLC, PLD : phospholipase A_2, C, D; G : GTP-binding regulatory protein; DAG : diacylglycerol; P-choline : phosphorylcholine.

REFERENCES

Blair, I. A., Barrow, S. E., Waddell, K. A., Lewis, P. J. and Dollery, C. T., 1982, Prostacyclin is not a circulating hormone in man, Prostaglandins, 23:579.

Bocckino, S. B., Blackmore, P.F., Wilson, P. B. and Exton, J. H., 1987, Phosphatidate accumulation in hormone-treated hepatocytes via a phospholipase D mechanisms, J. Biol. Chem. 262:15309

Boeynaems, J. M. and Galand, N., 1983, Stimulation of vascular prostacyclin by extracellular ADP and ATP, Biochem. Biophys. Res. Commun. 112:290.

Boeynaems, J. M., Galand, N. and Ketelbant, P., 1985, Prostacyclin production by the deendothelialized rabbit aorta, J. Clin. Invest. 76:7.

Boeynaems, J. M., Demolle, D. and Galand, N., 1987, Adrenergic stimulation of vascular prostacyclin : role of α_1-receptors in smooth muscle cells, Eur. J. Pharmacol. 144:193.

Burch, R. M., Luini, A. and Axelrod, J., 1986, Phospholipase A_2 and phospholipase C are activated by distinct GTP-binding proteins in response to $_1$-adrenergic stimulation in FRTL5 thyroid cells, Proc. Natl. Acad. Sci. USA 83:7201.

Burch, R. M. and Axelrod, J., 1987, Dissociation of bradykinin-induced prostaglandin formation from phosphatidylinositol turnover in Swiss 3T3 fibroblasts : evidence for G protein regulation of phospholipase A_2, Proc. Natl. Acad. Sci. USA, 84:6374.

Burnstock, G. and Kennedy, C., 1986, A dual function for adenosine triphosphate in the regulation of vascular tone, Circ. Res. 58: 319.

Chambard, J. C. and Pouysségur, J. 1983, Thrombin-induced phosphorylation in resting platelets and fibroblasts : evidence for common post-receptor molecular events, Biochem. Biophys. Res. Commun. 111:1034.

Chamley-Campbell, J. H. and Campbell, G. R., 1981, What controls smooth muscle phenotype, Atherosclerosis 40:347.

Charest, R., Blackmore, P. F. and Exton, J. H., 1985, Characterization of responses of isolated rat hepatocytes to ATP and ADP, J. Biol. Chem., 260:15789.

Colden-Stanfield, M., Schilling, W. P., Ritchie, A. K., Eskin, S. G., Navarro, L. T. and Kunze, D. L., 1987, Bradykinin-induced increases in cytosolic calcium and ionic currents in cultured bovine aortic endothelal cells, Circ. Res. 61:632.

Coughlin, S. R., Moskowitz, M. A., Antoniades, H. N. and Levine, L., 1981, Serotonin receptor-mediated stimulation of bovine smooth muscle cell prostacyclin synthesis and its modulation by platelet-derived growth factor, Proc. Natl. Acad. Sci. USA 78:7134.

Coughlin, S. R., Moskowitz, M. A. and Levine, L., 1984, Identification of a serotonin type 2 receptor linked to prostacyclin synthesis in vascular smooth muscle cells, Biochem. Pharmacol. 33:692.

Demolle, D. and Boeynaems, J. M., 1986, Prostacyclin production by the bovine aortic smooth muscle, Prostaglandins 32:155.

Demolle, D. and Boeynaems, J. M., 1988, Role of protein kinase C in the control of vascular prostacyclin : study of phorbol esters effects in bovine aortic endothelium and smooth muscle, Prostaglandins, in press.

Derian, C. K. and Moskowitz, M. A., 1986, Polyphosphoinositide hydrolysis in endothelial cells and carotid artery segments, J. Biol. Chem., 261:3831.

FitzGerald, G. A., Brash, A. R., Falardeau, P. and Oates J.A., 1981, Estimated rate of prostacyclin secretion into the circulation of normal man, J. Clin. Invest., 68:1272.

FitzGerald, G. A., Smith, B., Pedersen, A. K. and Brash, A. R., 1984, Increased prostacyclin biosynthesis in patients with severe atherosclerosis and platelet activation, New Engl. J. Med., 310:1065.

Fitzgerald, D. J., Roy, L., Catella, F. and FitzGerald, G. A., 1986, Platelet activation in unstable coronary disease, New Engl. J. Med. 315:983.

Forsberg, E. J., Feuerstein, G., Shohami, E. and Pollard, H. B., 1987, ATP stimulates inositol phospholipid metabolism and prostacyclin formation in adrenal medullary endothelial cells by means of P_2-purinergic receptors, Proc. Natl. Acad. Sci. USA 84:5630.

Hallam, T. J. and Pearson, J. D., 1986, Exogenous ATP raises cytoplasmic free calcium in fura-2 loaded piglet aortic endothelial cells, FEBS Lett. 207:95.

Hepburn, A., Demolle, D., Boeynaems, J. M., Fiers, W. and Dumont, J. E., 1988, Rapid phosphorylation of a 27 kD protein induced by tumor necrosis factor, FEBS lett. 227:175.

Hirafuji, M., Akiyama, Y. and Ogura, Y., 1987, Receptor-mediated stimulation of aortic prostacyclin release by 5-hydroxytryptamine, Eur. J. Pharmacol., 143:259.

Houston, D. A., Burnstock, G. and Vanhoutte, P. M., 1987, Different P_2-purinergic receptor subtypes on endothelium and smooth muscle in canine blood vessels, J. Pharm. Exp. Ther. 241:501.
Irving, H. R. and Exton, J. H., 1987, Phosphatidylcholine breakdown in rat liver plasma membranes, J. Biol. Chem., 262:3440.
Jaffe, E. A., Grulich, J., Weksler, B. B., Hampel, G. and Watanabe, K., 1987, Correlation between thrombin-induced prostacyclin production and inositol trisphosphate and cytosolic free calcium levels in cultured human endothelial cells, J. Biol. Chem., 262:8557.
Kokkas, B. and Boeynaems, J.M., 1988, Release of prostacyclin from the dog saphenous vein by 5-hydroxytryptamine, Eur. J. Pharmacol., in press.
Lambert, T. L., Kent, R.S. and Whorton, A. R., 1986, Bradykinin stimulation of inositol polyphosphate production in porcine aortic endothelial cells, J. Biol. Chem. 261:15288.
Lo, W. W. J., and Fan, T.-P. D., 1987, Histamine stimulates inositol phosphate accumulation via the H_1-receptor in cultured human endothelial cells, Biochem. Biophys. Res. Commun. 148:47.
Luckhoff, A. and Busse, R., 1986, Increased free calcium in endothelial cells under stimulation with adenine nucleotides, J. Cell. Physiol., 126:414.
Marcus, A. J., Weksler, B. B., Jaffe, E. A. and Broekman, M. J., 1980, Synthesis of prostacyclin from platelet-derived endoperoxides by cultured human endothelial cells, J. Clin. Invest. 66:979.
Morgan Boyd, R., Stewart, J. M., Vavrek, R. J. and Hassid, A., 1987, Effects of bradykinin and angiotensin II on intracellular Ca^{++} dynamics in endothelial cells, Am. J. Physiol., 253:C588.
Nairn, A. C. and Palfrey, M. C., 1987, Identification of the major Mr 100,000 substrate for calmodulin-dependent protein kinase III in mammalian cells as Elongation Factor-2, J. Biol. Chem. 262:17299.
Needham, L., Cusack, N. J., Pearson, J. D., and Gordon, J. L., 1987, Characteristics of the P_2 purinoceptor that mediates prostacyclin production by pig aortic endothelial cells, Eur. J. Pharmacol., 134:199.
Pearson, J. D., Slakey, L.L. and Gordon, J. L., 1983, Stimulation of prostaglandin production through purinoreceptors in cultured porcine endothelial cells, Biochem. J. 24:273.
Pirotton, S., Raspe, E., Demolle, D., Erneux, C., and Boeynaems, J. M., 1987a, Involvement of inositol 1,4,5-trisphosphate and calcium in the action of adenine nucleotides on aortic endothelial cells, J. Biol. Chem. 262:17461.
Pirotton, S., Erneux, C., and Boeynaems, J. M., 1987b, Dual role of GTP-binding proteins in the control of endothelial prostacyclin, Biochem. Biophys. Res. Commun, 147:1113.
Reilly, I.A.G., Roy, L. and FitzGerald, G. A., 1986, Biosynthesis of thromboxane in patients with systemic sclerosis and Raynaud's phenomenon, Br. Med. J. 292:1037.
Schafer, A. I., Crawford, D. D. and Gimbrone, M. A., 1984, Unidirectional transfer of prostaglandin endoperoxides between platelets and endothelial cells, J. Clin. Invest. 73:1105.
Van Coevorden, A. and Boeynaems, J. M., 1984, Physiological concentrations of ADP stimulate the release of prostacyclin from bovine aortic endothelial cells, Prostaglandins 27:615.
Weksler, B. B., Ley, C. W., and Jaffe, E. A., 1978, Stimulation of endothelial cells prostacyclin production by thrombin, trypsin and the ionophore A23187, J. Clin. Invest. 62:923.

PROSTACYCLIN AND ATHEROSCLEROSIS -

EXPERIMENTAL AND CLINICAL APPROACHES

R.J. Gryglewski, E. Kostka-Trąbka, A. Dembińska-Kieć and R. Korbut

Department of Pharmacology, Copernicus Academy of Medicine, 31-531 Cracow, Poland

INTRODUCTION

At the time of the discovery of prostacyclin(PGX,PGI_2) we also reported that its biosynthesis had been inhibited by 15-HPETE and by "a mixture of peroxides which were non-enzymically formed from arachidonic acid". A hypothesis was put forward that "a reduction in the amount of those lipid peroxides which inhibit the generation of PGX.... would be of considerable significance in preventing the development of atherosclerosis and arterial thrombosis"[1]. An experimental evidence was soon presented that a high lipid diet fed to rabbits had caused an increase in plasma lipid peroxides and a suppression of the generation of prostacyclin by arteries[2-4]. In human atherosclerotic lesions the generation of prostacyclin was also suppressed[5,6].

It may well be that the natural history of atherosclerosis is closely associated with a deficiency in both endothelial vasorelaxants, i.e.prostacyclin and "endothelium-derived relaxing factor" - EDRF[7,8] which are also inducers of thromboresistance and cytoprotective agents[9]. Indeed, in advanced experimental atherosclerosis the release and action of EDRF are suppressed[10,11]. Presently, we review our experimental and clinical data on the role of prostacyclin in atherosclerosis.

MATERIALS AND METHODS

Experimental animals

Male rabbits 2-3 kg of body weight were fed either a high-lipid diet (HLD) which consisted of 1 g cholesterol and 3 g olive oil per 100 g of rabbit pelet food daily for 1,2,4 or 12 weeks of a standard pelet rabbit diet.

Thromboresistance of rabbit aortas (RbAs)

A tubular segment of thoracic RbA of 5 cm in length was turned inside out and superfused with the rabbit citrated blood (2 ml/min) while its weight was continuously recorded. Basic properties of this assay system were as follows. Thromboresistance of the control RbAs was high and allowed to deposit only 74 ± 2 mg (n=14) of thrombi. Standardised incisions of the endothelial surface or the pretreatment of RbAs with aspirin (660 µM) decreased thromboresistance to 227 ± 25 mg (n=6) and 286 ± 30 mg (n=8), respectively. Prostacyclin (1 - 10 ng) dissipated blood thrombi. Nitric oxide given up to 30 µl of saturated aqueous solution of NO[10] had no such effect,although it potentiated the thrombolytic action of prostacyclin.

Bioassay of prostacyclin and PGE_2 generated by arterial slices

Arterial slices from RbA and rabbit mesenteric artery (RbMA) were incubated (250 mg/ml) in Krebs buffer for 10 min at room temperature and PGI_2-like activity was bioassayed by aggregometry[1] while PGE_2-like activity was bioassayed using superfused strips of rat stomach and rat colon.

Bioassay of TXA_2 generated during platelet aggregation

Platelet count in platelet rich plasma (PRP) was adjusted to 2×10^8 platelets/ml and 1 min after instillation of potassium arachidonate at a threshold proaggregatory concentration (50 - 100 µM) TXA_2 was bioassayed in 10-200 µl samples of PRP using superfused strip of RbMA as a bioassay organ and 11,9-epoxymethano analogue of PGH_2 (U 46619) as a stable TXA_2 agonist.

Activity of 12-lipoxygenase (12-LOX) in platelets

A suspension of washed platelets - 2×10^8 platelets/ml[12] - was incubated at 37°C with 1U/ml of thrombin for 10 min, and pH adjusted to 3.5. Platelets were spun down, 200 ng of PGE_2 was added, the supernatant extracted on Sep-Pak C 18 columns. These were eluted with methanol, washed with water, efluent evaporated and the residue subjected to the HPLC analysis on Altex C 18 column using a mixture of methanol: water: acetic acid (80:20:0.04) as a running solvent. UV absorbance was monitored at 234 nm. The activity of 12-LOX was quantified as an amount of 12-HETE generated by platelets.

Plasma malondialdehyde (MDA) levels

MDA was measured in plasma as a thiobarbituric acid (TBA)-reactive material[13].

Plasma fibrinolytic activity

Euglobulin clot lysis time (ECLT) was measured[14].

Statistics

The results are presented as mean ± standard deviation (M ± SD).Statistical significance was calculated by unpaired t Student test. For the evaluation of clinical scores paired t Student test and R Wilcoxon test were used.

Clinical trials

There were performed four placebo-controlled studies on the therapeutical effects of prostacyclin in 116 patients with completed ischaemic stroke (IS), central retinal vein occlusion (CRVO) and sudden unilateral deafness (SD). The IS study included two trials, a published one[15] in 26 patients and an unpublished trial in 30 patients (Fig.1). The unpublished CRVO and SD studies in 30 patients each were preceded by open clinical trials[16,17]. Patients were assigned at random to intravenous infusions of prostacyclin sodium salt (2-5 ng/kg/min) or placebo. The IS patients received no pharmacological treatment apart from adequate support of fluids,

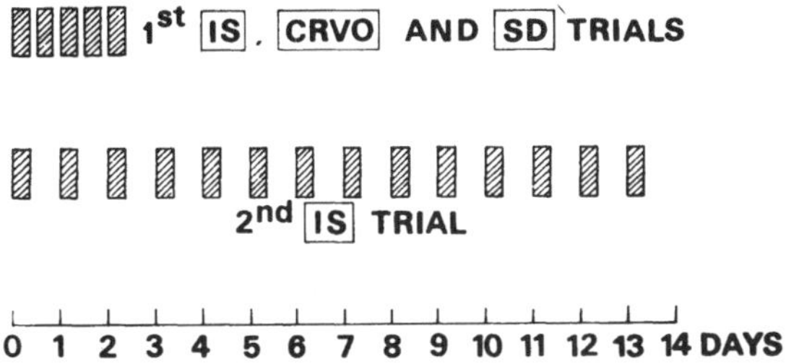

Fig. 1. The mode of 6-hrs infusions of PGI_2 or placebo in controlled clinical trials(IS,CRVO and SD).

electrolytes and glucose when necessary and, in addition, whatever was required to correct other than cerebral pathology (such as antibiotics, cardiac glycosides, antidyssrhythmic drugs). All the CRVO patients received a conventional therapy consisting of a mixture of rutin, ascorbic acid and calcium dobesilate. The criteria for the evaluation of clinical status of the patients were described previously[15-17]. All patients were admitted in the trial no later than 8 days after the symptoms of the disease had been diagnosed.

RESULTS

Thromboresistance of RbAs from rabbits fed HLD for a period of 4-12 weeks decreased significantly ($p<0.01$) from a control level of 74 ± 8 mg (n=14) of thrombi deposits to a level of 189 ± 21 mg (n=5). The spontaneous generation of a prostacyclin-like activity (PLA) by slices of RbAs of control animals was 180 ± 90 ng/g of wet tissue (n=18) and fell down to 88 ± 26 (n=7) and 85 ± 24 ng/g (n=5) in RbA slices of animals fed HLD for periods of 2 and 12 weeks, respectively. A difference in the release of PLA between control and HLD-fed animals was significant at a level of $p<0.02$. The spontaneous generation of a PGE_2-like activity by the control RbA slices was 176 ± 17 ng/g wet tissue (n=10) and did not change significantly after 2,4 and 12 weeks of feeding HLD. Contrary to RbAs rabbit mesenteric arteries (RbMAs) of control animals generated more of PLA than of a PGE_2-like activity (Fig.2), i.e. 1440 ± 200 ng/g wet tissue of PLA (n=18) as compared to 239 ± 85 ng/g of a PGE_2-like activity. This last did not change significantly in animals fed HLD up to 12 weeks of the observation period, whereas a dramatic fall in PLA was observed as early as 1 week of HLD, stayed for 12 weeks and was followed by the recovery of the release of PLA by RbMAs within next 12 weeks of feeding HLD (Fig.2).

Platelets in PRP of the control rabbits were aggregated by arachidonate at a threshold concentration of 77 ± 25 µM (n=20), and the corresponding amount of generated TXA_2 was 78 ± 47 ng equivalents of U 46619 per 2×10^8 platelets (n=20). Following 2,4 or 12 weeks of HLD neither of these two values were significantly changed. The lowest proaggregatory concentration of arachidonate and the highest amount of TXA_2 produced were observed following 12 weeks of HLD and they were 48 ± 31 µM of arachidonate and 112 ± 29 ng equivalents of U 46619 per 2×10^8 platelets (n=4). Washed platelets of control animals when stimulated by thrombin generated 358 ± 100 ng of 12-HETE per 2×10^8 platelets (n=15). No significant

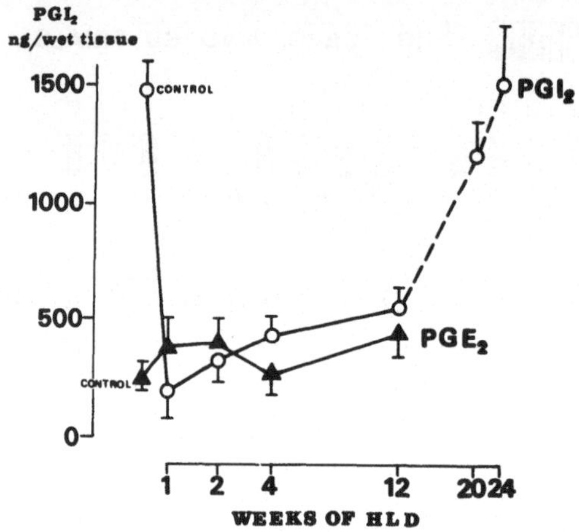

Fig. 2. The effect of duration of HLD on generation of prostacyclin(PGI_2) and PGE_2 by slices of rabbit mesenteric artery (RbMA). Dashed line denotes the results of experiments in which only PGI_2 but not PGE_2 was determined.

change in the amount of generated 12-HETE was observed after 2 and 4 weeks of HLD, however, after 12 weeks of HLD the amount of 12-HETE released by thrombin increased significantly up to 582 ± 75 ng per 2 x 10^8 platelets (n=4, $p < 0.02$).

In rabbits fed HLD plasma levels of a TBA-sensitive material were soon doubled (from 3.0 ± 0.5 μM, n=18 to 6.4 ± 1.8 μM, n=5) and remained so during the observation period. On the other hand, plasma fibrinolytic activity as measured by ECLT was suppressed up to 4 weeks of feeding animals with HLD (from 109 ± 45 min, n=20 to 191 ± 28 min, n=6 of ECLT), however, following 12 weeks of HLD the plasma fibrinolytic activity was increased and ECLT fell down to 70 ± 18 min (n=5).

Two weeks after prostacyclin had been administered to the IS patients their neurological deficit (mainly hemiparesis and aphasia) improved to a larger extent than in the placebo group, however, this difference reached statistical significance in neither of two controlled trials (Fig.3). Only during the initial 54 hours of the treatment alleviation of neurological status in the prostacyclin-treated group was on a border of statistical significance. Neither the elongation of the intermittent course of administration of prostacyclin up to two weeks nor extension of the observation period up to three weeks improved an appreciation of the therapeutical efficacy of prostacyclin (Fig.3).

In 30 CRVO patients 4 weeks after the commencement of the treatment an improvement was observed of pathological changes in eye fundus and in acuity of vision in 13 out of 15 patients of the prostacyclin group and in 12 out of 15 patients of the placebo group. The corresponding figures for the considerable improvement were 6/15 and 4/15. Neither of these two pairs were statistically different. A therapeutical effect of prostacyclin appeared only after 6 months when neovascularization was diagnosed in 4 out of 15 patients in the prostacyclin group and in 7 out of 15 patients in the placebo group. This difference was significant ($p < 0.05$ by Wilcoxon test).

Two weeks after the treatment of the SD patients with prostacyclin

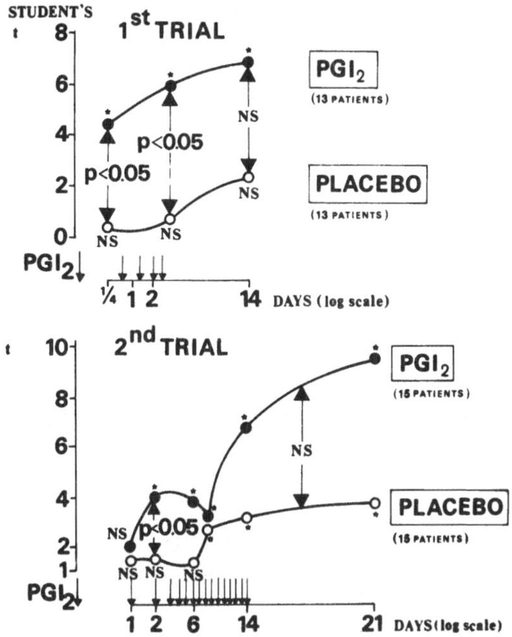

Fig. 3. Significance(t Student test) of the improvement in the neurological status of patients with ischaemic stroke (IS). Asterisks denote significant improvement as compared to initial number of scores. Arrows denote significance of difference between PGI_2-treated and placebo-treated groups.
NS - non-significant.

the improvement in hearing was found in 13 out of 15 patients as compared to 8 out of 15 patients in the placebo group. The highest rate of improvements in audiometric parameters was achieved by the prostacyclin therapy in patients whose initial impairement of hearing was only partial. In this prostacyclin-treated subgroup the complete recovery of hearing occurred in 6 out of 6 patients as compared to the placebo subgroup in which the improvement was detected in 2 out of 4 patients. The observed therapeutical effects of prostacyclin were statistically significant ($p < 0.05$ by Wilcoxon test).

DISCUSSION

A high-lipid diet (HLD) fed to rabbits for a period of 1 to 2 weeks caused an increase in plasma levels of a TBA-reactive material. A rise in levels of a TBA-reactive material need not be exclusively dependent on an accumulation of malondialdehyde (MDA) which derives from oxidation of lipids[13], however, since our experimental animals were fed non-physiological amounts of easily oxidisable lipids - it was likely to be so. An increase in plasma lipid peroxides is considered to be inhibitory for the biosynthesis of prostacyclin by vascular walls[1]. We have proposed that thus resulting deficiency in antithrombotic vasodilator, fibrinolytic and cytoprotective prostacyclin[11] triggers the development of atherosclerosis[2,3,18].

Presently, we have confirmed that a rise in plasma levels of a TBA-reactive material is associated with a selective suppression of the production of prostacyclin by aorta and mesenteric artery whereas the arterial biosynthetic capacity of PGE_2 remains unchanged. The generation of prostacyclin by arteries stays at a low level for the first three months of HLD and within this period a dramatic decrease in thromboresistance of endothelial surface has been observed. Following six months of HLD in parallel to the development of distinct morphological signs of atherosclerosis the arteries recover from the "lipid peroxide shock" and increase the production of prostacyclin[18]. A similar phenomenon has been observed in advanced human atherosclerosis[19]. We have noticed that during the first three months of HLD platelets have been hardly activated apart from an increased generation of 12-HETE at the end of the third month of HLD. A significant increase in generation of TXA_2 by platelets and low thresholds for aggregants appear only in advanced experimental and human atherosclerosis[18]. The suppression of plasma fibrinolytic activity which is typical for advanced human atherosclerosis and may be counteracted by exogenous prostacyclin[20] occurred also in rabbits fed HLD but only in the third month of HDL.

Recent studies in humans, monkeys and rabbits have shown that atherosclerosis is associated with an impaired response of arteries to endothelium-dependent vasorelaxants[10,11,21-24] and this phenomenon might be prevented by the treatment with calcium channel blockers[23] or by a dietary manipulation[24]. Although the above effect is usually ascribed to suppression of the release or action of EDRF[11], nonetheless, in some reports the authors have not differentiated between EDRF and prostacyclin[e.g.21,22,24] while both these endothelial products are released by the endothelium-dependent vasorelaxants[9].

We hypothesise that in atherosclerosis the release of prostacyclin and EDRF is suppressed in an unequal proportion (Fig.4). A deficiency in prostacyclin appears before any morphological damage is done[2] and recovers at the advanced stage of the disease[18,19] whereas the release and action of EDRF is inhibited only in advanced atherosclerosis[11,21] (Fig.4). Thus a deficiency in vascular prostacyclin is likely to play an initiative role in the development of atherosclerosis while the deficiency in EDRF seems to be a supportive factor for advancement of the disease.

This hypothesis could explain a relatively low therapeutical efficacy of prostacyclin in diseases which are associated with atherosclerosis apart from patients with sudden deafness (this paper) and peripheral vascular disease[25]. It may well be that in advanced atherosclerosis in contrast with the total mobilisation of endogenous prostacyclin, it cannot develop the vasoprotective action because of lack of the permissive effect of EDRF. A synergism has been recently reported[26] between prostacyclin and EDRF in platelets. Would this hypothesis be true, then patients with advanced atherosclerosis should benefit from a combined therapy with prostacyclin and molsidomine. This last is a prodrug for a nitroso-compound (SIN-1A) with a stimulatory action on guanylate cyclase[27] similar to that of EDRF or nitric oxide[8].

In advanced atherosclerosis SIN-1A(or a similar N-nitroso or S-nitroso compound) is expected to substitute a missing endogenous stimulator of guanylate cyclase. A pharmacologically-induced rise in c-GMP is likely to be a permissive factor for showing up the full-scale vasoprotective action of prostacyclin, both as an adenylate cyclase stimulator and a substrate for stable metabolites (6-keto-PGE_1,SMX) with fibrinolytic and cytoprotective actions[29].

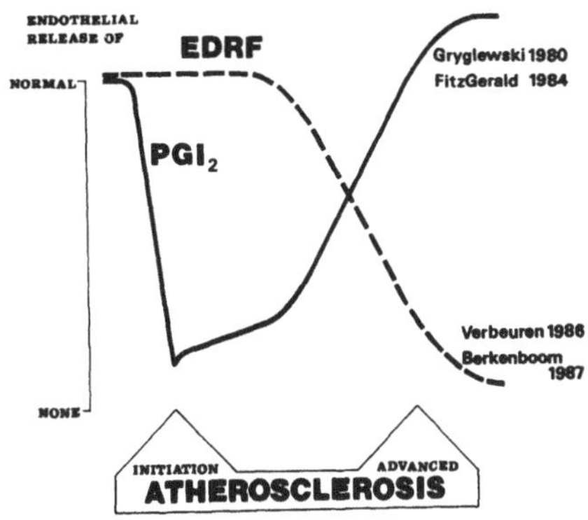

Fig. 4. A hypothesis on the role of endothelial PGI_2 and EDRF in the development of atherosclerosis. An initial drop in PGI_2-generating capacity of endothelium might be due to intoxication with plasma lipid peroxides[1] and derived from them hydroxyl radicals whereas the secondary deficiency in EDRF might reflect the overproduction of superoxide anions[28] by activated leukocytes and platelets which are sticking to the arterial wall.

REFERENCES

1. R. J. Gryglewski, S. Bunting, S. Moncada, R. J. Flower and J. R. Vane, Arterial walls are protected against deposition of platelet thrombi by a substance (prostaglandin X) which they make from prostaglandin endoperoxides, Prostaglandins, 12:685 (1976).
2. A. Dembińska-Kieć, T. Gryglewska, A. Żmuda, and R. J. Gryglewski, The generation of prostacyclin by arteries and by coronary vascular bed is reduced in experimental atherosclerosis in rabbits, Prostaglandins, 14:1025 (1977).
3. R. J. Gryglewski, A. Dembińska-Kieć, A. Żmuda, and T. Gryglewska, Prostacyclin and thromboxane A_2 biosynthesis capacities of heart, arteries and platelets at various stages of experimental atherosclerosis, Atherosclerosis, 31:385 (1978).
4. J. Larrue, M. Rigaud, D. Daret, J. Demond, J. Durand, and H. Bricaud, Prostacyclin production by cultured smooth muscle cells from atherosclerotic rabbit aorta, Nature, 285:480 (1980).
5. V. D'Angelo, M. Myśliwiec, M. B. Donati, G. De Gaetano, Defective fibrinolytic and prostacyclin-like activity in human atheromatous plaque, Thromb. Haemostas. 39:535 (1978).
6. H. Sinzinger, K. Silberbauer, W. Feige, W. Wagner, M. Witner, and A. Auerswald, Prostacyclin activity is diminished in differential types of morphologically controlled human atherosclerotic lesions, Thromb. Haemostas. 42:803 (1979).
7. R. F. Furchgott and J. Zawadzki, The obligatory role of endothelial cells in the relaxation of smooth muscle by acetylcholine, Nature, 288:373 (1980).
8. R. M. J. Palmer, A. G. Ferrige, and S. Moncada, Release of nitric oxide accounts for the biological activity of endothelium-derived relaxing factor, Nature, 327:524 (1987).

9. J. R. Vane, R. J. Gryglewski, and R. M. Botting, The endothelial cell as a metabolic and endocrine organ, TIPS 8:491 (1987).
10. R. L. Jayakody, M. P. J. Senartane, A. B. R. Thomson, and C. T. Kappagoda, Cholesterol feeding impairs endothelium-dependent relaxation of rabbit aorta, Can. J. Physiol. Pharmacol. 63:1206 (1985).
11. T. J. Verbeuren, F. H. Jordaens, A. E. Van Hoydonck, and A. G. Herman, Release and relaxation induced by endothelium-derived relaxing factor in atherosclerotic rabbit aorta, Proc. Brit. Pharmacol. Soc. Abstracts, C 141 (1987).
12. M. Radomski and S. Moncada, An improved method for washing of human platelets with prostacyclin, Thromb. Res. 30:383 (1983).
13. T. Shimizu, K. Kondo, and O. Hayashi, Role of prostaglandin endoperoxides in the serum thiobarbituric acid reaction, Arch. Biochem. Biophys. 206:217 (1981).
14. K. N. von Kaulla and R. L. Schulz, Methods for evaluation of human fibrinolysis: studies with two combined techniques, Ann. J. Clin. Pathol. 29:104 (1958).
15. J. Huczyński, E. Kostka-Trąbka, W. Sotowska, K. Bieroń, L. Grodzińska, A. Dembińska-Kieć, E. Pykosz-Mazur, E. Pęczak, and R. J. Gryglewski, Prostacyclin in patients with completed ischaemic stroke.A controlled trial, in:"Prostacyclin-Clinical Trials", R. J. Gryglewski, A. Szczeklik, and J. C. McGiff, eds, Raven Press, New York (1985).
16. H. Żygulska-Mach, E. Kostka-Trąbka, L. Grodzińska, K. Bieroń, E. Telesz, and R. J. Gryglewski, Prostacyclin in the therapy of central retinal vein occlusion, in:"Prostacyclin - Clinical Trials", R. J. Gryglewski, A. Szczeklik, and J. C. McGiff, eds, Raven Press, New York (1985).
17. E. Olszewski, J. Sekula, E. Kostka-Trąbka, L. Grodzińska, A. Dembińska-Kieć, K. Bieroń, M. Basista, A. Kędzior, and R. J. Gryglewski, Prostacyclin in the treatment of sudden deafness, in:"Prostacyclin-Clinical Trials", R. J. Gryglewski, A. Szczeklik, and J. C. McGiff, eds, Raven Press, New York (1985).
18. R. J. Gryglewski, Prostaglandins, platelets and atherosclerosis, in: CRC Crit.Rev.Biochemistry, G. D. Fasman,ed., CRC Press Inc.(1980).
19. G. A. Fitzgerald, B. Smith, A. K. Pedersen, and A. R. Brash, Increased prostacyclin biosynthesis in patients with severe atherosclerosis and platelet activation, New Engl. J. Med. 310:1060 (1984).
20. A. Dembińska-Kieć, E. Kostka-Trąbka, and R. J. Gryglewski, Effect of prostacyclin on fibrinolytic activity in patients with arteriosclerosis obliterans, Thromb. Haemostas. 47:190 (1982).
21. G. Berkenboom, M. Depierreux, and J. Fontaine, The influence of atherosclerosis on the mechanical responses of human isolated coronary arteries to substance P, isoprenaline and noradrenaline, Brit. J. Pharmacol. 92:113 (1987).
22. P. C. Freiman, G. G. Mitchell, D. D. Heistad, M. L. Armstrong, and D. G. Harrison, Atherosclerosis impairs endothelium-dependent vascular relaxation to acetylcholine and thrombin in primates, Circulation Res. 58:783 (1986).
23. J. B. Habib, C. Bossaler, S. Wells, C. Williams, J. D. Morrisett, and P. D. Henry, Preservation of endothelium-dependent vascular relaxation in cholesterol-fed rabbit by treatment with the calcium blocker PN 200110, Circulation Res. 58:305 (1986).
24. D. G. Harrison, M. L. Armstrong, P. C. Freiman, and D. D. Heistad, Restoration of endothelium-dependent relaxation by dietary treatment of atherosclerosis, J. Clin. Invest. 80:1808 (1987).
25. J. J. F. Belch, B. McArdle, J. G. Pollock, C. D. Forbes, A. McKay, P. Leiberman, G. D. O. Lowe, and C. R. M. Prentice, Epoprostenol (prostacyclin) and severe arterial disease: a double blind trial, Lancet 1:315 (1983).

26. M. W. Radomski, R. M. J. Palmer, and S. Moncada, Comparative pharmacology of endothelium-derived relaxing factor, nitric oxide and prostacyclin in platelets, Brit. J. Pharmacol. 92:181 (1987).
27. M. Nishikawa, M. Kanamori, and H. Hidaka, Inhibition of platelet aggregation and stimulation of guanylate cyclase by an antianginal agent Molsidomine and its metabolites, J. Pharmacol. Exp. Ther. 220:183 (1982).
28. R. J. Gryglewski, R. M. J. Palmer, and S. Moncada, Superoxide anion is involved in the breakdown of endothelium-derived vascular relaxing factor, Nature 320:454 (1986).
29. R. J. Gryglewski, The impact of prostacyclin studies on the development of its stable analogues, in:"Prostacyclin and its stable analogue Iloprost", R. J. Gryglewski and G. Stock, eds, Springer Verlag, Berlin-Heidelberg-New York-London-Paris-Tokyo (1987).

PLATELET-NEUTROPHIL INTERACTIONS IN THE EICOSANOID PATHWAY

Aaron J. Marcus, Lenore B. Safier, Harris L. Ullman, Naziba Islam, M. Johan Broekman, J.R. Falck, Sven Fischer and Clemens von Schacky

Divisions of Hematology-Oncology, Departments of Medicine, New York Veterans Administration Medical Center, and the Specialized Center of Thrombosis Research, Cornell University Medical College, New York, NY 10010 (A.J.M., L.B.S., H.L.U., N.I., M.J.B.), Department of Molecular Genetics, University of Texas Health Science Center, Dallas, Texas (J.R.F.), and the Medizinische Klinik Innenstadt der Universitat Munchen, F.R.G. (S.F., C.v.S.)

INTRODUCTION

Trauma to blood vessels or degenerative changes in atherosclerotic plaques result in exposure of structures in the subendothelium, such as collagen, which in its native state is stimulatory for platelet adhesion and subsequent activation (Marcus, 1988). Platelet activation is also accompanied by initiation of the blood coagulation cascade, culminating in the formation of thrombin. Collagen and thrombin are capable of stimulating the oxygenation of free arachidonate, thereby activating the eicosanoid pathway in both platelets and endothelial cells. Many of the eicosanoids which form are actually autacoids with significant biological properties.

As the above cited processes evolve, cells are brought into close contact, thereby providing the opportunity for biochemical interchange of eicosanoid precursors, intermediates and end products (Marcus et al., 1980, Marcus et al., 1984, Marcus et al., 1987).

In stimulated human platelets, 12-hydroxyeicosatetraenoic acid (12-HETE) is the lipoxygenase product produced. In previous research of the metabolism of this eicosanoid, we found that unstimulated neutrophils metabolized 12-HETE released from platelets to 12,20-dihydroxyeicosatetraenoic acid (12,20-DiHETE). The latter reaction occurred via an omega-hydroxylation process of the cytochrome P-450 type (Marcus et al., 1987). Recently, we demonstrated further metabolism of 12,20-DiHETE by an NAD-dependent dehydrogenase in unstimulated neutrophils to a new and previously unidentified eicosanoid 12-hydroxyeicosatetraen-1,20-dioic acid (12-HETE-1,20-dioic acid) (Marcus et al., 1988).

RESULTS

During our initial studies of the metabolism of platelet 12-HETE to 12,20-DiHETE by unstimulated neutrophils, we observed a more polar product which eluted prior to 12,20-DiHETE when this was studied by reversed-phase high performance liquid chromatography (RP-HPLC). Utilizing several different types of biochemical systems, we identified the new metabolite as 12-HETE-1,20-dioic acid. When incubation times were extended to periods such as 10 min the new metabolite increased 4-fold. The product also formed when stimulated platelets were utilized as an endogenous source of 12-HETE.

We then carried out detailed time-course studies in which purified 12,20-DiHETE was incubated with neutrophils for periods up to 60 min. Production of 12-HETE-1,20-dioic acid increased with time, and this increase was accompanied by a decrease in the quantity of remaining 12,20-DiHETE. As a control, 12,20-DiHETE was incubated in a comparable manner in the absence of cells. Under these conditions, no other products were detected, indicating that the new product did not result from nonspecific conversion of 12,20-DiHETE.

Initially, the new metabolite was studied by UV absorption spectrophotometry. The spectrum exhibited an absorption maximum at 236 nm, which did not differ significantly from that of 12-HETE or 12,20-DiHETE. Therefore, the basic double bond structure was similar to that of compounds preceding the new product in the 12-HETE-neutrophil metabolic sequence.

An observation of interest was that initial spectral studies carried out on the RP-HPLC samples demonstrated a downward shift in the maximum of the UV spectrum. Subsequently, we determined that this downward shift in the UV spectrum was due to residual acetic acid from the eluting solvent which had been removed from the samples by evaporation. When we added NH_4OH to an approximate concentration of 0.4 mM, the absorption maximum was restored to 236 nm. The above changes were thought to be due to possible lactonization of the compound, which is a reversible phenomenon reported to occur with hydroxy acids (Morrison and Boyd, 1983).

For identification of 12-HETE-1,20-dioic acid by HPLC, we used chemically synthesized material as the standard. The compound was more stable when converted to the dimethyl ester form and, therefore, the biological samples of the new metabolite obtained from incubations of 12,20-DiHETE with neutrophils were also methylated with diazomethane.

We discerned that chemically synthesized 12-HETE-1,20-dioic acid in its dimethyl ester form chromatographed as a single peak with a retention time of 29.97 min in methanol/water/acetic acid (75:25:0.01). The methylated biological metabolite had a retention time of 29.85 min, which was virtually identical to that of the chemically synthesized standard.

The HPLC identification of 12-HETE-1,20-dioic acid was corroborated by thin-layer chromatographic studies (TLC). The methyl ester of the biosynthesized compound migrated with an R_f value of 0.57 which was identical to that of the chemically synthesized product. For gas chromatography-mass spectrometry, the biosynthesized product, purified by RP-HPLC was converted to the methylesters-trimethylsilylether. The compound was somewhat thermally unstable as demonstrated by a broad tailing peak on gas chromatography. A mass-spectrum taken in the electron impact (EI) mode included an ion representing C1-C12 which corresponded to similar ions in the spectra of the methylester-trimethylsilylethers of 12-HETE and 12,20-DiHETE. This indicated that the C1-C12 portion of the molecule remained intact in the metabolite. The hydrogenated derivative was then subjected to mass spectrometry in the EI mode. This demonstrated the presence of three double bonds between C1 and C12 and a single double bond between C12 and C20. Spectra taken in the positive chemical ionization mode (+CI) confirmed unequivocally that the metabolite was 12-HETE-1,20-dioic acid.

We found that 12-HETE-1,20-dioic acid formation had an absolute requirement for NAD. Prior to disruption of the neutrophil preparations, they were pretreated with diisopropylfluorophosphate (DFP) and NADPH was added before assay. These conditions were already known to be necessary for the omega-hydroxylation of 12-HETE by disrupted neutrophils (Marcus et al., 1987). Nevertheless, sonication of DFP-pretreated neutrophils resulted in loss of 12-HETE-1,20-dioic acid production. It, therefore, appeared that the co-factor requirement of the 12,20-DiHETE dehydrogenase which catalyzed 12-HETE-1,20-dioic acid formation was different from that of the 12-HETE-omega-hydroxylase. Addition of NAD restored 12-HETE-1,20-dioic acid formation lost during cell disruption by 77%. This requirement for NAD was specific, since NADP was a poor substitute and little effectiveness was noted following addition of NADPH or NADH.

Subcellular localization of 12,20-DiHETE dehydrogenase activity was carried out as follows: neutrophils were initially pretreated with DFP and then sonicated. Subcellular granule, microsomal, and cytosol fractions were prepared by standard differential ultracentrifugation techniques. All fractions were assayed in the presence of 1 mM NAD. Total activity was recovered in the 10,000 x g supernatant (cytosol and membranes). When the latter supernatant was ultracentrifuged at 100,000 x g, the resulting cytosol and microsomal fractions unexpectedly contained only 12% and 13% of the starting total activity, respectively. Since 75% of the starting enzyme activity was lost when the cytosol and microsomal fractions were separated, recombination experiments were carried out. When the cytosol and membrane fractions were combined, 82% recovery of 12,20-DiHETE dehydrogenase activity resulted. If the cytosol was boiled and the supernatant of the precipitated protein recombined with the membrane fraction, the cytosolic contribution of the recombined activity was lost. Activity was equivalent only to that of the membranes alone. In the converse situation, when boiled membranes were added to unboiled cytosol, only 22% of the starting total enzyme activity was recovered. Results of these subcellular experiments indicated that the neutrophil 12,20-DiHETE dehydrogenase enzyme system had subcellular heat-labile components in both the cytosol and membrane fractions.

DISCUSSION

Stimulated platelets lipoxygenate free arachidonic acid in the C-12 position in the presence or absence of aspirin or other non-steroidal anti-inflammatory agents. Release of the product 12-HETE occurs, and this eicosanoid can be metabolized by unstimulated neutrophils in close proximity. The resulting product 12,20-DiHETE forms via a cytochrome P-450 omega-hydroxylation mechanism (Marcus et al., 1987). Time course studies carried out in this work with purified 12,20-DiHETE demonstrated that the latter is further oxidized by unstimulated neutrophils to 12-HETE-1,20-dioic acid. These reactions are summarized in Figure 1.

It was necessary to make certain comparisons between 12-HETE-1,20-dioic acid formation and metabolism of leukotriene B_4

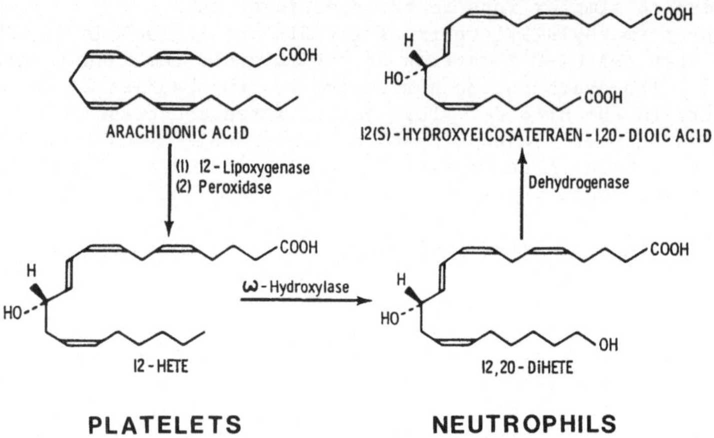

Fig. 1. **Metabolism of 12-HETE released from stimulated platelets in the presence of unstimulated neutrophils.** Cytochrome P-450-mediated omega-hydroxylation at C20 is followed by further oxidation at C20 to form 12-HETE-1,20-dioic acid.

(LTB_4). When neutrophils are stimulated LTB_4 is produced and is regarded as the most important proinflammatory eicosanoid synthesized by these cells. We endeavored to determine whether the enzyme system responsible for catalyzing conversion of 12,20-DiHETE to 12-HETE-1,20-dioic acid might also be involved in the metabolism of LTB_4. The trihydroxy acid, 20-hydroxy-LTB_4 (the omega-hydroxylated metabolite of LTB_4) is further processed to the dioic acid, 20-carboxy LTB_4 by human neutrophils (Hansson et al., 1981, Jubiz et al., 1982, Shak and Goldstein, 1984, Powell, 1984). We, therefore, added 20-hydroxy-LTB_4 to incubates of neutrophils and 12,20-DiHETE in order to observe effects on 12-HETE-1,20-dioic acid formation and ascertained that partial inhibition had occurred. These results suggested that the two substrates (20-OH-LTB_4 and 12,20-DiHETE) were actually competing for the same enzyme. We had reported earlier that there was a competition by LTB_4 for the 12-HETE-omega-hydroxylase in unstimulated neutrophils (Marcus et al., 1987). This suggests that platelet 12-HETE and its metabolites can exert a modulating influence on the metabolism of neutrophil-derived LTB_4.

Since NAD was required as a cofactor for conversion of 12,20-DiHETE to 12-HETE-1,20-dioic acid, there was a clear-cut distinction between the enzyme(s) involved and the neutrophil 12-HETE-omega-hydroxylase. The omega-hydroxylase is a cytochrome P-450 enzyme which is typically NADPH-dependent. The absence of conversion of 20-hydroxy-LTB_4 to the 20-carboxy-LTB_4 form in neutrophil sonicates, despite the presence of NADPH (Soberman et al., 1985, Shak and Goldstein, 1985), might indicate that if the 20-OH-LTB_4 enzyme system is similar to that which processes 12,20-DiHETE, the reported loss of activity may be explained by a requirement for NAD as a cofactor. It should be mentioned that omega-hydroxy fatty acid dehydrogenases with a specific requirement for NAD as cofactor have been reported in other tissues, such as mammalian liver (Mitz and Heinrikson, 1961, Wakabayashi and Shimazono, 1961, Wakabayashi et al., 1962, Kamei et al., 1964, Bjorkhem and Danielsson, 1970). Some of these differed from the enzyme we have studied in neutrophils in that specificities for omega-hydroxy-acids were for those of medium molecular weight (C9-C11). Dehydrogenases in rat liver homogenates catalyzing oxidation of long chain 16-hydroxypalmitic and

18-hydroxystearic acids to corresponding dicarboxylic acids have been demonstrated (Bjorkhem and Danielsson, 1970).

In the reaction proceeding from 12,20-DiHETE to the dioic acid, it is possible that an aldehyde intermediate may form. This has been postulated by Mitz and Heinrikson (1961) and demonstrated by Kamei et al. (1964). In the analytical procedures utilized by us, we did not detect evidence for an aldehyde intermediate. However, this may have been due to rapid conversion of this hypothetical intermediate to the dioic acid form (Marcus et al., 1988).

The reactions we have described in this paper represent a novel extension of previous research in which it was demonstrated that eicosanoid pathway interactions can occur between different cell types, such as platelets, neutrophils and endothelial cells (Marcus, 1986). When stimulated platelets produce 12-HETE, it is initially transformed by unstimulated neutrophils to 12,20-DiHETE - a new eicosanoid which neither cell can synthesize alone. In a recent classification of cell-cell interactions which we have devised, this category of reactions would be classified as Type IIB. Now we have found that the neutrophil continues to metabolize 12,20-DiHETE to 12-HETE-1,20-dioic acid. It is also known that platelets and neutrophils are brought into apposition during thrombosis and the inflammatory response. The metabolic interchange of biochemical substances generated by these cells may serve to modulate host defense mechanisms (Marcus, 1988).

REFERENCES

Bjorkhem, I., and Danielsson, H., 1970, Omega- and (omega-1)-oxidation of fatty acids by rat liver microsomes, Eur. J. Biochem., 17:450.

Hansson, G., Lindgren, J. A., Dahlen, S.-E., Hedqvist, P., and Samuelsson, B., 1981, Identification and biological activity of novel omega-oxidized metabolites of leukotriene B_4 from human leukocytes, FEBS Lett., 130:107.

Jubiz, W., Radmark, O., Malmsten, C., Hansson, G., Lindgren, J. A., Palmblad, J., Uden, A.-M., and Samuelsson, B., 1982, A novel leukotriene produced by stimulation of leukocytes with formylmethionylleucylphenylalanine, J. Biol. Chem., 257:6106.

Kamei, S., Wakabayashi, K., and Shimazono, N., 1964, Studies on omega-oxidation of fatty acids in vitro, J. Biochem. (Tokyo), 56:72.

Marcus, A. J., Weksler, B. B., Jaffe, E. A., and Broekman, M. J., 1980, Synthesis of prostacyclin from platelet-derived endoperoxides by cultured human endothelial cells, J. Clin. Invest., 66:979.

Marcus, A. J., Safier, L. B., Ullman, H. L., Broekman, M. J., Islam, N., Oglesby, T. D., and Gorman, R. R., 1984, 12S,20-Dihydroxyicosatetraenoic acid: a new icosanoid synthesized by neutrophils from 12S-hydroxyicosatetraenoic acid produced by thrombin- or collagen- stimulated platelets, Proc. Natl. Acad. Sci. U.S.A., 81:903.

Marcus, A. J., 1986, Transcellular metabolism of eicosanoids, in: "Progress in Hemostasis and Thrombosis," B. S. Coller, ed., Grune & Stratton, New York.

Marcus, A. J., Safier, L. B., Ullman, H. L., Islam, N., Broekman, M.J. and von Schacky, C., 1987, Studies on the mechanism of omega-hydroxylation of platelet 12-hydroxyeicosatetraenoic acid (12-HETE) by unstimulated neutrophils, J. Clin. Invest., 79:179.

Marcus, A. J., Safier, L. B., Ullman, H. L., Islam, N., Broekman, M. J., Falck, J. R., Fischer, S., and von Schacky, C., 1988, Platelet-neutrophil interactions. 12S-hydroxyeicosatetraen-1,20-dioic acid: a new eicosanoid synthesized by unstimulated neutrophils from 12S,20-dihydroxyeicosatetraenoic acid, J. Biol. Chem., 263:2223.

Marcus, A. J., 1988, Hemorrhagic disorders: abnormalities of platelet and vascular function, in: "Cecil Textbook of Medicine, 18th Edition," J. B. Wyngaarden and L. H. Smith, Jr., eds., Saunders, Philadelphia.

Marcus, A. J., 1988, Eicosanoids: transcellular metabolism, in: "Inflammation: basic principles and clinical correlates," J. I. Gallin, I. M. Goldstein, and R. Snyderman, eds., Raven Press, New York.

Mitz, M. A., and Heinrikson, R.L., 1961, Omega hydroxy fatty acid dehydrogenase, Biochem. Biophys. Acta, 46:45.

Morrison, R. T., and Boyd, R. N., 1983, "Organic Chemistry, 4th Edition," Allyn and Bacon, Inc., Boston.

Powell, W. S., 1984, Properties of leukotriene B_4 20-hydroxylase from polymorphonuclear leukocytes, J. Biol. Chem., 259:3082.

Shak, S., and Goldstein, I. M., 1984, Omega-oxidation is the major pathway for the catabolism of leukotriene B_4 in human polymorphonuclear leukocytes, J. Biol. Chem., 259:10181.

Shak, S., and Goldstein, I. M., 1985, Leukotriene B_4 omega-hydroxylase in human polymorphonuclear leukocytes, J. Clin. Invest., 76:1218.

Soberman, R. J., Harper, T. W., Murphy, R. C., and Austen, K. F., 1985, Identification and functional characterization of leukotriene B_4 20-hydroxylase of human polymorphonuclear leukocytes, Proc. Natl. Acad. Sci. U.S.A., 82:2292.

Wakabayashi, K., and Shimazono, N., 1961, Studies in vitro on the mechanism of omega-oxidation of fatty acids, Biochim. Biophys. Acta, 48:615.

Wakabayashi, K., Kamei, S., Murakami, H., and Shimazono, N., 1962, On the presence of omega-hydroxyfatty acid: NAD oxidoreductase, J. Biochem. (Tokyo), 52:464.

This work was supported by grants from the Veterans Administration, National Institutes of Health HL-18828-12 SCOR (to A. J. M., M. J. B., and C. v. S.), HL-29034 (to M. J. B.), GM-31278 (to J. R. F.), the Edward Gruenstein Fund, the Sallie Wichman Fund, and SM Louis Fund (to A. J. M.), and the Deutsche Forschungsgemeinschaft (to S. F. and C. v. S.).

ARTERIAL CELL INTERACTIONS: MECHANISTIC STUDIES RELATED TO EICOSANOID AND GROWTH FACTOR-INDUCED ALTERATIONS IN CHOLESTEROL METABOLISM

David P. Hajjar[*], Aaron J. Marcus[°+], Kenneth B. Pomerantz[°], and Katherine A. Hajjar[°x]

Departments of Biochemistry[*], Medicine[°], Pathology[*] and Pediatrics[x], Cornell University Medical College and the New York Veterans Administration Medical Center[+], New York, NY

Intracellular cholesterol accumulation in the arterial wall and the proliferation of intimal smooth muscle cells are the hallmarks of atherogenesis. Humoral fluid-phase interactions between arterial endothelial and smooth muscle cells have been studied *in vivo* and *in vitro* in an attempt to delineate whether these cells communicate in the regulation of intimal hyperplasia and cholesterol and lipoprotein metabolism in arterial smooth muscle cells.

Recently, we demonstrated that arterial endothelial cells alter cholesterol metabolism in co-cultured arterial smooth muscle cells (1). These results were similar to previous animal studies where we demonstrated that the neointima of re-endothelialized arteries accumulated free and esterified cholesterol due to altered cholesteryl ester (CE) metabolism (2). This tissue also produced less prostacyclin during diet-induced hypercholesterolemia (3). These findings, subsequently corroborated by other investigators (4,5), suggested that endothelial cells, which are a rich source of eicosanoids (6), modulate cholesterol metabolism in smooth muscle cells. However, in the major studies cited above, the mechanisms involved in the interaction of these cell types were not defined.

Similarly, there have been few studies done to date which have clarified the pathophysiological effects of reduced prostacyclin (PGI_2) and PGE_2 synthesis in atherosclerotic tissue derived from humans ans animals. In particular, it has been shown that endothelial and smooth muscle cells derived from human atherosclerotic tissue have a reduced capacity to synthesize PGI_2 and PGE_2 from either exogenous or endogenous arachidonic acid (7,8). Eicosanoid metabolism is a function of substrate availability as well as the activities of arachidonyl CoA - lysophospholipid acyltransferases and eicosanoid synthetic enzymes. The mechanisms by which atherogenic stimuli

TABLE 1

EFFECTS OF ENDOTHELIAL CELLS ON SMOOTH MUSCLE CELL CHOLESTEROL METABOLISM

	CE hydrolytic activity (pmol/hr/mg prot)	6-keto $PGF_{1\alpha}$ ($ng/10^6$ cells)
SMC alone	747 ± 145[abcdgh]	5.2 ± 0.2[i]
EC/SMC	995 ± 86[a]	47.8 ± 0.6[i]
EC(ASA)/SMC	716 ± 125	
EC(ETYA)/SMC	566 ± 55	
SMC + PGI_2	1627 ± 127[ce]	
SMC + 12-HETE	1005 ± 88[de]	
SMC + $C_{20:4}$	872 ± 28	
ECCM-I/SMC	1520 ± 188[bf]	
SMC-CM/EC(ETYA) -> ECCM-II/SMC	944 ± 51[f]	
SMC-CM/SMC	745 ± 62	
SMC + PDGF	1629 ± 224[g]	62.2 ± 0.4[i]
EC + PDGF/SMC	1960 ± 111[h]	86.5 ± 0.3[i]
SMC(ASA)	825 ± 116	0.4 ± 0.1[i]
SMC(ASA) + PDGF	937 ± 94	0.7 ± 0.2
EC(ASA) + PDGF/SMC(ASA)	962 ± 210	0.4 ± 0.1

All data shown are from one of four separate experiments. Mean ± SD; N=6. Final concentrations are 0.1 mM ASA, 0.1 mM ETYA, 100 nM PGI_2, 100nM 12-HETE, 100 nM $C_{20:4}$, and 10 ng/ml PDGF. Values with the same superscript are significantly different ($p<0.05$). From ref. (14).

Abbreviations include:

ASA (cyclooxygenase inhibitor) : aspirin
$C_{20:4}$: arachidonic acid
ECCM : endothelial cell-conditioned medium
ETYA (cyclooxygenase and lipoxygenase inhibitor): 5,8,11,14 - eicosatetraynoic acid
PGI_2 : prostacyclin
PDGF : platelet-derived growth factor
SMC-CM: smooth muscle cell-conditioned medium

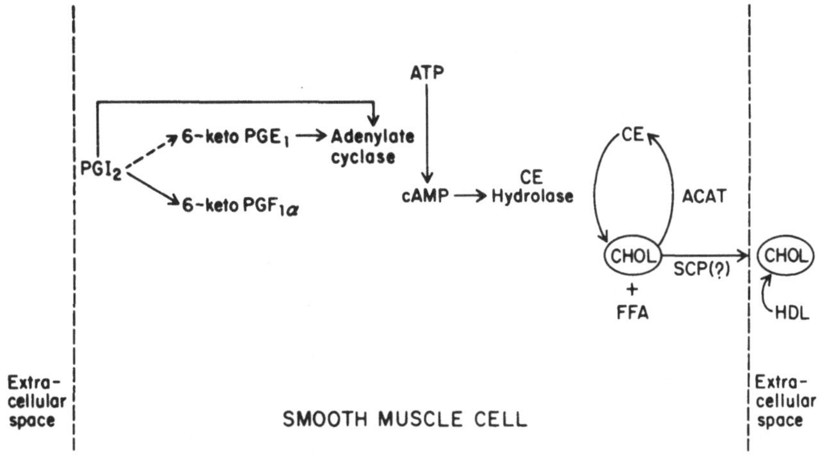

Fig. 1. Hypothetical model describing the physiological action of PGI_2 in the activation of the CE hydrolases via cyclic AMP levels with the subsequent enhancement of cholesterol egress from the arterial smooth muscle cell. (From ref. 21).

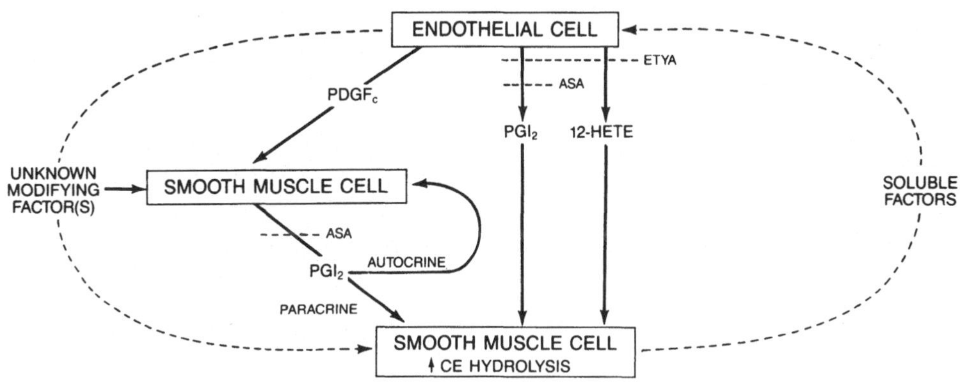

Fig. 2. Hypothetical model depicting arterial endothelial cell (EC) modulation of cholesteryl ester (CE) catabolism in co-cultured arterial smooth muscle cells (SMC). Endothelial cells secrete PDGF and eicosanoids (PGI_2 and 12-HETE) which increase CE hydrolysis in the co-cultured smooth muscle cells. Enhancement by eicosanoids is abolished by preincubation of arterial endothelial cells with aspirin (a cyclooxygenase inhibitor) or ETYA (a cyclooxygenase and lipoxygenase inhibitor). Furthermore, smooth muscle cells secrete unknown factor(s) modified by the co-cultured endothelial cells which also enhance CE catabolism in the smooth muscle cell. (From ref. 14).

such as lipid accretion alter vascular PGI_2 production via these metabolic pathways also have not been fully explained.

Since we and others have demonstrated that:

1. Endothelial cells synthesize eicosanoids and growth factors (6,9);
2. Endothelial cell-derived growth factors stimulate smooth muscle cell proliferation (9,10);
3. Exogenous growth factors such as platelet-derived growth factor can stimulate PGI_2 production (11);
4. Eicosanoids (PGI_2) can modulate CE metabolism in arterial cells (12,13)

We investigated the role of PDGF and arterial endothelial cell-derived eicosanoids in the modulation of CE hydrolysis in smooth muscle cells when co-cultured directly with endothelial cells on their apical surface (14). Furthermore, we studied the effects of lipid-loading on eicosanoid production by arterial smooth muscle cells in order to define the relationship between cellular lipid accretion and eicosanoid metabolism (15). Such a relationship has been suggested repeatedly in the literature by the observation that ingestion of aspirin contributes to the reduction of the incidence and severity of a second myocardial infarction in men (16,17).

The major eicosanoids synthesized by arterial endothelial cells included PGI_2, 12-hydroxy-5,8,10,14-eicosatetraenoic acid (12-HETE) and, to a lesser extent, prostaglandin E_2 (14). Exogenously added PGI_2 and 12-HETE stimulated CE hydrolytic activity in smooth muscle cells by 49 and 35%, respectively, when co-cultured with aspirin-treated endothelial cells (14) (Table 1). Aspirin-treated endothelial cells, when co-cultured with smooth muscle cells did not stimulate CE hydrolytic activity in smooth muscle cells, which was also the case with non-aspirinated endothelial cells. These findings suggest a role for eicosanoids in the regulation of cholesterol metabolism (14).

However, such a cellular mechanism is not parsimonious for the vessel wall since small amounts of PGI_2 produced by the endothelium is used not only to regulate hemostasis and prevent a thrombogenic event but also promotes intracellular lipolytic activation. Other humoral agents derived from endothelial cells such as $PDGF_c$ stimulated CE hydrolytic activity almost 2-fold in smooth muscle cells cultured alone or co-cultured with endothelial cells (14) (Table 1). Aspirin-treated endothelial cells, when incubated with 10 ng/ml PDGF, did not stimulate CE hydrolytic activity in co-cultured smooth muscle cells (14). Therefore, we postulated that growth promoting activity may enhance CE catabolism via the PGI_2-cyclic AMP-CE hydrolysis cascade (Fig 1; ref 14). This hypothesis is further supported by our recent observations that PDGF stimulates PGI_2 production in smooth muscle cells (14) (Table 1). Our results suggest that the regulation of cholesterol metabolism in smooth muscle cells can involve, at least in part, growth factors and endothelial cell derived eicosanoids (Fig 2; ref 14). These effectors may play a central role in the regulation of cholesterol metabolism, hemostasis, and the inflammatory response.

TABLE 2

EFFECT OF CE ACCUMULATION ON PGI_2 RELEASE FROM AORTIC SMC: ROLE OF ENDOGENOUS VS EXOGENOUS SUBSTRATE

A. Endogenous substrate: effect of A-23187

TREATMENT	CONTROL	A-23187
NORMAL CELLS	0.2 ± 0.1	32.4 ± 4.0*
CE-ENRICHED CELLS	2.3 ± 1.8	6.4 ± 0.5*#

B. Exogenous substrate: effect of AA

TREATMENT	0 hours	24 hours	48 hours
NORMAL CELLS	10.8 ± 0.7	20.5 ± 0.8	14.4 ± 0.5
CE-ENRICHED CELLS	1.5 ± 0.7*	13.3 ± 0.7*	13.2 ± 0.7

(ng 6-keto-$PGF_{1\alpha}$ /mg protein, mean ± SD, * = $p < 0.01$ DMEM vs A-23187, # = $p < 0.05$ normal vs CE-enriched cells). From ref. 15.

A) SMC incubated with DMEM with or without cLDL (100 ug/ml) for six days were then exposed to DMEM with or without A-23187 (5.0 uM). N = 3.

B) Normal and CE-enriched SMC as described above were exposed to AA, (10 uM). SMC were then incubated for an additional time intervals up to 48 hours in DME in the absence of additional lipid-loading prior to re-exposure to AA (10 uM, 30 minutes). Supernatants were assayed for 6-keto-$PGF_{1\alpha}$ by RIA. N = 4.

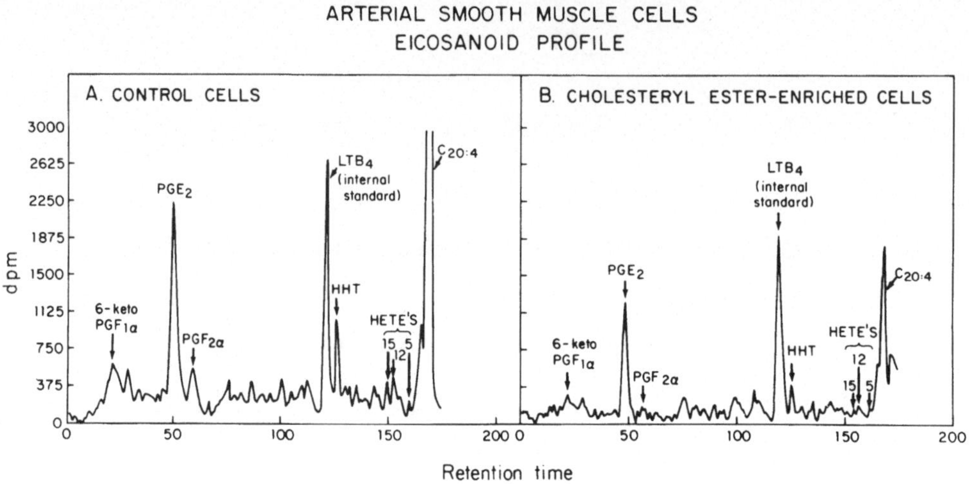

Fig. 3. Representative HPLC profiles of eicosanoid products synthesized from normal (Panel A) and cholesterol-enriched (Panel B) arterial SMC incubated with [^{3}H] arachidonic acid (AA) and exposures to ionophore A-23187. After addition of [^{3}H]-LTB_4 to monitor recovery, supernatants were diluted, acidified to pH 3.5, and extracted on reversed phase C-18 columns (15). Eicosanoids were eluted and reconstituted into 30% methanol. Identity of each eicosanoid was determined by its equivalent retention times using authentic eicosanoid standards. CE-enrichment resulted in a generalized decrease in arachidonate release and conversion to other eicosanoids. (From ref. 15).

These findings have supported our recent hypothesis based on related studies done in our laboratory, i.e. that lipid-loaded cells have a reduced capacity to synthesize PGI_2, which in turn, can cause down-regulation of the CE hydrolases in the cell, predisposing to further lipid deposition. In these studies, we found that CE-enriched arterial smooth muscle cells had a substantial decrease in the percentage of phospholipid-derived arachidonate relative to linoleate as compared to control cultures, suggesting that linoleate competitively inhibits arachidonate release and subsequent conversion to eicosanoids (15). In addition, the lipid-laden cells synthesize 50% less eicosanoids (Fig 3), particularly PGI_2, than control cells when exposed to the calcium ionophore, A-23187, or to exogenous arachidonate (15) (Table 2). A-23187 released equivalent amounts of arachidonate from all phospholipids, suggesting specificity of uptake but not release of arachidonate by cellular phospholipases (15). Potential mechanisms for reduced eicosanoid production in lipid-laden cells include direct inhibition of phospholipase A_2 (Table 3) or cyclooxygenase activities by cholesterol, or competitive inhibition of eicosanoid production by linoleate derived from LDL. We have

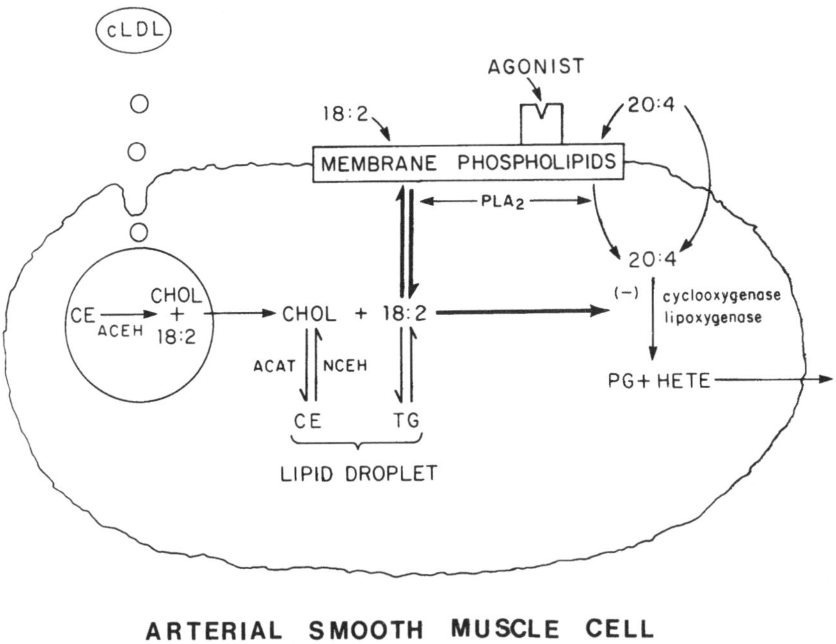

Fig. 4. A potential mechanism of reduced eicosanoid synthetic capacity by CE-enriched SMC. Cholesteryl esters derived from cationized LDL (cLDL), consist principally of cholesteryl linoleate. Cholesteryl esters are hydrolyzed in the lysosome by acid CE hydrolase (ACEH). Free linoleic acid may then be incorporated into cellular phospholipid for phospholipase A_2 (PLA_2)-induced hydrolysis. Alternatively, linoleate may also act directly as a competitive inhibitor of AA conversion to other eicosanoids. Excess cholesterol and free fatty acids may also accumulate in esterified form as CE and triglycerides (From ref. 15).

summarized these results in a hypothetical model shown in Fig 4. These findings may have pathophysiological significance in that a reduction in PGI_2 synthetic capacity of arterial cells will exacerbate CE deposition, since PGI_2 is known to promote hydrolysis of intracellular CE via cyclic AMP-dependent protein kinase (18,19). These studies are further supported by the observations that the stable PGI_2 analogue, carbocyclin, reduces cellular triglyceride, cholesterol, and CE content in intimal smooth muscle cells derived from atherosclerotic plaque areas (20). Therefore, eicosanoids, derived in part from endothelial cells, may mediate cholesterol efflux from lipid-laden foam cells during atherogenesis (14), and may provide a cellular mechanism by which these bioactive lipids regulate intracellular cholesterol levels in the arterial wall during atherosclerosis.

TABLE 3

EFFECT OF CE ENRICHMENT ON PHOSPHOLIPASE-A_2 ACTIVITY

TREATMENT	6-keto-$PGF_{1\alpha}$	PGE_2	AA ($\times 10^4$)
Control	147 ± 4.7	188 ± 8.7	3.27 ± 0.21
A-23187	875 ± 106*	630 ± 42*	20.70 ± 0.15*
cLDL	125 ± 12	156 ± 14	2.50 ± 0.45
cLDL/A23187	483 ± 30*#	478 ± 32*#	14.82 ± 0.04*#

Data is expressed as dpm/mg protein (mean ± SEM, * - $p < 0.05$ from control, # - $p < 0.05$ cLDL vs control group). N = 4.

Arterial smooth muscle cells were exposed to 10 uM A-23187 in the presence of fatty acid free bovine serum albumin (0.2%) for 30 minutes at 37 °C. Supernatants were assayed for cyclooxygenase products by TLC

References

1. D. P. Hajjar, D. J. Falcone, J. B. Amberson, and J. M. Hefton, Interaction of arterial cells. I. Endothelial cells alter cholesterol metabolism in co-cultured smooth muscle cells, J. Lipid Res., 26:1212 (1985).
2. D. P. Hajjar, D. J. Falcone, S. D. Fowler, and C. R. Minick, Endothelium modifies the altered metabolism of the injured aortic wall, Am. J. Pathol., 102:28 (1981).
3. A. Eldor, D. J. Falcone, D. P. Hajjar, C. R. Minick, and B. B. Weksler, Diet-induced hypercholesterolemia inhibits the recovery of prostacyclin production by injured rabbit aorta, Am. J. Pathol., 107:186 (1982).
4. R. S. Rosenfeld, L. Drouet, J. Cintron, I. Paul, J. Won, and T. H. Spaet, In vitro synthesis of DNA, protein, and lipids by the de-endothelialized rabbit aorta, Arteriosclerosis, 1:418 (1981).
5. S. Moore, L. W. Belbeck, M. Richardson, and W. Taylor, Lipid accumulation in the neointima formed in normally fed rabbits in response to one or six removals of the aortic endothelium, Lab. Invest, 47:37 (1982).
6. A. J. Marcus, B. B. Weksler, E. A. Jaffe, and M. B. Broekman, Synthesis of prostacyclin from platelet-derived endoperoxides by cultured human endothelial cells, J. Clin. Invest., 66:979 (1980).
7. A. Dembińska-Kieć, T. Gryglewska, A. Zmuda, and R. J. Gryglewski, The generation of prostacyclin by arteries and by the coronary vascular bed is reduced in experimental atherosclerosis in rabbits, Prostaglandins, 14:1025 (1977).

8. R. J. Gryglewski, A. Dembińska-Kièć, A. Zmuda, and T. Gryglewska, Prostacyclin and thromboxane A_2 biosynthesis capacities of heart, arteries and platelets at various stages of experimental atherosclerosis in rabbits, Atherosclerosis, 31:385 (1978).
9. P. E. Di Corleto, and D. F. Bowen-Pope, Cultured endothelial cells produce a platelet-derived growth factor-like protein, Proc. Natl. Acad. Sci. (USA), 80:1919 (1983).
10. K. A. Hajjar, D. P. Hajjar, R. L. Silverstein, and R. L. Nachman, Tumor necrosis factor-mediated release of platelet-derived growth factor from cultured endothelial cells, J. Exp. Med., 166:235 (1987).
11. S. R. Coughlin, M. A. Moskowitz, B. R. Zetter, H. N. Antoniades, and L. Levine, Platelet-dependent stimulation of prostacyclin synthesis by platelet-derived growth factor, Nature, 288:600 (1980).
12. D. P. Hajjar, B. B. Weksler, D. J. Falcone, J. M. Hefton, K. Tack-Goldman, and C. R. Minick, Prostacyclin modulates cholesteryl ester hydrolic activity by its effects on cyclic adenosine monophosphate in rabbit aortic smooth muscle cells, J. Clin. Invest., 70:479 (1982).
13. D. P. Hajjar, and B. B. Weksler, Metabolic activity of cholesteryl esters in aortic smooth muscle cells is altered by prostaglandins I_2 and E_2, J. Lipid Res., 24:1176 (1983).
14. D. P. Hajjar, A. J. Marcus, and K. A. Hajjar, Interactions of arterial cells. Studies on the mechanisms of endothelial cell modulation of cholesterol metabolism in co-cultured smooth muscle cells, J. Biol. Chem., 262:6976 (1987).
15. K. B. Pomerantz, and D. P. Hajjar, J. Biol. Chem., in press, (1988).
16. The Steering Committee of the Physician's Health Study Research Group, Preliminary report: findings from the aspirin component of the ongoing physician's health study, New Engl. J. Med., 318:262 (1988).
17. The Coronay Drug Project Research Group: Aspirin in coronary artery disease, J. Chronic Dis., 29:625 (1976).
18. D. P. Hajjar, Regulation of neutral Cholesteryl esterase in arterial smooth muscle cells: stimulation by agonists of adenylate cyclase and cyclic AMP-dependent protein kinase, Arch. Biochem. Biophys., 247:49, (1986).
19. D. P. Hajjar, Herpes virus infection prevents activation of cytoplasmic cholesteryl esterase in arterial smooth muscle cells, J. Biol. Chem., 261:7611 (1986).
20. A. N. Orekhov, V. V. Tertov, S. A. Kudryashov, Kh. A. Khashimov, and V. N. Smirnov, Primary culture of human aortic intima cells as a model for testing antiatherosclerotic drugs, Atherosclerosis, 60:101 (1986).
21. D. P. Hajjar, Prostaglandins and cyclic nucleotides. Modulators of arterial cholesterol metabolism, Biochem. Pharmacol., 34:295 (1985).

EFFECTS OF PLASMA LIPOPROTEINS ON EICOSANOID METABOLISM BY CULTURED VASCULAR SMOOTH MUSCLE CELLS

Jacky Larrue

Unité de Recherches de Cardiologie U8 INSERM

33600 Pessac, France

Prostacyclin (PGI_2) inhibits platelet aggregation and is an arterial vasodilator [1]. PGI_2 is synthetized by vascular endothelial and smooth muscle (SMC)[2,3], principally from arachidonate (AA) derived from cellular lipids [4] and its production by cultured SMC is stimulated by exposure to various agonists or drugs (see 5 for a review). In plasma, AA exists in low concentrations as free fatty acid namely in an albumin-bound form or in esterified form in lipoproteins [6]. When investigated in vitro on the microsomal fraction of aorta, a linear negative correlation between the amount of LDL cholesterol and the PGI_2 synthetase activity was found by contrast to the positive correlation observed with HDL cholesterol [7-9]. The observed inactivation of PGI_2 synthetase has been proposed to be closely associated with lipid peroxyde, nevertheless, the fact that serum lipoproteins were succeptible to peroxidation during the preparation procedures leads to a quite obscure situation [9]. Recently, by contrast to the previous results, it was demonstrated that either HDL or LDL are able to stimulate prostaglandin synthesis in several intact cell systems [10-13]. The mechanisms of their stimulation may be complex, involving both the lipid and the protein moieties of the lipoproteins, since HDL and LDL provide AA to cellular cyclooxygenase [11-13], and stimulate the release of AA from endogenous cellular stores for prostaglandin synthesis [12], both mechanisms compatible with the receptor-mediated effect of LDL on PGE_2 synthesis recently described in fibroblasts [14].

The present study was designed to test the influence of ApoB and ApoA containing lipoproteins upon the metabolism of AA and the synthesis of PGI_2 by vascular SMC grown in tissu culture. The results confirmed that lipoproteins non selectively act by providing AA to cell and, that, in addition, indirect competitive and/or inhibitory effects occur.

MATERIAL AND METHOD

Preparation of plasma lipoproteins

Plasma lipoproteins were prepared by ultracentrifugation of human pooled plasma samples of different total lipid concentrations essentially as described by Havel et al [15]. The ApoB containing fractions were isolated at densities $d \leq 1.070$ (LDL + VLDL). HDL 2b were isolated at densities between 1.070 and

1.100 g/ml then subfractions with d $\leq$ 1.125 g/ml were discarded and the HDL_3 were isolated at densities between 1.125 and 1.21 g/ml. For ApoA1 preparation, a (HDL_{2b} + HDL_3) fraction was treated with guanidium chloride according to Nichols et al [16]. Prior to incubation with cells, the lipoproteins were dialysed twice against 100 volumes of 5mM ammonium bicarbonate buffer and once against 40 volumes of HAM F10 medium containing penicillin (100 U/ml), streptomycin (100 μg/ml) then filtered through 0.22 μm filter. The total lipoprotein content was determined by the method of Lowry [17] and total cholesterol content by the method of Zlatkis and Zak [18].

Incubations

Cells were plated into 12 wells cluster plates (Costar) in HAM F10 medium containing 10 % FCS and 1 % penicillin-streptomycin and incubated at 37°C (5.10^4 cells/well) in an atmosphere of 5 % CO_2. Prior to study, the just confluent cells were washed twice in HAM F10 medium. The control culture received 1.0 ml of HAM F 10 without FCS and experimental cultures received 1.0 ml of HAM F 10 in which different amounts of filtered lipoproteins were added. When activators or inhibitors of eicosanoid synthesis were used, they were added at desired concentration in the same medium.

Radioimmuno-assay

After 24 hrs of incubation at 37°C, the media were removed and assayed for 6 keto $PGF_{1\alpha}$, the stable hydrolysis product of prostacyclin, by radioimmuno assay using a specific antibody to 6 keto $PGF_{1\alpha}$, ^{125}I RIA KIT (NEN) with the modification that HAM F10 medium was used in preparation of the standard curves. Data are reported as mean $\pm$ SD of at least 5 different experiments. Cell number in each well was determined in a heamocytometer after trypsination.

Effects of LPS on AA metalism

In order to test the effects of isolated LPS or Apolipoproteins on AA metabolism, just confluent cells were washed twice to remove unincorporated labelled material. Prelabelled cells received 2.5 ml HAM F 10 medium (with 5 % FCS or not - control experiments -) and the desired amount of dialysed lipoproteins or apoproteins. After 24 hrs incubation, media were collected and analysed as follow : the total radioactivity (RA) released into the incubation medium was measured in 50 μl aliquots by scintillation counting. The remaining media were then acidified to pH : 3 using HCl (1.0 N) then extracted twice with 3 vol. ethyl acetate. The extracts were evaporated under N_2 then concentrated in ethyl acetate and analysed either by thin layer chromatography (Silica gel G) or RP-HPLC (μ Bondapack 10 μm). The solvants used were the organic phase of ethyl acetate, iso-octane, acetic acid, water (110, 50, 20, 100 v/v) and methanol-water (70-30 v/v) for TLC and HPLC respectively.

RESULTS

LDL and HDL lipoproteins stimulate PGI_2 synthesis by vascular smooth muscle cells.

When purified ApoB and ApoA lipoproteins were incubated at different final concentrations from 25 to 400 μg/cholesterol/ml (Fig. 1), the stimulation of the the 6 keto $PGF_{1\alpha}$ release in the incubation medium appeared dose dependant. Control cells (incubated in media alone) release 58.8 $\pm$ 9.2 μg 6 keto-$PGF_{1\alpha}$ 10^5 cells/24 h. ApoB containing lipoproteins (LDL + VLDL) induce a significant increase of 6 keto-$PGF_{1\alpha}$ production only for concentration over 50 μg cholesterol/ml. Highest doses (400 μg cholesterol/ml induced a 6 fold increase of 6 keto $PGF_{1\alpha}$ production to 292 $\pm$ 62 pg/10^5 cells. PGI_2 synthesis was increased by this concentration of HDL_3, however, the stimulative effect of HDL_3 was less than that of LDL (220 $\pm$ 53 pg/10^5 cells). The concentrations of 6 keto PGF_{1a} found in incubations containing HDL_3 were 1.5 to 4 times those of the control cultures.

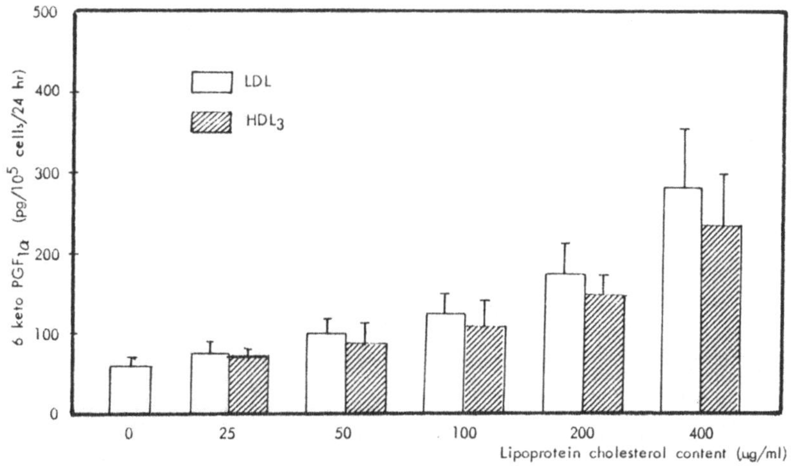

Figure 1 : Dose response effect of LDL and HDL_3 on 6 keto PGF_{1a} release by SMC
mean of 6 exp. $\pm$ SD

The effects of lipoproteins were compared as a function of the original plasma cholesterol concentrations. The composition of the different experimental pools were the following : Total cholesterol and triglycerides exhibit a wide range of concentrations ranging from 182 to 341 and 75 to 236 mg cholesterol/dl respectively. LDL and HDL vary from 114 to 237 (LDL) and 44 to 70 (HDL)mg cholesterol/dl. These values lead to ApoB/ApoA1 ratio ranging from 1.27 to 0.81 (table 1).

Table 1 - Lipids and apolipoproteins in plasma

Experimental Series	Plasma concentration (mg/dl)							
	Total Chol.	Tri-glyc	HDL Chol	LDL Chol.	VLDL Chol.	$ApoA_1$	ApoB	$ApoB/ApoA_1$
1.	182	120	44	114	24	118	99	0.84
2.	192	75	54	123	15	123	100	0.81
3.	209	249	42	117	50	121	114	0.94
4.	213	99	60	133	20	164	134	0.82
5.	240	155	53	156	31	146	146	1.00
6.	258	215	44	171	43	158	176	1.11
7.	302	179	51	215	36	128	162	1.27
8.	325	128	70	229	26	190	190	1.00
9.	341	236	57	237	47	154	186	1.21

Figure 2 demonstrates that for a given concentration of 150 µg cholesterol/ml, the ability of LDL to promote 6 keto PGF_{1a} release did not depend on the original plasma cholesterol concentrations. By contrast, HDL_3 extracted from plasma of hypercholesterolemic patients appeared less efficient than HDL_3 originated from normocholesterolemic pools. These results suggest that LDL and HDL_3

may have a differential mechanism of action in stimulating PGI_2 synthesis by SMC.

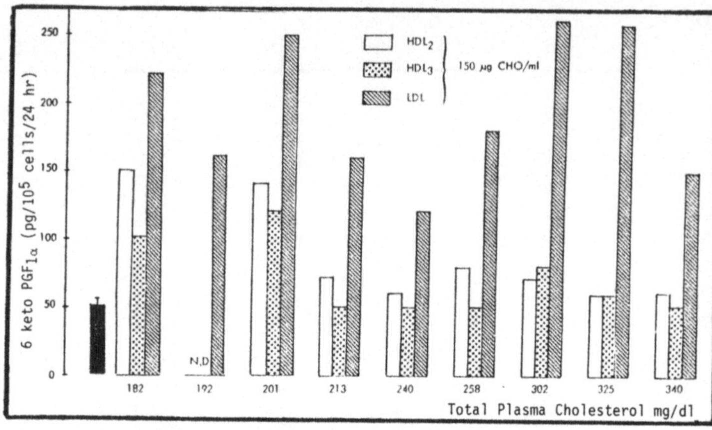

Figure 2 - Effects of HDL_3 and LDL from different plasma pools on 6 keto PGF_{1a} release by SMCs.

The effects of agents which modulate phospholipase A_2 activities in the SMC were evaluated. Human vascular SMC were incubated with LDL or HDL_3 at an identical protein concentration of 350 µg/ml in the presence of either activators (Bradykinin (BK) $5.10^{-6}M$, Angiotensin II (Ag II) $5.10^{-6}M$) or inhibitors (Dexamethazone (Dx) - $10^{-6}M$) of phospholipase A_2 activities. Figure 3 summarizes the results obtained showing that BK and Ag II produced a synergistic (BK) or additive (Ag II) effect on LDL stimulated PGI_2 secretion. By contrast, neither BK nor Ag II modify HDL_3 induced PGI_2 secretion.

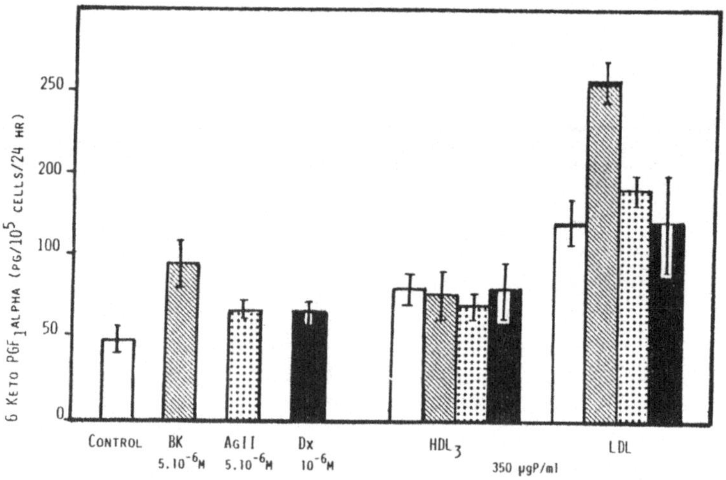

Figure 3 - Influence of lipoproteins on agonist induced 6 keto PGF_{1a} release by human SMC.

To investigate more directly the mechanism of PGI_2 induced generation by ApoB and ApoA containing lipoproteins, vSMCs were prelabelled with [14C] AA then incubated with different concentrations of either HDL_3 or LDL.

Figure 4 shows that after 24 hrs incubation increasing concentrations of HDL_3 and LDL dose-dependently stimulated the release of radioactivity from endogenous cellular stores. The media was removed, acidified extracted and analysed by TLC and RP-HPLC. Figure 5 indicate that total cyclooxygenase derived products (CO dP : prostaglandins + HHT + II HETE) dose dependently decreased in the incubation medium when concentration of LDL or HDL_3 increased. Accordingly, the accumulation of [14C] 6 keto PGF_{1a} also decreased with increasing lipoproteins concentrations. These results indicated some lipoprotein-dependent regulating mechanisms at the level of cyclooxygenase rather than PGI_2 synthetase activities since the relative concentration of 6 keto $PGF_{1}a$ increase by comparison of total cyclooxygenase derived products.

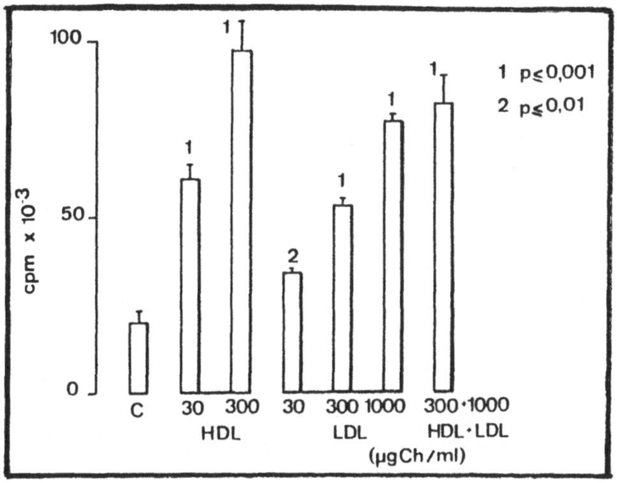

Figure 4 - Effects of ApoA and Apo B containing LPS on the release of radioactivity by [14C] AA prelabelled vSMC.

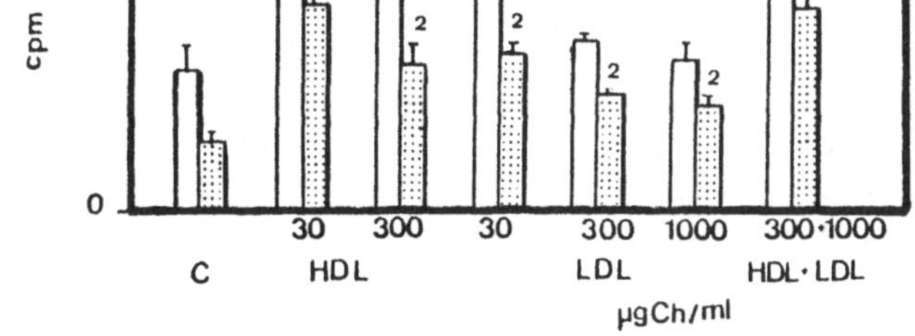

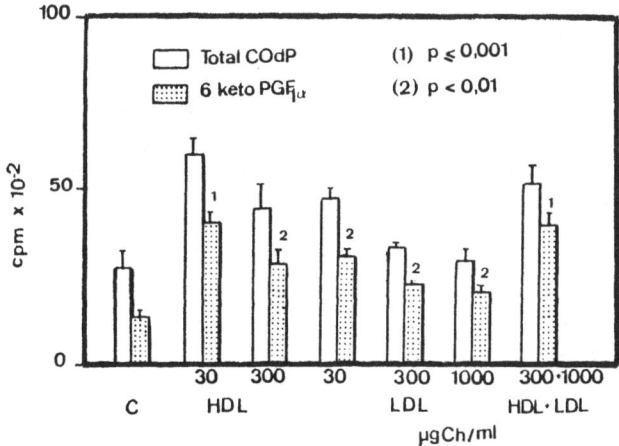

Figure 5 - Effects of ApoA and ApoB containing LPS on the synthesis of [14 C] cyclooxygenase derived products (COdP) and 6 keto PGF_{1a} from endogenous sources in [14C] AA prelabelled SMC.

The inhibitory capacities of HDL_3 and LDL on the oxydative pathways of AA metabolism was further confirmed by studying the transformation of [14C] AA exogenously added to vSMCs in the presence of growing amounts of VLDL, LDL and HDL. Figure 6 demonstrate that Apo A1 and ApoB containing lipoproteins dose dependently inhibited the transformation of [14C] AA. HDL appeared more efficient than LDL or vLDL in this process. Moreover, lipoproteins were also efficient in inhibiting HETEs formation (12 and 15 HETE) but at a lesser extend.

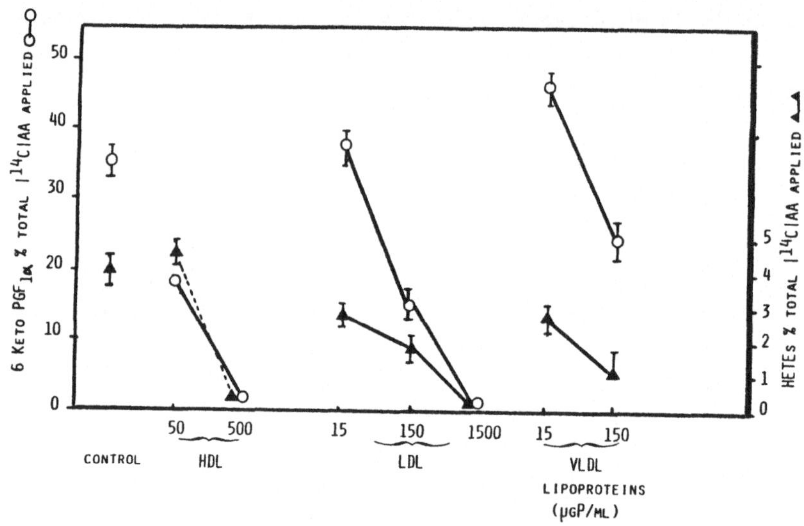

Figure 6 - Effects of HDL, LDL and VLDL on exogenous [14C] AA transformation by cultured vSMC.

DISCUSSION

These results indicate that plasma high and low density lipoproteins stimulate the formation of PGI_2 from vascular smooth muscle cells grown in tissue culture. Incubation of the cells with HDL_3 or LDL produce a dose dependent accumulation of 6 keto PGF_{1a} in the culture media. The magnitude of the release induced by HDL_3 was lesser than that produced by LDL. The in vitro stimulation of prostacyclin synthesis produced by LDL did not vary with the origin of the serum (normocholesterolemic vs hypercholesterolemic) by contrast to effect of HDL_3 which was lesser when the $ApoA_1$ containing lipoprotein come from hypercholesterolemic sera. These results suggest some differential stimulating mechanism between HDL and LDL. Both lipoproteins provide AA to intracellular pools available for cyclooxygenase but the differential effects of phospholipase A_2 modulation on HDL_3 and LDL induced 6 keto PGF_{1a} release suggest that these pools may be different.

In addition, the cyclooxygenase activity appears as a second enzymatic step of regulation of lipoprotein induced PGI_2 secretion. The inhibitory effect might depend on direct or indirect mechanism linked to fatty acid accumulation in cells incubated with HDL or LDL and needs further experiment. The fact that HDL_3 have been shown to produce 2 times higher accumulation of linoleic acid than LDL in the phospholipid fraction of SMCs [13] may partly explain the differences observed since linoleic acid may be a substrate for cyclooxygenase in vascular wall [19] and cultured SMC (Larrue et al unpublished results) leading to a competitive mechanism on cyclooxygenase activity. However, it is important to consider that a possible regulative mechanism in prostaglandin formation by vSMC under lipoproteins treatment is that the mobilisation by lipoproteins of high quantities of free AA may reduce directly or by the way of peroxyde desactivation, the cyclooxygenase activities and regulate the level of PGI_2 production.

While further experimentation is required in order to more directly demonstrate the mechanism underlying the modulatory action of LDL and HDL_3 on eicosanoid synthesis, our data provide evidence that both LDL and HDL contribute to the regulation of eicosanoid production in vascular SMC and suggest that this process involved both the lipid and the apoprotein moieties.

REFERENCES

1. S. Moncada, R. Gryglewski, S. Bunting and J.R. Vane. An enzyme isolated from arteries transforms prostaglandin endoperoxides to an unstable substance that inhibits platelet aggregation. Nature, 263 : 663 (1976).

2. B.B. Weksler, C.W. Ley and E.A. Jaffe. Stimulation of endothelial cell prostacyclin production by thrombin, trypsin and the Ionophore A23187. J. Clin. Invest. 62 : 923 (1978).

3. N.L. Baenzinger, P.R. Becherer and P.W. Majerus. Characterization of prostacyclin synthesis in cultured human arterial smooth muscle cells, venous endothelial cells and skin fibroblasts. Cell, 16 : 967 (1979).

4. P.A. Ramwell, E. Leouey, A. Sintelos. Regulation of the arachidonic acid cascade. Biol. Reprod. 16 : 70 (1977).

5. J. Larrue, B. Dorian, J. Bonnet and H. Bricaud. Eicosanoid production by vascular smooth muscle cells in culture. In "Vascular Smooth Muscle in tissue culture". R. Campbell and J. Campbell Ed. Acad. Press., 2 (1987).

6. V.P. Dole, A.T. James, J.P.W. Webb, M.A. Rizack, M.F. Sturman. The fatty acid and patterns of plasma lipids during alimentary lipemia. J. Clin. Invest. 38 : 1554 (1959).

7. J. Bietz and W. Forster. Influence of human low density and high density lipoprotein cholesterol on the in vitro prostaglandin I_2-synthetase activity. Biochim. Biophys. Acta, 620 : 352 (1980).

8. J. Bietz, M. Panse, S. Fischer, C. Mora and W. Forster. Inhibition of prostaglandin I_2 (PGI_2) formation by LDL-cholesterol or LDL-peroxides. Prostaglandins, 26 : 885 (1983).

9. A. Szczeklik, R.J. Gryglewski, B. Domagala, A. Zmuda, J. Hartwich, E. Wozny, M. Grzywacz, J. Madej and T. Gryglewska. Serum lipoproteins, lipid peroxides and prostacyclin biosynthesis in patients with coronary heart disease. Prostaglandins 22 : 795 (1981).

10. L. Fleisher, A.R. Tall, L. Witte, R. Miller and P. Cannon. Stimulation of arterial endothelial cell prostacyclin synthesis by high density lipoproteins. J. Biol. Chem. 257 : 6653 (1982).

11. K.B. Pomerantz, A R. Tall, S.J. Feinmark and P.J. Cannon. Stimulation of vascular smooth muscle cell prostacyclin and prostaglandin E synthesis by plasma high and low density lipoproteins. Circ. Res. 84 : 554 (1984).

12. J. Larrue, J. Crockett, D. Daret, J. Demond-Henri, H. Bricaud. Modulation of vascular smooth muscle cells, prostacyclin synthesis by plasma lipoproteins . Eur. Heart J., 5 (S1) : 159, (1984).

13. A.A. Spector, A.M. Scanu, T.L. Kaduce, P.H. Figard, G.M. Fless, R.L. Czevionke. Effect of human plasma lipoproteins on prostacyclin production by cultured endothelial cells. J. Lipid Res. 26 : 288 (1985).

14. A.J. Habenicht, H.A. Dresel, H. Goerg, J.A. Weber, M. Stoehr, J.A. Glomset, R. Ross and G. Schettler. Low density lipoprotein receptor dependant prostaglandin synthesis in Swiss 3T3 cells stimulated by platelet derived growth factor. Proc. Natl. Acad. Sci., 83 : 1344 (1986).

15. R.J. Havel, H. Eder and J. Bragdon. The distribution and chemical preparation of ultracentrifugally separated lipoproteins in human serum. J. Clin. Invest. 34 : 1345 (1958).

16. A.V. Nichols, E.L. Gong, P.J. Blanche, T.M. Forte and D.W. Anderson. Effects of guanidine hydrochloride on human plasma high density lipoproteins. Biochim. Biophys. Acta,446 : 226 (1976).

17. O. Lowry, N. Rosebrough, A. Farr and J. Randall. Protein measurement with the Folin phenol reagent. J. Biol. Chem. 193 : 265 (1951).

18. A. Zlatkis and B. Zak. Study of a new cholesterol reagent. Anal. Biochem. 29 : 143 (1969).

19. C.D. Funk and W.S Powell. Metabolism of linoleic acid by prostaglandin endoperoxide synthetase from adult and fetal blood vessel. Biochim. Biophys. Acta, 754 : 57 , (1983).

ACKNOWLEDGEMENTS

This work has been realized in part through a collaborative Research with Drs N. Perova, V. Metelskaya and Y. Tchakov from the Institute of Preventive Cardiology, Department of Biochemistry, Cardiology Research Center, Moscow (USSR). This work was supported in part by Grants from Fournier Laboratories and the University of Bordeaux II.

EICOSANOID SYNTHESIS IN PLATELET-DERIVED GROWTH FACTOR-STIMULATED FIBROBLASTS

Andreas J.R. Habenicht, Peter Salbach, and Matthias Goerig

University of Heidelberg, Medical School

Bergheimer Str. 58, D-6900 Heidelberg, FRG

Morphological studies have shown that the "proliferative lesion" - the early stage of arteriosclerosis - consists of endothelial cells, differentiating macrophages, activated T-lymphocytes, and proliferating intimal smooth muscle cells (1,2). These cells - under appropriate conditions - synthesize and release many biologically potent molecules including growth factors (1), arachidonic acid (AA) metabolites (3,6), interleukins (12), colony stimulating factors (13), and others (20).

Among the biologically potent molecules released by the cells that have been identified in proliferative lesions are metabolites of AA- termed eicosanoids (9,10). Each of the cell types has a characteristic eicosanoid synthesis pattern. For example, arterial smooth muscle cells preferentially synthesize prostaglandin (PG) E_2, macrophages preferentially form thromboxane A_2 and leukotriene B_4 (3), and the major product of AA of endothelial cells is prostacyclin (PGI_2). On the other hand, each of the eicosanoids have distinct - and in some instances opposite - biological activities on their target cells: nM concentrations of PGE_2 suppress the activation of T-lymphocytes (18), similarly low concentrations of PGI_2 exert strong antiaggregatory effects on platelets and influence cholesterol homeostasis in smooth muscle cells (11; see also article by D. Hajjar, this volume), and nM concentrations of leukotriene B_4 have strong chemotactic effects towards leukocytes (17). Therefore, the fact that cells found in the arterial wall both synthesize and respond to eicosanoids might indicate that these molecules play a role in the modulation of the arterial wall in vivo.

One of the molecules that has been suggested to induce the proliferation of smooth muscle cells in the arterial wall is platelet-derived growth factor (PDGF) (1). More recently it has been shown that PDGF-like molecules are formed and secreted by both macrophages and endothelial cells (1). These findings increased the interest in the role of PDGF in atherogenesis since it now became conceivable that PDGF might be formed by macrophages and endothelial cells and potentially has effects on cell growth within the arterial wall in vivo.

We have been interested in changes in lipid metabolism induced by PDGF (4-8). We found that PDGF activates a phosphatidylinositol specific phospholipase C and induces release of AA from cellular phospholipids (5). In addition the growth factor activates the key enzyme of PG, prostacyclin (PGI_2

and thromboxane biosynthesis, the PGH synthase (6). This dual effect of PDGF results in greatly stimulated PG synthesis (6). However, stimulation of AA release and activation of PGH synthase is only part of the complicated network of changes in lipid metabolism induced by the growth factor. Mammalian cells must take up unsaturated fatty acids from extracellular sources before endogenous eicosanoid synthesis can occur. Therefore we and others have hypothesized that physiological carriers of unsaturated fatty acids, the plasma lipoproteins, might be involved in the regulation of eicosanoid synthesis (8,14,15). We have reported previously that low density lipoprotein (LDL) stimulates PG synthesis in PDGF stimulated Swiss 3T3 cells through a process that seems to depend on the LDL pathway (8).

The conclusion that the LDL pathway was involved in the stimulatory effect of LDL on PG synthesis was based on the following lines of evidence:
1) Modification of the lysine residues of Apolipoprotein B by acetylation prevented the stimulatory effect of LDL on PG synthesis; 2) PDGF stimulated expression of functional LDL receptors; 3) PDGF stimulated uptake of LDL;
4) chloroquine, an inhibitor of lysosomal degradation, inhibited the stimulatory effect of native LDL. These previous studies raised an important question: does LDL stimulate PG synthesis by providing AA to the PGH synthase pathway?

To answer this question, we reconstituted human plasma LDL with different species of cholesterylester, namely cholesteryloleate and cholesterylarachidonate (19). The reconstitution protocol results in LDL particles with an altered fatty acid composition but unaltered biological activity (i.e. the reconstituted LDL species retain the ability to bind to the LDL receptor and to regulate HMG-CoA reductase activity, ACAT activity, and the expression of the LDL receptor) (19). Gas liquid chromatography was used to show that the reconstitution protocol drastically altered the fatty acid composition of the native LDL (Table 1). PDGF stimulated cells were then incubated with the two reconstituted LDL species and the amounts of PGE_2 that were formed in response to PDGF and LDL were determined. In addition we preincubated parallel cultures with chloroquine in order to test the possibility that the reconstituted LDL species enter the cells through the LDL pathway. We found that LDL that had been reconstituted with cholesterylarachidonate was significantly more active than LDL reconstituted with cholesteryloleate. However, the cholesteryloleate reconstituted LDL was still biologically active (Table 2). It should be born in mind that the reconstitution protocol does not significantly alter the phospholipid composition of the LDL and that the phospholipids of LDL contain approximately 50 % of the total AA content (results not shown). The experiment shown in Table 2 also demonstrates that chloroquine blocks both the effect of native LDL and reconstituted LDL on PGE_2 synthesis in PDGF stimulated cells.

Taken together these preliminary results lend further support to the hypothesis that LDL stimulates PG synthesis by providing AA to the PGH synthase pathway and that this effect of LDL depends on the LDL pathway.

Table 1. Total fatty acid composition (major fatty acid species) of native LDL and LDL reconstituted with cholesteryloleate (chol. oleate-LDL) or cholesterylarachidonate (chol. arach.-LDL).

Native human plasma LDL was prepared as described (8) and reconstituted according to Krieger et al. (19) using different cholesterylester species as indicated in the Table. Total fatty acid analysis was done by by gas liquid chromatography.

Fatty acid species	native LDL	Chol. oleate-LDL (ug fatty acid/ug protein)	Chol. arach.-LDL
Palmitic acid (16:0)	0.28	0.22	0.33
Oleic acid (18:1)	0.21	0.57	0.05
Linoleic acid (18:2)	0.60	0.13	0.14
Arachidonic acid (20:4)	0.11	0.05	0.83

Table 2. Effects of reconstituted LDL species on PG synthesis in PDGF stimulated Swiss 3T3 cells.

Experiment was performed (8) and LDL was reconstituted as described (19). 20 ug LDL-protein/ml of each LDL species was used. 12 h after addition of PDGF, parallel cultures were incubated with LDL for 6 h and the concentration of PGE_2 was determined thereafter as described (6). Numbers represent means of duplicate dishes ± S.D.

	Prostaglandin E_2 ($ng/10^6$ cells)
PDGF	64 ± 8
PDGF + native LDL	109 ± 7
PDGF + Chol.oleate-LDL	97 ± 18
PDGF + Chol.arach.-LDL	173 ± 19
PDGF + native LDL + chloro	52 ± 10
PDGF + Chol.oleate-LDL + chloro	46 ± 16
PDGF + Chol.arach.-LDL + chloro	58 ± 13

Acknowledgements: This work was supported by the Deutsche Forschungsgemeinschaft and the Forschungsrat Rauchen und Gesundheit.

References

1. Ross, R., 1986, The pathogenesis of arteriosclerosis - an update. N. Engl. J. Med. 314:488.

2. Jonasson, L., Holm, J., Skalli, O., Bondjers, G., and Hansson, G.K., 1986, Regional accumulation of T cells, macrophages, and smooth muscle cells in the human atherosclerotic plaque. Arteriosclerosis 6: 131

3. Goerig, M., Habenicht, A.J.R., Heitz, R., Zeh, W., Katus, H., Kommerell, B., Ziegler, R., and Glomset, J.A., 1987, Sn-1,2-diacylglycerols and phorbol diesters stimulate thromboxane synthesis by de novo synthesis of prostaglandin H synthase in human promyelocytic leukemia cells. J. Clin. Invest. 79:903.

4. Habenicht, A.J.R., Glomset, J.A., and Ross, R., 1980, Relation of cholesterol and mevalonic acid metabolism to the cell cycle in smooth muscle and Swiss 3T3 cells stimulated to divide by platelet-derived growth factor. J. Biol. Chem. 255:5134.

5. Habenicht, A.J.R., Glomset, J.A., King, W.C., Nist, C., Mitchell, C.D., and Ross, R., 1981, Early changes in phosphatidylinositol and arachidonic acid metabolism in Swiss 3T3 cells stimulated to divide by plateled-derived growth factor. J. Biol. Chem. 256: 12329.

6. Habenicht, A.J.R., Goerig, M., Grulich, J., Rothe, D., Gronwald, R., Loth, U., Schettler, G., Kommerell, B., and Ross, R., 1985, Human platelet-derived growth factor stimulates prostaglandin synthesis by activation and rapid de novo synthesis of cyclooxygenase. J. Clin. Invest. 75: 1381.

7. Habenicht, A.J.R., Glomset, J.A., Goerig, M., Grulich, J., Rothe, D., Gronwald, R., Schettler, G., and Kommerell, B., 1985, Cell cycle dependent changes in arachidonic acid and glycerol metabolism in Swiss 3T3 cells stimulated by platelet-derived growth factor. J. Biol. Chem. 260: 1370.

8. Habenicht, A.J.R., Dresel, H.A., Goerig, M., Weber, J.A., Stoehr, M., Glomset, J.A., Ross, R., and Schettler, G., 1986, Low density lipoprotein receptor-dependent prostaglandin synthesis in Swiss 3T3 cells stimulated by platelet-derived growth factor. Proc. Natl. Acad. Sci. USA 83: 1344.

9. Majerus, P.W., 1983, Arachidonate metabolism in vascular disorders. J. Clin. Invest. 72: 1521.

10. Needleman, P., Turk, J., Jakschik, B.A., Morrison, A.R., and Lewkowith, J.B., 1986, Arachidonic acid metabolism. In: Ann. Rev. Biochem. 55: 69-102.

11. Hajjar, D.P., Weksler, B.B., Falcone, D.J., Hefton, J.M., Goldman, K.T., and Minick, C.R., 1982, Prostacyclin modulates cholesterylester hydrolytic activity by its effects on cyclic adenosine monophosphate in rabbit aortic smooth muscle cells.
J. Clin. Invest. 70: 479.

12. Dinarello, C.A., and Mier, J.W., 1987, Lymphokines.
N. Engl. J. Med. 317: 940.

13. Clark, S.C., and Kamen, R., 1987, The human hematopoetic colony-stimulating factors.
Science 236: 1229.

14. Pomerantz, K.B., Fleisher, L.N., Tall, A.R., and Cannon, P.J., 1985, Enrichment of endothelial cell arachidonate by lipid transfer from high density lipoproteins: relationship to prostaglandin I_2 synthesis.
J. Lipid Res. 26: 1269.

15. Fleisher, L., Tall, A.R., Witte, L., Miller, R., and Cannon, P.J., 1982, Stimulation of arterial endothelial prostacyclin synthesis by high density lipoproteins.
J. Biol. Chem. 257: 6653.

16. Sporn, M.B., and Roberts, A.B., 1987, Peptide growth factors, tissue repair, and cancer.
J. Clin. Invest. 78: 329.

17. Ford-Hutchinson, A.W., Bray, M.A., Doig, M.V., Shipley, M.E., and Smith, M.J.H., 1980, Leukotriene B_4, a potent chemokinetic and aggregating substance released from polymorphonuclear leukocytes.
Nature 286: 264.

18. Chouaib, S., Robb, R.J., Welte, K., and Dupont, B., 1987, Analysis of prostaglandin E_2 effect on T- lymphocyte activation.
J. Clin. Invest. 80: 333.

19. Krieger, M., Brown, M.S., Faust, J.R., and Goldstein, J.L., 1978, Replacement of endogenous cholesteryl esters of low density lipoprotein with exogenous cholesteryl linoleate.
J. Biol. Chem. 253: 4093.

20. Nathan, C.F., 1987, Secretory products of macrophages.
J. Clin. Invest. 79: 319.

GENETIC HDL DEFICIENCY STATES

J.M. Ordovas, D.C. King, and E.J. Schaefer

Lipid Metabolism Laboratory, USDA Human Nutrition Research Center on Aging at Tufts University
Boston, MA 02111 U.S.A.

INTRODUCTION

Barr et al (1) were the first investigators to document a relationship between decreased high density lipoprotein (HDL) cholesterol concentrations and coronary artery disease (CAD). Subsequently in 1966 Gofman et al. in a prospective study (2) found a similar association of plasma HDL levels with disease. This concept was not properly recognized until Miller and Miller published their work in 1975 (3). In 1977 two more studies, Framingham (4) and Tromso (5) provided further evidence of the inverse relationship between levels of HDL in plasma and risk of CAD.

HDL particles are comprised of approximately 50% protein, 25% phospholipid, 20% cholesterol and 5% triglyceride (weight %) (6). Two major subclasses are identified based on analytical ultracentrifugation: HDL2 (d 1.063-1.125 g/ml) and HDL3 (1.125-1.200 g/ml) (7). Utilizing analytical ultracentrifugation and polyacrylamide gel electrophoresis further heterogeneity is noted in each one of these fractions (8). This heterogeneity may be due to the various pathways of HDL synthesis. Constituents of HDL are derived by direct synthesis of HDL by the liver or the intestine or, as a result of chylomicron and very low density lipoprotein (VLDL) metabolism. HDL can be further modified by uptake of lipids, specifically free cholesterol, from peripheral cells. Both the liver and the kidney appear to play an important role in the catabolism of HDL particles (9). Apolipoproteins (apo) A-I and A-II are the major proteins of HDL. Other minor constituents found in this density region are apoB, apoE, apoC-I, apoC-II, apoC-III, apoLp(a), apoD, apoE and apoG. Besides the possible structural role of these apolipoproteins, apoA-I and apoC-I have been reported to activate LCAT, while apoA-II enhances hepatic lipase (HL) activity (10-12). A specific receptor for apoA-I and apoA-II has been reported on the surface of skin fibroblasts and arterial smooth muscle cells (13,14). ApoC-II activates lipoprotein lipase, while apoC-III inhibits hepatic chylomicron remnant uptake (15,16). ApoE's presence in HDL may be associated with the removal of these particles by the liver. The role of other minor components in HDL is unknown.

It is our purpose to review current knowledge of the clinical, genetic, and biochemical features of familial HDL deficiency syndromes, occuring in the absence of other disorders, such as severe hypertriglyceridemia or lecithin:cholesterol acyltransferase deficiency. Disease entities falling within this category include Tangier Disease (17,18). HDL deficiency with planas xanthomas (19), apoA-I Milano (20), apolipoprotein A-I and C-III deficiency (21), apolipoprotein A-I, C-III and A-IV deficiency (22), and familial hypoalphalipoproteinemia (23).

TANGIER DISEASE

Tangier disease was the first HDL deficiency to be described (17,18). The original two probands for this disease were siblings, aged 5 and 6 years, from Tangier Island, Virginia, in the southern Chesapeake Bay area of the United States. They had enlarged yellow-orange tonsils, mild hepatosplenomegaly, and lymphadenopathy. Both subjects had their tonsils removed, and pathology showed macrophages containing increased amounts of cholesterol ester. Similar findings have been noted in the bone marrow, skin, rectal mucosa, liver, spleen, lymph nodes, omentum and conjuctiva of Tangier patients. Both probands developed mild transient peripheral neuropathy during adolescence, and also mild diffuse corneal opacification. No xanthomas have been observed, and they do not have evidence of CAD. At least 20 other kindreds have been described inthe original probands (24,25). No CAD has been detected in homozygotes under age 40 years. However premature CAD has been detected in these subjects under the age of 60 (24). Homozygotes for Tangier disease have mild hypertriglyceridemia, increased VLDL cholesterol, decreased LDL cholesterol, and marked HDL deficiency. Plasma HDL cholesterol, apoA-I and apoA-II values are 4%, 1% and 9% of normal, while other apolipoproteins are present in normal or slightly reduced concentrations (26,27). Heterozygotes for this disorder have plasma HDL cholesterol, apoA-I and apoA-II that are 50% of normal.

The molecular basis of this defect remains to be elucidated. In these subjects most of the plasma apoA-I is found in the d 1.21 fraction, while apoA-II is found mainly in the region of density 1.063, suggesting that apoA-I and apoA-II are not on the same particle. Moreover, Tangier homozygotes have an increased proportion of proapoA-I in their plasma (28). The sequence of the apoA-I gene in one of the homozygotes did not reveal any mutation altering the primary structure of the protein (29). It has been reported that these subjects have hypercatabolism of HDL components, especially apoA-I (30,31). Abnormalities in LDL, LCAT, or HL have not been reported. It has been suggested that the defect is due to an abnormality in the uptake and processing of HDL by macrophages (32).

HDL DEFICIENCY WITH PLANAR XANTHOMAS

The proband for this disorder was a 48 year old white Swedish woman with a history of widespread yellow skin discoloration around the eyes and in the groin since early childhood (19). Both upper eyelids were thickened and infiltrated with small firm nodules. In addition she had a history of angina. Histochemical tests revealed intracellular histiocytic deposition of free and esterified cholesterol in skin lesions and rectal mucosa. Moderate hepatomegaly was noted in this patient, and her tonsils were normal. She had elevated plasma triglyceride and VLDL cholesterol values, normal LDL cholesterol level, and HDL cholesterol level which was 6% of normal. Her plasma apoA-I and apoA-II levels were 1% and 19% of normal, respectively (19). All the other apolipoproteins were present in normal concentrations. The biochemical defect underlying this disorder remains to be defined.

APOA-I MILANO

The proband was a 49 year old Italian man originally from the region of Lake Garda in the northern part of Italy (20). He did not have evidence of coronary artery disease, and he was heterozygous for this abnormality. No homozygotes have been reported, and the affected individuals are characterized by moderate hypertriglyceridemia, and HDL cholesterol levels that were 20% of normal. On isoelectric focusing apoA-I bands were present in both normal and abnormal position due to the heterozygosity of all the subjects analyzed. Sequence analysis of apoA-I Milano indicates the presence of cysteine instead of arginine at residue 173 (33). Kinetic data indicate that apoA-I Milano is catabolized at a faster rate than normal apoA-I, and that this hypercatabolism accounts for the deficiency of apoA-I observed in these subjects. No premature CAD has been reported in any of the carriers of this mutation (34).

FAMIALIAL APOLIPOPROTEIN A-I AND C-III DEFICIENCY

The probands for this disorder were two sisters from Detroit, Michigan (USA) (21). They had severe premature CAD in their early thirties, mild diffuse corneal opacification and yellow-orange plaques on the trunk, neck, eye lids, chest, arms and back. Tonsils in these patients were normal. The two homozygous had marked HDL deficiency, and decreased levels of apoA-I were detectable and apoC-III was undetectable in plasma. ApoA-II was 50% of normal in homozygotes. All other apolipoproteins were present. Infusion of normal plasma in the homozygotes yielded a plasma half life for HDL protein of 2-3 days, in contrast with the hypercatabolism of the HDL constituents in Tangier disease. The molecular defect in this disorder has been recently characterized (35), and is due to a DNA rearrangement affecting the adjacent apoA-I and apoC-III genes located on the long arm of chromosome 11. This rearrangement is due to a DNA inversion containing portions of the 3' end of both genes, more specifically between the fourth exon of the apoA-I gene, and the first intron of the apoC-III gene. This inversion results in the absence of normal apoA-I and apoC-III mRNA and consequently the lack of both proteins in plasma. Heterozygotes for this disorder had apoA-I and apoC-III levels in plasma that were 50% of normal. The defect in this condition, therefore is due to the inability to synthesize both apoA-I and apoC-III.

FAMILIAL APOLIPOPROTEIN A-I, C-III AND A-IV DEFICIENCY

This disorder was originally described as plasma apoA-I absence (36), and later as familial apoA-I and C-III deficiency (variant II) (22). The proband was a 45 year old white female from Alabama (USA) who developed angina at age 42, and died shortly after coronary artery bypass surgery. She had mild corneal opacification, but no signs of xanthomas or abnormal tonsils. Of the 38 members of this kindred that were tested, 18 were heterozygotes. None of them developed premature coronary artery disease before the age of 40. Two of them developed CAD before 60 years and one of them died at 56 of a myocardial infarction. The homozygous proband had marked HDL deficiency and undetectable plasma levels of apoA-I and apoC-III, while apoA-II values were 11% of normal. Heterozygotes had HDL cholesterol, apoA-I, apoC-III and apoA-IV levels that were 54%, 53% 83% and 54% of normal, respectively. In addition vitamin E and linoleic acid deficiencies were also noted in the proband. A series of restriction fragment length polymorphisms (RFLP) 5' of the apoA-I gene have been identified in heterozygotes for this disorder, but not in the unaffected members of this kindred or in the general population. Further analysis of the abnormal allele indicates that the molecular defect is due to a complete deletion of the entire apoA-I, apoC-III and apoA-IV gene complex, resulting in an inability to synthesize these three apolipoproteins.

FAMILIAL HYPOALPHALIPOPROTEINEMIA

Familial hypoalphalipoproteinemia appears to be the most common genetic disorder associated with HDL deficiency (HDL cholesterol levels less than 10th percentile of normal values). This disorder has an autosomal dominant mode of inheritance, and it has been associated with premature CAD as well as stroke (23). No xanthomas, ocular abnormalities or enlargement of the liver or spleen, or neurological disorders have been noted. Kinetic data suggest that these subjects have decreased synthesis of HDL apoA-I (37). The molecular defect is not known. We have recently found an RFLP that is more common in probands of kindreds with familial hypoalphalipoproteinemia than in the normal population. This RFLP is detected, after restriction analysis of genomic DNA with PstI, on the 3' region of the apoA-I gene (38). Further family studies to more fully understand this finding are in progress.

CONCLUSION

With the exception of familial hypoalphalipoproteinemia which may be the most common genetic defect associated with premature CAD, all the HDL deficient states reported are extremely rare. However, the findings in these disorders indicate that there is an independent role for HDL deficiency in the pathogenesis of CAD, particularly in kindreds with decreased production of HDL apoA-I.

REFERENCES

1. D.P. Barr, E.M. Russ, and H.A. Eder (1951) Amer. J. Med. 11:480-493.
2. J.W. Gofman, W. Young, and R. Tandy (1966) Circulation 34:679-697.
3. G.J. Miller and N.F. Miller (1975) Lancet 1:16-20.
4. T. Gordon, W.P. Castelli, M.C. Hjortland et al. (1977) Amer. J. Med. 62:707-714.
5. N.E. Miller, O.H. Forde, D.S. Thelle et al. (1977) Lancet 1:965-968.
6. R.J. Havel, H.A. Eder, J.H. Bragdon (1955) J. Clin. Invest. 34:1345-1353.
7. D.W. Anderson, A.V. Nichols, S.S. Pan, and F.T. Lindgren (1978) Atherosclerosis 29:161-179.
8. A.V. Nichols, P.J. Blanche, and E.L. Gong (1983) in: "CRC Handbook of Electrophoresis", Vol. III, Lewis, L. and Opplt, J eds., CRC Press, Boca Raton, pp. 29-47.
9. C.K. Glass, R.C. Pittman, G.A. Keller, and D. Steinberg (1983) J. Biol. Chem. 258:7161-7167.
10. C.J. Fielding, V.G. Shore, and P.E. Fielding (1972) Biochem. Biophys. Res. Commun. 46:1493-1498.
11. A.K. Soutar, G.W. Garner, G.N. Baker, et al. (1975) Biochemistry 14:3057-3064.
12. C.E. Jahn, J.O. Osborne, E.J. Schaefer, and H.B. Brewer Jr. (1983) Eur. J. Biochem. 131:25-29.
13. R. Biesbroeck, J.F. Oram, J.J. Albers, and E.L. Bierman (1983) J. Clin. Invest. 71:525-539.
14. E.L. Bierman, E.A. Brinton, M. Oppenheimer, and J.F. Oram (1986) in: Atherosclerosis VII. N.J. Fidge and P.J. Nestel, editors Elsevier pp. 195-197.
15. J.C. LaRosa, R.I. Levy, P.N. Herbert, S.E. Lux, and D.S. Fredrickson (1970) Biochem. Biophys. Res. Commun. 41:57-62.
16. F. Shelburne, J. Hanks, W. Myers, and S. Quarfordt (1980) J. Clin. Invest. 65:652-658.
17. D.S. Fredrickson, P.H. Altrocchi, and L.C. Avioli (1961) Ann. Intern. Med. 55:1016-1031.

18. D.S. Fredrickson (1964) J. Clin. Invest. 43:228-236.
19. A. Gustafson, W. McConathy, P.Alaupovic, M.D. Curry, and B. Persson (1979) Scand. J. Clin. Lab. Invest. 39:377-388.
20. G. Franceschini, C.R. Sirtori, A. Capurso, K.H. Weisgraber, and R.W. Mahley (1980) J. Clin. Invest. 66:982-990.
21. R.A. Norum, J.B. Lakier, S. Goldstein, A. Angel, R.B. Goldberg, W.D. Black, D.K. Noffze, P.J. Dolphin, J. Edelglass, D.D. Borograd, and P. Alaupovic (1983) N. Engl. J. Med. 306:1513-1519.
22. E.J. Schaefer, J.M. Ordovas, S. Law, G. Ghiselli, M.L. Kashyap, L.S. Srivastava, W.H. Heaton, J.J. Albers, W.E. Conners, and H.B. Brewer Jr. (1985) J. Lipid Res. 26:1089-1101.
23. C. Vergani, and A. Bettale (1981) Clin. Chim. Acta 114:45-52.
24. E.J. Schaefer, L.A. Zech, D.S. Schwartz, and H.B. Brewer Jr. (1980) Ann. Inter. Med. 93:261-266.
25. P.N. Herbert, G. Assmann, A.M. Gotto Jr., D.S. Fredrickson (1982) in: The Metabolic Basis of Inherited Disease. J.B. Wyngaarden, D.S. Fredrickson, J. Goldstein, and M. Brown eds. 5th ed. McGraw-Hill, New York. pp. 589-621.
26. L.O. Henderson, P.N. Herbert, D.S. Fredrickson, and R.J. Heinen (1978) Metabolism 27:165-174.
27. P. Alaupovic, E.J. Schaefer, W.J. McConathy, J.S. Fesmire, and H.B. Brewer Jr. (1981) Metabolism 30:809-816
28. V.I. Zannis, A.M. Lees, R.S. Lees and J.L. Breslow (1982) J. Biol. Chem. 257:4978-4986.
29. V.I. Zannis, J.L. Breslow, J.M. Ordovas, S.K. Karathanasis (1984) Circulation 70:29.
30. E.J. Schaefer, D.W. Anderson, L.A. Zech, F.T. Lindgren, T.J. Bronzert, E.A. Rubalcaba, and H.B. Brewer Jr. (1978) J. Lipid Res. 22:217-218.
31. E.J. Schaefer, C.B. Blum, R.I. Levy, L.L. Jenkins, P. Alaupovic, D.M. Foster, and H.B. Brewer Jr. (1978) N. Engl. J. Med. 299: 905-910.
32. G. Schmitz, G. Assmann, B. Brennhausen, and H.J. Schaefer (1987) J. Lipid Res. 28:87-99.
33. K.H. Weisgraber, T.P. Bersot, R.W. Mahley, G. Franceschini, C.R. Sirtori (1980) J. Clin. Invest. 66:901-909.
34. G. Franceschini, G. Vecchio, G. Gianfranceschi, D. Magani, and C. Sirtori (1985) J. Biol. Chem. 260:16321-16325.
35. S.K. Karathanasis, E. Ferris, and I.A. Haddad (1987) Proc. Natl. Acad. Sci. USA 84:7198-7202.
36. E.J. Schaefer, W.H. Heaton, M.G. Wetzel, H.B. Brewer Jr. (1982) Arteriosclerosis 2:16-26.
37. N. Le, and H.N. Ginsberg (1987) Circulation 76:1294.
38. J.M. Ordovas, E.J. Schaefer, D. Salem, R.H. Ward, C.J. Glueck, C. Vergani, P.W.F. Wilson, and S.K. Karathanasis (1986) N. Engl. J. Med. 314:671-677.

APOLIPOPROTEIN A-I: DEFICIENCY IN TANGIER DISEASE

Marie-France Dumon, Monique Clerc and Michel Clerc

Laboratoire de Biochimie Médicale A, Université de Bordeaux II

146, rue Léo Saignat, 33076 - Bordeaux Cedex

INTRODUCTION

The near absence of apolipoprotein A-I (apo A-I) characterizes Tangier disease. However, no unequivocal evidence of a molecular defect in apo A-I or apo A-I gene has been established (1-13). It could signify that the molecular defect is not necessarily the same in every patient.

Nevertheless none of the 29 cases reported in the literature seems affected with a premature atherosclerosis (14, 15). In spite of the apo A-I related HDL deficiency which contrasts with the classical association between cardiovascular disease and HDL deficiency (16), the absence of atherosclerosis in Tangier homozygotes is quite consistent with the lack of cholesterol deposition in the arterial wall and on the contrary the massive storage of cholesteryl esters in histiocytes (17).

At present it is generally admitted that the apo A-I deficiency seems better to correspond to an enhanced catabolic rate of Tangier HDL (HDL_T) than to a lowered synthesis. In some cases indeed a decreased plasma residence time of HDL (6) as well as the predominance of pro-forms of Apo A-I (4, 18, 19) or a 6 to 10 fold enhanced HDL binding to Tangier monocytes secondly degraded in the lysosomal compartment of these monocytes (20) have been shown.

We here report upon studies of the plasma lipoproteins and HDL uptake by monocyte-derived macrophages from a control and a Tangier homozygous patient discovered two years ago in our laboratory.

MATERIALS AND METHODS

Case report

The patient, a 40 years old man, has been tonsillectomized at the end of his first year and does not present any other major affection. Two years ago a pruritus anal leads to examine the digestive tract and revealed the presence of small yellowish-gray plaques on the intestinal mucosa (containing numerous foamy histiocytes). The patient presented also splenomegaly, recurrent multineuropathy and flocculent infiltration throughout the corneal stroma (evident only by Slit-lamp), tonsil area were enlarged despite the previous tonsillectomy but no lymphadenopathy or hepatomegaly was detected. The electrophoresis of plasma lipoproteins showed a severe deficiency of HDL, well correlated to the low levels of HDL-Cholesterol

(0.05 mmol/L) and of total cholesterol (1.34 mmol/l) whereas triglyceridemia was slightly enhanced (2.62 mmol/L). These signs, completely different of those observed in Fish-eye syndrom and the lack of the main secondary causes of HDL deficiency (hepatic, digestive, renal, thyroïd), suggest the diagnosis of Tangier disease, according to the early description of Fredickson (17). They prompted us to examine his family : two sons, a girl and his wife. The latter and the girl being normals the two sons were shown heterozygous, presenting lowered values of apo A-I (0,6 g/l) and HDL-Cholesterol (respectively 12 % and 15 %) as well as decreased percentages of HDL in discontinuous polyacrylamide gel electrophoresis (respectively 11 % and 16 % for 35 % to 42 % in normals).

Materials

Serum anti-alpha and beta lipoproteins were purchased from Behring (Mannheim, G.F.R.) ; monospecific serum anti apo A-I, A-II, B from IMMUNO-A.G. (Vienne, Austria) ; anti apo C-II, C-III and E from Daiichi (Tokyo, Japan) ; agarose from I.B.F. (Clichy, France); ampholytes, molecular weight markers and isoelectric-focusing (I.E.F.) markers from Pharmacia (Uppsala, Sweden) ; all the chemicals of analytical grade were from Sigma (Saint Louis, Mo, U.S.A.)
Electrophoresis equipments were from L.K.B. (Bromma, Sweden). Veinous Blood samples from 12 h fasting patient and controls were collected on EDTA, centrifuged at 3000 rpm and plasma was immediately used or frozen at -20°C.

Methods

Electrophoresis and immunochemical derived methods. Electrophoresis in discontinuous polyacrylamide gel(D.P.A.G.E.),immuno-electrophoresis(I.E.A.), immunofixation (I.F.A.), Laurell's two-dimensional electrophoresis to detail the distribution of apoproteins within lipoproteins (2 D.E.I.A.) and 2 dimensional electrophoresis of lipoprotein apoproteins (2 D.E.L.A.) which consists of three steps (separation of crude plasma lipoproteins in P.A.G.E., delipidation of these lipoproteins in the first gel, separation of the lipoprotein's apoproteins in a second P.A.G. : T = 15 %, C = 2.6 %) have been previously described (21).

Two-dimensional I.E.F. immunoelectrophoresis (2 D.I.E.F.E.). The methodology described by Marcel et al (22, 23) has been applied to total plasma in the ampholytes pH 3.5 - 9.5 range and first dimension I.E.F. narrow strips had been interfaced with 1.5 % agarose plates containing appropriate anti apo A-I or B concentrations.

Isolation of LDL and HDL. Fresh blood plasma from control and patient were used to isolate LDL and HDL by stepwise centrifugal flottation according to Havel et al. (24). LDL and HDL were respectively obtained between the densities 1.006 - 1.063 g/ml and 1.063 -1.210 g/ml. In these conditions Tangier patient's HDL were quite entirely contained in the infranatant at 1.210 g/ml. To isolate these HDL it was necessary to dialyze the infranatant against 10 L of 10 mM Tris-HCl, pH 7.4 at 4°C during 48 hours with gentle stiring and lyophilize it.

Fluorescence quenching of LDL_2 by iodide ions. It was carried out according to Lehrer as described by Dang et al. (25).

RESULTS AND DISCUSSION

In our patient apo A-I was always drastically decreased (from 20 to 50 mg/l) while apo A-II was heavily lowered (from 40 to 70 mg/l) and apo B was often lowered (from 400 to 900 mg/l). These variations were apparently diet dependent. The HDL in D.P.A.G.E. as well as in agarose gel were undetected and the LDL (LDL_T) showed a higher mobility than controls. The increased anodic mobility of LDL_T support the hypothesis of their higher anionic charge (agarose gel) and their smaller size (PAG).

The immunochemical analysis revealed that HDL and apo A-I were not completely absent in the patient's plasma but abnormaly distributed among the lipoproteins. I.E.A., with a fivefold increase of Tangier plasma employed, showed a thin α lipoprotein arc of low mobility in the forefront of the β lipoprotein region. The immunoprints corresponding to I.F.A. demonstrated numerous differences between normal and Tangier plasma : apo A-I appeared as a very light band in the LDL region and was absent in the normal HDL region (HDL_2, HDL_3), apo A-II appeared as a single band in the LDL region of normal serum but was distributed as A-I and A-II in the LDL_T and cathodic post LDL_T region. The 2 D.E.I.A. exhibited great heterogeneity in Tangier plasma : a single apo B component appeared in the normal HDL region but 3 peaks were observed in the fast moving LDL region LDL_T (containing apo A-I and apo A-II) and another small peak in the forefront of the HDL_T region (containing apo A-I and apo A-II). These results are in accordance with previous data concerning other Tangier patients (1, 26).

The 2 D.E.L.A. (fig. 1) showed the apoproteins from delipidated lipoproteins that have already been electrophoretically fractionated. A small quantity of apo A-I was observed for the Tangier plasma in normal HDL_3 region of the lipidogram, but the main proportion of this component was present in the abnormal pre-LDL region (HDL_T region) where the 2 D.E.I.A. (fig. 2) revealed two superposed peaks (probably apo $A\text{-}I_0$ and apo $A\text{-}I_{+2}$ or proapo A-I, see below). Following the immunochemical assays, the 2 D.E.L.A. demonstrates that the distribution of major apoproteins in the electrophoretic track of normal or Tangier plasma is characteristic of each lipoprotein and reveals various combinations which are distinct from classical lipoproteins (that one appear more heterogeneous than predicted). This is in perfect accordance with the new concept of lipoprotein particles (27) which correspond to all the types of associations of apoproteins which are not visible on traditional electrophoresis. In the case of HDL_T, the 2 D.E.L.A. showed that apo B, apo $A\text{-}I_T$, apo A-II may be free or combined diversely : B + A-I, B + A-II, B + A-I and A-II. Using affinity chromatography lipoproteins containing apo A-II, B, CII, CIII and E but no apo A-I have been isolated from the plasma of our patient by Dr. P. Puchois (personal communication). By another way using ultracentrifugally isolated HDL_T and LDL_T as compared to normal lipoproteins (HDL_N and LDL_N) differences have been observed in agarose plates of Ouchterlony : the ratio B/A-I was high in HDL_T and low in HDL_N, on the contrary it was low in LDL_T and very high in LDL_N (fig. 3).

The above data support the hypothesis of lipoprotein particles in which apo $A\text{-}I_T$ could be bound to apo B and other apo like AII, CII, CIII and/or E. From the three following experiments other arguments have been obtained which showed that apo B and $A\text{-}I_T$ may be bound. At first the 2.D.I.E.F.E. described by Marcel et al. (22, 23) was applied to normal and Tangier total plasma in ampholytes pH gradient from 3.5 to 9.5. It was observed that the prevailing apo B particles from Tangier plasma were shifted toward the acidic edge of the gel as apo $A\text{-}I_T$ and compared to normal plasma. Furthermore the two dimensional electrophoresis (2 DE) showed the predominance of basic isoforms (AI_{+2} and AI_0) in Tangier plasma as compared to the normal one containing : $A\text{-}I_{+2}$ $A\text{-}I_{+1}$ as minor components and AI_0, AI_{-1} AI_{-2} as main components (18, 19). The basicity of the pro-forms of Tangier apo A-I is consistent with their combination to acidic residues of apo B. Finally the fluorescence quenching by iodide ions of Tangier LDL_2 which as been measured higher than the controls one enhances the chances of Tangier LDL_2 to externalize apo B residues as it reveals a smaller lipid core in Tangier LDL_2 and a shallower embedding of TRY residues that are mainly responsible for apoprotein fluorescence.

CONCLUSION

In a new case of homozygote Tangier disease the data obtained using electrophoretic and immunoelectrophoretic methods with the aid of ampholytes, detergents and agarose or P.A.G. suggest the presence of lipoprotein particles which associate mainly apo B, apo A-I_T and/or apo A-II. This observation support the hypothesis that such particles which are nearly absent in normal plasma could explain their uptake by monocyte-derived macrophages and through the apo B fraction their hydrolysis in lysosomes as indicated by Schmitz et al. (20).

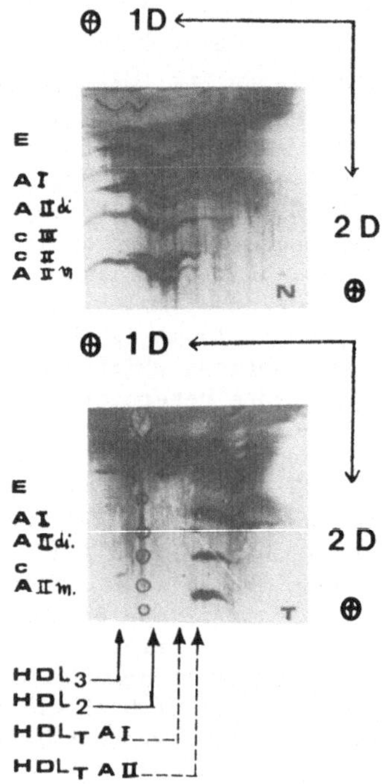

Fig. 1 : 2 D.E.L.A. - 1st dimension : Plasma lipoproteins are separated in discontinous PAG (3% anodic) then are delipidated in the gel.
2nd dimension : The apolipoproteins are separated vertically (anode at bottom) in 15% PAG. In Tangier plasma (T) apo A-I_T appears in the LDL region as A-II. A-I_T is slightly shifted on the left, A-II on the right. It shows that particles with A-I, A-II, A-I and A-II may be predicted in the apo B region.

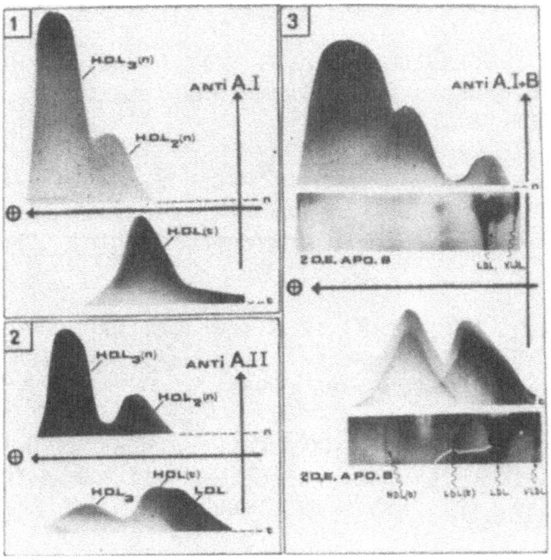

Fig. 2 : 2 D.E.I.A. The anode is on the left (1st dimension) and on the top (2nd dimension) Normal (n) and Tangier (t) serum are compared in anti A-I, A-II and AI + B (2nd dimension).

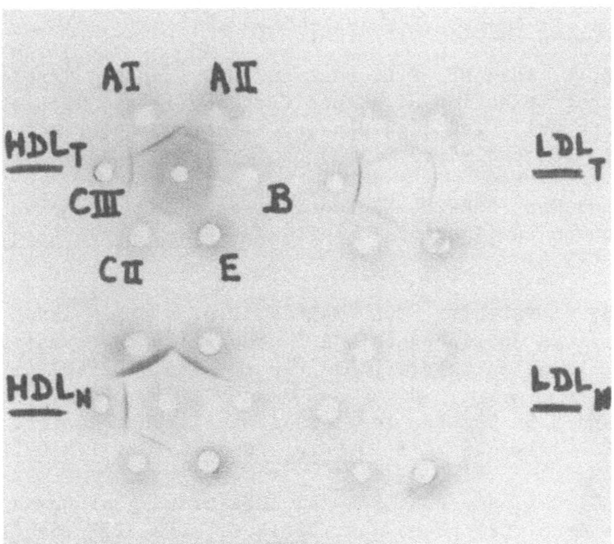

Fig. 3 : Double-immunodiffusion of OUCHTERLONY of normal and Tangier HDL or LDL ultracentrifugally isolated and dialyzed.

REFERENCES

1. G. ASSMANN, E. SMOOTZ, K. ADLER, A. CAPURSO and K. OETTE, The lipoprotein abnormality in Tangier disease : quantitation of A apoproteins, J. Clin. Invest., 59, 565-575 (1977)

2. V.I. ZANNIS, J.L. BRESLOW and A.J. KOTZ, Isoproteins of human apolipoprotein A-I demonstrated in plasma and intestinal organ culture, J. Biol. Chem., 255, 8612-8617 (1980)

3. V.I. ZANNIS, D.M. KURNIT and J.L. BRESLOW, Hepatic apo A-I and apo E and intestinal apo A-I are synthesized in precursor isoprotein forms by organ cultures of human foetal tissues, J. Biol. Chem., 257, 536-544 (1982)

4. V.I. ZANNIS, A.M. LEES, R.S. LEES and J.L. BRESLOW, Abnormal apolipoprotein A-I isoprotein composition in patients with Tangier disease, J. Biol. Chem., 257, 4978-4986 (1982)

5. L.L. KAY, R.M. RONAN, E.J. SCHAEFFER and H.B. BREWER, Tangier disease : a structural defect in apolipoproteins A-I (Apo $A\text{-}I_{Tangier}$), Proc. Natl. Acad. Sci. (USA), 79, 2485-2489 (1982)

6. E.J. SCHAEFFER, L.L. KAY, L.A. ZECH and B.H. BREWER, Tangier disease : HDL deficiency due to defective metabolism of an abnormal apo A-I, J. Clin. Invest., 70, 934-945 (1982)

7. H.B. BREWER, T. FAIRWELL, M. MENG, L. KAY and R. RONAN, Human proapo $A\text{-}I_{Tangier}$: isolation of proapo $A\text{-}I_{Tangier}$ and amino acid sequence of the propeptide, B.B.R.C., 113, 934-940 (1983)

8. J.I. GORDON, H.F. SINS, S.R. LENTZ, C. EDELSTEIN, A. SCANU and A.W. STRAUSS, Proteolytic processing of Human preproapolipoprotein A-I. A proposed defect in the conversion of Pro A-I to A-I in Tangier disease, J. Biol. Chem., 258, 4037-4044 (1983)

9. G. SCHMITZ, G. ASSMANN, S.C. RALL and R.W. MAHLEY, Tangier disease : defective recombination of a specific Tangier apolipoprotein A-I isoform (proapo A-I) with high density lipoproteins, Proc. Natl. Acad. Sci. (USA), 80, 6081-6085 (1983)

10. S.N. LAW and H.B. BREWER, Nucleotide sequence and the encoded amino acids of human apolipoprotein A-I $_{m}$RNA, Proc. Natl. Acad. Sci. (USA), 81, 66-70 (1984)

11. P.K. WEECH, J. FROHLICH, Y.L. MARCEL, T.D. N'GUYEN and R.W. MILNE, Tangier disease apolipoprotein A-I compared with normal plasma A-I using monoclonal antibodies, BBA report, 835, 402-407 (1985)

12. S.N. LAW and H.B. BREWER, Tangier disease : the complete $_{m}$RNA sequence encoding for preproapo A-I, J. Biol. Chem., 260, 12810-12814 (1985)

13. S.N. LAW, G. GRAY and H.B. BREWER, cDNA cloning of Human apo A-I : Amino acid sequence of preproapo A-I, B.B.R.C., 112, 257-264 (1983)

14. E.J. SCHAEFFER, L.A. ZECH, D.E. SCHWARTZ and H.B. BREWER, Coronary heart disease prevalence and other clinical features in familial high density lipoprotein deficiency (Tangier disease), Ann. Intern. Med., 93, 261-266 (1980)

15. T.J.C. JASAWANT, Etude clinique, biochimique et génétique d'un nouveau cas de maladie de Tangier, Thèse Doct. Med., Bordeaux, 25 (1985)

16. G.J. MILLER and N.F. MILLER, Plasma HDL concentration and development of ischemic heart disease, Lancet, 1, 16-20 (1975)

17. P.N. HERBERT, G. ASSMANN, A.M. GOTTO and D.S. FREDRICKSON, Familial lipoprotein deficiency, in : "Metabolic basis of inherited disease", M.S. Brown, Eds, Mac Graw Hill, New York, 589-621 (1983)

18. S. VISVIKIS, M.F. DUMON, J. STEINMETZ, T. MANABE, M.M. GALTEAU, M. CLERC and G. SIEST, Plasma apolipoproteins in Tangier disease as studied with two-dimensional electrophoresis, Clin. Chem., 33, 120-122 (1987)

19. T. MANABE, S. VISVIKIS, M.F. DUMON, M. CLERC and G. SIEST, Evaluation of lipoproteins and apolipoproteins in serum of a Tangier Patient by micro-scale two dimensional electrophoresis, Clin. Chem., 33, 468, 472 (1987)

20. G. SCHMITZ, G. ASSMANN, H. ROBENEK and B. BRENNHAUSEN, Tangier disease : A disorder of intracellular membrane traffic, Proc. Natl. Acad Sci. (USA), 82, 6305-6309 (1985)

21. M.F. DUMON, S. VISVIKIS, T. MANABE and M. CLERC, Immunochemical study of the plasma low and high density lipoproteins in Tangier disease, F.E.B.S. Letters, 201, 163-167 (1986)

22. Y.L. MARCEL, P.K. WEECH, T.D. N'GUYEN, R.W. MILNE and W.J. McCONATHY, Apolipoproteins as the basis for heterogeneity in high density lipoprotein 2 and high density lipoprotein 3, Eur. J. BIochem., 143, 467-476 (1984)

23. P.K. WEECH, J. FROHLICH, Y.L. MARCEL, T.D. N'GUYEN and R.W. MILNE, Tangier disease apolipoprotein A-I compared with normal plasma A-I using monoclonal antibodies, B.B.A., 835, 402-407 (1985)

24. R.J. HAVEL, H.A. EDER and J.H. BRAGDON, The distribution and chemical composition of ultracentrifugally separated lipoproteins in human serum, J. Clin. Invest., 34, 1345-1353 (1955)

25. M.F. DUMON, M. FRENEIX-CLERC, N. DOUSSET, Q.Q. DANG, A. MARET, R. SALVAYRE, M. CLERC and L. DOUSTE-BLAZY, Physical alterations of plasma lipoproteins in Tangier disease and their hypothetic involvement in pathogenesis, Lipid Storage disorders, N.A.T.O.-INSERM Coll., Plenum press Ed., (sous presse) (1988)

26. G. ASSMANN, O. SIMANTKE, H.E. SCHAEFFER and E. SMOOTZ, Characterization of High density lipoproteins in patients heterozygous for Tangier disease, J. Clin. Invest., 60, 1025-1035 (1977)

27. P. PUCHOIS, P. ALAUPOVIC and J.C FRUCHART, Mise au point sur les classifications des lipoproteines plasmatiques, Ann. Biol. Clin., 43, 831-840 (1985)

APOLIPOPROTEIN C-II DEFICIENCIES: IN VIVO MODELS FOR ASSESSING THE SIGNIFICANCE OF DEFECTIVE LIPOLYSIS ON LIPOPROTEIN METABOLISM

W. Carl Breckenridge

Department of Biochemistry
Dalhousie University
Halifax, Nova Scotia, B3H 4H7

INTRODUCTION

Plasma lipolytic processes are a major driving force in normal lipoprotein metabolism. The formation of both atherogenic and "protective" lipoproteins is dependent on the maintenance of proper biological activity of lipoprotein lipase. It is now apparent that maintenance of sufficient biological activity of lipoprotein lipase requires the synthesis of an active form of the enzyme in tissues that take up fatty acids as well as the secretion of biologically active apolipoprotein (apo) C-II so that proper activation of lipoprotein lipase is accomplished (1,2). As further characterization of genetic defects in lipolytic enzymes and apo C-II are elucidated there is increasing evidence which indicates that a number of subtle defects in apo C-II are responsible for a variety of lipoprotein abnormalities. Apo C-II deficiency may be defined very broadly as insufficient biological activating potential for lipoprotein lipase to allow for adequate lipolysis and maintenance of normal triacylglycerol concentrations under normal dietary fat loads. Most examples of apo C-II deficiency have been due to an almost complete deficiency of biological activity. However there is preliminary information to suggest that some defects may limit lipolytic activity and cause serious hyperlipidemia especially when combined with other defects.

APOLIPOPROTEIN C-II ABNORMALITIES

Originally the hallmark features of apolipoprotein C-II deficiency (1,2) included a deficiency of apolipoprotein C-II by immunoassay, by electrophoretic characterization and by biological activating potential for lipoprotein lipase. Plasma post heparin lipolytic activity was low but could be stimulated by the addition of normal plasma. Infusion of normal plasma (1), apo C preparations (2) or synthetic fragments of apo C-II (4) resulted in a rapid decrease of plasma triacylglycerol concentrations. However as further cases have been described it is apparent that apo C-II deficiency may be divided into several types: a) a complete absence (5), b) frameshifts in the apo C-II gene which result in large accumulations of variants which are inactive (6,7), single amino acid substitutions resulting in variants which are active (8) as well as other variants which are present in minor amounts (4,9). Two recent publications are excellent examples of frameshifts (6,7). The original kindred for apo C-II deficiency possesses a variant apo C-II which does not yield immunoprecipitations with polyclonal anti apo C-II. There is a large charge shift for the variant which may be identified on isoelectric focussing gels by Western blotting technques with the same polyclonal antibody that does not precipitate the variant apo C-II (10). The variant appears to be due to a frameshift in the gene such that there is an altered amino acid composition from residue 69 and truncation at residue 75 by a new stop codon (6). A second variant has a frameshift such that the amino acid sequence is altered from residue 70, along with an additional sequence up to 96 residues (7). It is postulated that during translation the original stop codon is not detected. These variants indicate that there may be potentially a large number of apo C-II variants since appropriate frameshifts in the region of the gene responsible for the terminal 19 amino acids, which are important for biological activity, may yield truncated or extended apo C-II variants (2).

A single substitution of glutamine at lysine 55 in apo C-II does not result in loss of biological activity (8). Two other variants with slight changes in electrophoretic mobility on two dimensional electrophoresis are present in extremely small quantities (4,9). They may be inactive or present in insufficient quantities for normal activation. A new apo C-II variant has been detected in large quantities in a hyperlipidemic subject (11). Although this variant, has normal mobility in two dimensional electrophoresis, and is immunoreactive with anti apo C-II it has low biological activity. It is possible that this variant may also be due to a single amino acid substitution in an area of the apolipoprotein which is crucial for activity.

Three combined defects have been described. A combined defect of lipoprotein lipase deficiency, apo C-II deficiency and E2,E2 phenotype was reported (12) while a combined deficiency of apo C-I and apo C-II yielded the usual characteristics of apo C-II deficiency (13). A new defect involving a combination of an apo C-II variant with low biological activity, and apo E deficiency (11) is particularily interesting and will be described below.

Thus it is clear that one of the most crucial assays for elucidation of biological defects in apo C-II is a careful assessment of the activation potential of apo C-II for lipoprotein lipase. Defective apolipoproteins may have normal electrophoretic or immunologic properties but be non functional.

LIPOPROTEIN METABOLISM IN APO C-II DEFICIENCY

The process of catabolism of triacylglycerol-rich lipoproteins has been studied extensively over the past decade (14). There is clear evidence that lipoprotein lipase hydrolyzes triacylglycerols in VLDL and chylomicrons resulting in a reduction in particle size and a progressive enrichment in cholesteryl esters but a loss of surface lipids and apo C and E. A substantial portion of the phospholipid and cholesterol lost from human VLDL, in vitro, is in the form of vesicles with a cholesterol/phospholipid ratio of 1/1 (15). These vesicles are not readily disrupted by HDL once formed. However if sufficient HDL is present in the incubation the formation of the vesicles is largely prevented. In the absence of HDL a major portion of apo C-II and C-III is lost in association with a spherical HDL-sized particle which also contains apo E (15,16). These particle undergo rearrangement in the presence of HDL to resemble plasma HDL (14,17). This process is interesting since it suggests a potential mechanism for the loss of excess cholesteryl ester and apo E from VLDL (14). Further studies,in vivo and in vitro in the presence of physiological proportions of HDL, have established that most of the apo C-II and C-III leave a VLDL particle during lipolysis, before apo E is lost. Despite a number of studies involving lipoprotein lipase, hepatic lipase and various amounts of HDL, the maximum loss of apo E from VLDL is approximately 35% and the resulting IDL-like particle is larger than plasma LDL and contains excess cholesteryl ester (Murdoch and Breckenridge, unpublished). While the loss of apo E to HDL is associated with some loss of cholesteryl ester into the region of large HDL at present this loss is insufficient to account for the complete conversion of all VLDL particles to LDL.

In apo C-II deficiency there is a severe accumulation of triglyceride-rich lipoproteins. Administration of plasma, apo C peptides or fragments of apo C-II results in a rapid clearance of triacylglycerols and remnant lipoproteins (1,2,3). In our experience there is relatively little accumulation of LDL or HDL during the treatment. Since most studies indicate that the chylomicrons and VLDL of apo C-II deficient subjects are markedly enriched in apo E it is probable that the lipoproteins are removed as remnants following loss of apo C-III. A recent case of apo C-II deficiency raises a number of important questions (11). In addition to a non-functioning apo C-II this patient has apo E deficiency. The major lipoprotein abnormality is the accumulation of a cholesteryl ester-rich chylomicron remnant which contains apo B-100, apo B-48, apo A-I, and apo A-IV. These characteristics are quite similar to a previous report of apo E deficiency (18) with the exception that this patient has severe elevations of chylomicron remnants in contrast to the previous studies. The data from this case of combined apo C-II/apo E deficiency is consistent with the concept that there is extremely slow lipolysis of the chylomicrons. It is postulated that with slow lipolysis and the action of lipid transfer protein (19) there is a progressive enrichment of the chylomicrons with cholesteryl ester. Since the individual lacks apo E it is postulated that there is very inadequate clearance of the chylomicron remnants. If this mechanism proves correct it is possible that subtle defects in apo C-II which result in reduced activation potential, may delay the clearance of triacylglycerol-rich lipoproteins from plasma.

Marked enrichment of triacylglycerols in LDL and HDL has been noted in hypertriglyceridemia (3,20). Furthermore it has been calculated that large VLDL subfractions from some hypertriglyceridemic patients contain larger proportions of cholesteryl ester in relation to apo B than is found for LDL (20). These alterations may be a function of the prolonged half lives of triacylglycerol-rich lipoproteins and the lipid transfer protein which allows for an extensive enrichment of the VLDL with cholesteryl esters and LDL and HDL with triacylglycerols. In our studies of lipolysis it seems that insufficient cholesteryl ester is lost by some triacylglycerol-rich lipoprotein,during lipolysis, to allow for conversion to LDL. Such remnants may be removed by receptors prior to conversion to LDL. If the extent of enrichment of chylomicrons and VLDL with cholesteryl ester is a function of the rate of lipolysis versus the rate of lipid transfer activity it is possible that factors such as apo C-II activation potential or ratios of apo C-II/C-III may influence the extent to which triacylglycerol-rich lipoproteins may accumulate cholesteryl ester before conversion to remnants for removal or further conversion to LDL.

SUMMARY

There are now several types of apo C-II deficiency. The accumulating evidence suggests that more variants will be discovered which may be inactive or partially active. It seems important to establish accurate assays for assessing optimal properties of lipoproteins for lipolysis in order to continue to diagnose these defects and to study whether rates of lipolysis may influence the accumulation of cholesteryl esters in triacylglycerol-rich lipoproteins.

REFERENCES

1. W. C. Breckenridge, J. A. Little, G. Steiner, A. Chow and M. Poapst, Hypertriglyceridemia associated with a deficiency of apolipoprotein C-II, N. Engl. J. Med. 298:1265 (1978)
2. W. C. Breckenridge, Apolipoprotein C-II deficiency, in: Lipoprotein Deficiency Syndromes, A. Angel and J. Frohlich, Plenum , New York, (1968).
3. J. A. Little, D. W. Cox, W. C. Breckenridge, V. M. McGuire, Familial apo C-II Deficiency, in: Atherosclerosis V, pp 671-675 Springer Verlag, New York (1980).
4. G. Baggio., E. Manzato, C. Gabelli, R. Fellin, S. Martini, G. B. Enzi, F. Veriato, M. R. Baiocchi, D. L. Sprecher, M. L. Kashyap, H. B. Brewer and G. Crepaldi, Apolipoprotein C-II Deficiency Syndrome, J. Clin. Invest. 77:520 (1986)
5. K. Saku, C. Cedres, B. McDonald, B. A. Hynd, B. W. Liu, L. S. Srivastava and M. L. Kashyap, C-II anapolipoproteiemia and severe hyperglyceridemia. Am. J. Med. 77:457 (1984).
6. P. W. Connelly, G. F. Maguire, T. Hofmann and J. A. Little, Structure of apolipoprotein C-II Toronto, a nonfunctional human apolipoprotein, Proc. Natl. Acad. U.S.A. 84:270 (1987).
7. P. W. Connelly, G. F. Maguire and J. A. Little, Apolipoprotein C-II St. Michael. J. Clin. Invest. 80:1597 (1987).
8. H. J. Menzel, J. P. Kane, M. J. Malloy and R. J. Havel. A variant primary structure of apolipoprotein C-II in individuals of African descent, J. Clin. Invest. 77:595 (1986).
9. D. L. Sprecher, I. Taam, R. E. Gregg, S. S. Fojo, D. M. Wilson, M. L. Kashyap and H. B. Brewer Jr., Identification of an apo C-II variant (apo C-II Bethesda) in a kindred with apo C-II deficiency and type I hyperlipoproteinemia, J. Lipid Res. 29:273 (1988)
10. G. F. Maguire, J. A. Little, G. Kakis and W. C. Breckenridge, Apolipoprotein C-II deficiency associated with nonfunctional mutant forms of apolipoprotein C-II, Can. J. Biochem. Cell Biol. 622:847 (1984).
11. R. Rebourcet, W. C. Breckenridge, J. L. Bresson, F. Rey, R. Benbrahem and J. Rey, Hyperlipoproteinemia associated with apo C-II variants and apo E deficiency, in Proceedings of the 4th Colloquim, Fondation de Recherche sur L'Atherosclerose, Brussells (1988).
12. A. F. H. Stalenhoef, A. F. Casparie, P. N. M. Demacker, J. T. J. Stouten, J. A. Lutterman and A. van't Laar, Combined deficiency of apolipoprotein C-II and lipoprotein lipase in familial hyperchylomicronemia, Metabolism 30:919 (1981).
13. M. F. Dumon and M. Clerc, Preliminary report on a case of apolipoproteins C-I and C-II deficiency, Clin. Chim. Acta 157:239 (1986).
14. W. C. Breckenridge, The catabolism of very low density lipoproteins, Can. J. Biochem. Cell Biol. 63:890 (1985).

15. S. P. Tam and W. C. Breckenridge, Apolipoprotein and lipid distribution between vesicles and HDL-like particles formed during lipolysis of human very low density lipoproteins by perfused rat heart, J. Lipid Res. 24:1343 (1983).
16. S. P Tam, P. L. Dory and D. Rubinstein, Fate of apolipoproteins C-II, C-III and E during lipolysis of human very low density lipoproteins in vitro, J. Lipid Res. 22:641 (1981).
17. S. P. Tam and W. C. Breckenridge, The interaction of lipolysis products of very low density lipoprotein with plasma high density lipoprotein (HDL): perfusate HDL with plasma HDL subfractions, Biochem. Cell Biol. 65:252 (1987).
18. G. Ghiselli, E. J. Schaefer, P. Gascon and H. B. Brewer Jr., Type III hyperlipoproteinemia associated with apolipoprotein E deficiency, Science 214:1239 (1981).
19. Y. L. Marcel, Lecithin cholesteryl acyl transferase and intravascular cholesterol transport, Adv. Lipid Res. 19:117 (1982).
20. S. Eisenberg, High density lipoprotein metabolism, J. Lipid Res. 25:1017 (1984).

GENETIC VARIATION IN THE APOLIPOPROTEINS C-II AND C-III

R.E. Ferrell[1], M.I. Kamboh[1], B.S. Sepehrnia[1]
L.L. Adams-Campbell[2], and K.M. Weiss[3]

[1]Human Genetics Division and [2]Department of Epidemiology
University of Pittsburgh, Pittsburgh, PA 15261
[3]Department of Anthropology, Pennsylvania State University
University Park, PA 16802

INTRODUCTION

The C group apolipoproteins are primarily associated with triglyceride-rich lipoprotein particles (1). They are important in triglyceride metabolism where they act as allosteric affectors of lipoprotein lipase. Apolipoprotein C-II is an activator of lipoprotein lipase while C-III is an inhibitor of both lipoprotein lipase and hepatic lipase. Apolipoprotein C-III may also be involved in the receptor mediated uptake of triglyceride-rich lipoproteins. Genetically determined deficiency of C-II is associated with a functional deficiency of lipoprotein lipase and severe hypertriglyceridemia (2). Menzel, et al (3) have reported a protein polymorphism in APO C-II in U.S. Blacks that involved the substitution of glutamine for lysine at residue 55 of the C-II polypeptide chain. They reported a frequency of 12% for this variant in a sample of 50 normolipidemic U.S. Blacks. Sepehrnia, et al (4) confirmed the presence of this polymorphism in U.S. Blacks, demonstrated its presence in Nigerians, and verified an autosomal codominant pattern of segregation in families. Reports of genetic variation in apolipoprotein C-III are restricted to two types of rare families. One having combined apolipoprotein A-I/C-III deficiency due to a sequence inversion involving both the APO A-I and C-III genes (5). Deficient family members have abnormalities of triglyceride metabolism that are corrected by the infusion of normal C-III (6). The second variant is recognized by unusually high levels of unglycosylated APO C-III (APO C-III$_0$) due to the replacement of threonine by alanine at position 74 of the polypetide chain, which prevents O-glycosylation (7). The latter mutation is not assocociated with gross alteration in lipoprotein levels.

In the present study we have used an isoelectric focusing-immunoblotting procedure (4, 8) to determine the occurrence and frequency of genetic variation in apolipoproteins C-II and C-III in the general population, and have initiated studies of the impact of such variation on lipoprotein levels.

MATERIALS AND METHODS

Plasma samples from U.S. Caucasians and Blacks, native Africans, and American Indians from North and Central America are available in the Human

Genetics Laboratory. These samples were stored at -80°C prior to typing. Lipid and lipoprotein analyses were performed by standard Lipid Research Clinics methods (9). Apolipoprotein C-III quantitation was performed by single radial immunodiffusion (10) using commercially available plates (Daiichi, Tokyo). Isoelectric focusing of plasma samples was carried out over the pH range 3.5-5 for C-II and C-III followed by immunoblotting using goat anti-human apolipoprotein C-II or C-III (Diichi, Tokyo) as the primary antibody and rabbit anti-goat IgG conjugated with alkaline phosphatase as the second antibody. Isoprotein bands were visualized by histochemical staining of alkaline phosphatase (4, 8). The effect of APO C-II genotype on lipoprotein and lipid levels was estimated after adjustment of quantitative variables for the effect of age, height and weight by linear regression analysis. The allelic effect of the APO C2*2 allele was estimated by the method of Boerwinkle, et al (11). The distribution of apolipoprotein C-III levels was resolved into its components by the method of Bhattacharya (12).

RESULTS

Apolipoprotein C-II. Among all populations screened only U.S. and African Blacks were found to be polymorphic for APO C-II. Table 1 presents the APO C-II phenotypes and allele frequencies in these two groups. U.S. Blacks exhibit two primary alleles, APO C2*1 and APO C2*2, with frequencies of 0.975 and 0.025, respectively. The Nigerian population is polymorphic for the same alleles, and is characterized by the presence of two additional alleles, APO C2*3 and APO C2*4, at low frequency. The frequency of the APO C2*2 allele in Africans is approximately twice that seen in U.S. Blacks. This is consistent with historical admixture between blacks and whites in the U.S., but may also reflect population heterogeneity among African populations. The isoprotein patterns for each of the observed APO C2 phenotypes are shown in Figure 1. All heterozygous, phenotypes are consistent with autosomal codominant expression of two alleles coding for a monomeric protein.

To estimate the effect of the APO C-II polymorphism on the quantitative levels of total cholesterol, total HDL-cholesterol, LDL-cholesterol, HDL subfractions and triglycerides, each quantitative variable was tested for the effect of the concomitant variables age, $(age)^2$, height, $(height)^2$, weight and body mass index by stepwise regression analysis using the statistical package SAS. After adjustment for significant concomitant variables individuals were classified by APO C-II phenotype and the effect of phenotype on quantitative variables were estimated by two-way multivariate analysis of variance. The APO C-II 2-1 phenotype is associated with lower mean total cholesterol levels compared with the 1-1 phenotype, and the cholesterol lowering effect is reflected in lower mean values for both LDL- and total HDL-cholesterol. The 2-1 phenotype was also associated with an increase in plasma triglycerides compared to the 1-1 phenotype. These results are consistent with the proposed role of apolipoprotein C-II in lipoprotein metabolism. To avoid the problem of compounding environmental and genetic effects, the analysis of the phenotypic effect of C-II variation was restricted to the African population sample. The frequency of the APO C2*2 allele is such that only two individuals of the 2-2 phenotype were observed. This number is not sufficient to determine whether the phenotypic effects observed show a clear gene dosage effect.

Apolipoprotein C-III. Screening of native plasma from a variety of populations for genetically determined qualitative polymorphism did not reveal any interindividual variation, although intraindividual heterogeneity due to variation in glycosylation of APO C-III was clearly resolved. However, striking quantitative differences between individuals were seen on

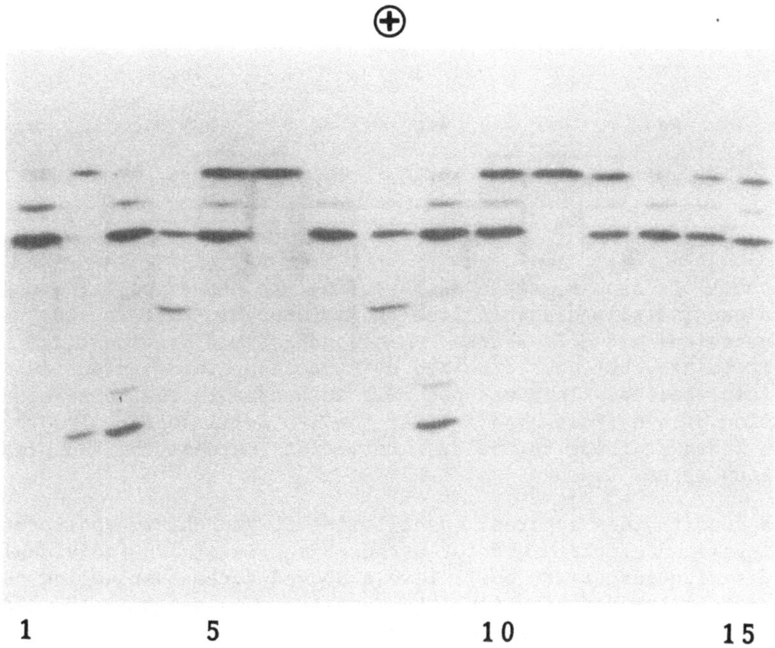

Figure 1. Apolipoprotein C-II phenotype patterns. APO C-II 1-1, lanes 1, 7 and 13; APO C-II 2-1, lanes 5, 10, 12, 14 and 15; APO C-II 2-2, lanes 6 and 11; APO C-II 3-2, lane 2; APO C-II 3-1, lanes 3 and 9; APO C-II 4-1, lanes 4 and 8.

Table 1. Phenotype counts and gene frequencies of apolipoprotein C-II alleles in populations of African ancestry.

Population	Number Tested	Phenotypes 1-1	2-1	2-2	3-1	4-1	3-2	Allele Frequency
U.S. Blacks	137	130	7	0	0	0	0	APO C2*1=0.975
		(130)	(7)	-	-	-	-	APO C2*2=0.025
Nigerians	361	320	35	2	2	1	1	APO C2*1=0.939
		(318)	(37)	(1)	(3)	(1)	(0)	APO C2*2=0.055
								APO C2*3=0.004
								APO C2*4=0.002

Numbers in parentheses are the expected number of observations assuming Hardy-Weinberg equilibrium

Table 2. Presumed genotypes and mean apolipoprotein C-III levels in Mayan Indians.

Presumed Genotypes	Mean C-III (mg/dl)	Standard Deviation
NN	14.1	5.92
Nd	7.8	3.92
dd	2.6	2.12

IEF/immunoblots from a sample of Mayan Indians of Mexico. The most common pattern was indistinguishable from the monomorphic pattern observed in U.S. Caucasians and U.S. Blacks. A second pattern was identical in isoelectric points, but characterized by reduced staining intensity of all isoprotein species. This was presumed to be due to the homozygous expression of a deficiency allele at the APO C-III locus. The third pattern had a staining intensity intermediate between the two presumed homozygous types.

To confirm this apparent quantitative polymorphism, apoliprotein C-III levels were determined for a random sample of 123 individuals. The total distribution of APO C-III levels showed three overlapping peaks. Resolution of the distribution into Gaussian components by the procedure of Bhattacharya (12) confirmed the trimodal distribution. Table 2 presents the means and standard deviations of APO C-III levels for each component and the presumed genotypes associated with each class. The quantitative data are consistent with the segregation of two alleles with identical isoelectric points, but with the less frequent allele associated with reduced levels of circulating APO C-III. The mean of the presumed heterozygous class is intermediate between the means of the two homozygous classes. Unfortunately, the available family data was insufficient for a formal segregation analysis.

Because of the geographical remoteness of the Mayan population, detailed lipid parameters are not available. The deficiency phenotype is associated with lower mean cholesterol levels, but the data to address the more interesting question of its impact on triglyceride levels are not available. The molecular basis of the deficiency phenotype is still under investigation. Analysis of the Xmn I, Pst I, Sst I, Pvu IIA and Pvu IIB restriction fragment length polymorphisms (13) in the APO A-I/C-III region of the gene cluster revealed no major differences between individuals carrying the common or the deficiency allele at APO C-III locus. Thus the APO C-III gene in the deficient individuals seems grossly intact.

DISCUSSION

Genetically determined variation in the plasma apoliproteins has provided unique insights into many aspects of lipid metabolism. This has been particularly true in the case for those rare mutations that lead to the aggregation of premature coronary artery disease in families. Our efforts are directed toward the detection of common polymorphism at these loci, which may not be associated with any clinical phenotype, but which may impact in a more subtle way on lipid metabolism. Combinations of these clinically benign alleles may play an important role in determining lipoprotein and lipid levels in the general population, and thus influence the risk of cardiovascular disease in the population.

ACKNOWLEDGMENT

The authors wish to thank Dr. P. P. Majumder for assistance in the analysis of the quantitative data, Dr. Steve Humphries for providing the DNA probes used to analyze the A-I/C-III/A-IV cluster, and Dr. C. F. Sing for helpful discussions of the general strategies of the measured genotype approach to data analysis. This work supported in part by N.I.H. Grant H.L. 39107.

REFERENCES

1. J. L. Breslow, Apolipoprotein genetic variation and human disease, Physiological Rev. 68:85 (1988).
2. W. C. Breckenridge, J. A. Little, G. Steiner, A. Chow, and M. Poapst, Hypertriglyceridemia associated with deficiency of apolipoprotein C-II, N. Eng. J. Med. 298:1265 (1978).
3. H. J. Menzel, J. P. Kane, M. J. Malloy, and R. J. Havel, A variant primary structure of apolipoprotein C-II in individuals of African descent, J. Clin, Invest. 77:595 (1986).
4. B. Sepehrnia, M. I. Kamboh, and R. E. Ferrell, Genetic studies of human apolipoproteins. III. Polymorphism of apolipoprotein C-II, Hum. Hered. (1988). In press.
5. S. K. Karathanasis, DNA inversion inactivates both the apo AI and apo CIII gene in patients with combined apo AI and apo CIII deficiency and premature coronary artery disease. Circulation (Suppl. II) 74:157 (1986).
6. R. A. Narum, J. B. Lakier, S. Goldstein, A. Angel, R. B. Goldberg, W. D. Block, D. K. Noffze, P. J. Dolphin, J. Edelglass, D. D. Bogorad, and P. Alaupovic, Familial deficiency of apolipoproteins A-I and C-III and precocious coronary-artery disease, N. Engl. J. Med. 306:1513 (1982).
7. H. Maeda, R. K. Hashimoto, T. Ogura, S. Hiraga, and H. Uzawa, Molecular cloning of a human APO C-III variant: Thr 74 --- Ala 74 mutation prevents O-glycosylation, J. Lipid Res. 28:1404 (1987).
8. M. I. Kamboh, R. E. Ferrell, and B. Sepehrnia, Genetic studies of human apolipoproteins. II. A. rapid one-dimensional isoelectric focusing technique to characterize apolipoproteins A-I, A-II, A-IV and C-II of unfractionated plasma, Electrophoresis 8:355 (1987).
9. Lipid Research Clinics Program. Manual of Laboratory Operations. Lipid and Lipoprotein Analysis (DHEW Publication No. M.I.H. 75-628) National Institutes of Health, Bethesda, MD. (1974).
10. G. Mancini, A. O. Carbonara, and J. F. Heremans, Immunochemical quantiation of antigens by single radial immunodiffusion, Immunoche 2:235 (1965).
11. E. Boerwinkle, R. Chakraborty, and C. F. Sing, The use of measured genotype information in the analysis of quantitative phenotypes in man. I. Models and analytical methods. Ann. Hum. Genet. 50:181 (1986).
12. C. G. Bhattacharya, A. simiple method of resolution of a distribution into Gaussian components, Biometrics 23:115 (1967).
13. R. A. Hegele, and J. L. Breslow, Apolipoprotein genetic variation in the assessment of atherosclerosis susceptibility. Genet. Epidemiol. 4:163 (1987).

THE EFFECT OF APOLIPOPROTEIN E ALLELE SUBSTITUTIONS ON PLASMA LIPID AND APOLIPOPROTEIN LEVELS

Louis M. Havekes[1], Peter de Knijff[1], Marijke Smit[2] and Rune R. Frants[2]

[1]Gaubius Institute TNO, Leiden, and
[2]Dept. of Human Genetics, Sylvius Laboratories
University of Leiden, The Netherlands

INTRODUCTION

The apolipoprotein E (apo E) present on very low density lipoprotein (VLDL-) and chylomicron-remnants, plays a central role in the hepatic metabolism of these particles as this apolipoprotein is recognized with high affinity by hepatic lipoprotein receptors (1,2).

Human apo E can be separated by isoelectric focusing into three major isoforms i.e. E2, E3 and E4 which differ in pI by a single charge unit, apo E4 being the most basic and E2 the most acidic. This heterogeneity is the result of three different apo E alleles, E*4, E*3 and E*2 at one single genetic locus (3,4).

Apo E3 is the most commonly occurring or wild type form. Apo E4 is derived from E3 by a Cys → Arg substitution at position 112 and is designated as E4(Cys_{112} → Arg). Apo E2 is derived from E3 by an Arg → Cys substitution at position 158 and is designated as E2(Arg_{158} → Cys).

From population studies several hypotheses have been elaborated about the influence of the apo E polymorphism on plasma lipid levels in a normal population. It has been postulated that the E*2 allele is associated with subnormal plasma and LDL-cholesterol levels whereas the E*4 allele seems to be associated with elevated plasma cholesterol levels (5-8).

In this paper we report the apo E phenotype and gene frequencies together with plasma levels of cholesterol, triglyceride, apo B and apo E in 2,018 35-year old male individuals randomly selected from the Dutch population.

MATERIALS AND METHODS

EDTA plasmas were obtained by venapuncture and stored at -20°C until the assays were performed. Apo E phenotyping was performed using a recently developed rapid micromethod which is based on isoelectric focusing of delipidated plasma samples followed by immunoblotting (9) using a polyclonal anti-apo E antiserum as first antibodies. This method is especially suitable for a large scale screening.

Plasma cholesterol and triglycerides were measured enzymatically using Boehringer test-kits (cholesterol CHOD-PAP and triglyceride GPO-PAP, respectively). Apo B concentrations were measured by immunonephelometric assay (INA) as described by Rosseneu et al. (10). Plasma apo E levels were measured by enzyme-linked immunosorbent assay (ELISA) as described by Bury et al. (11).

Differences in apo E phenotype distribution between different population samples were determined by chi-square analysis. Differences in plasma cholesterol, triglycerides, apo B and apo E levels between groups of different apo E phenotypes were evaluated by Student's t-test for unpaired samples. The average effects of the three apo E alleles on the plasma cholesterol apo B and apo E concentrations were estimated exactly according to the method of Sing and Davignon (12).

RESULTS AND DISCUSSION

Figure 1 represents an immunoblot of an isoelectric focusing slab gel (pH range 5-7) with delipidated plasma samples from 19 normolipidemic individuals. The relatively weak bands acidic to the major bands originate from partial sialylation and/or deamidation of the major isoforms and thus are not taken into consideration for the phenotyping.

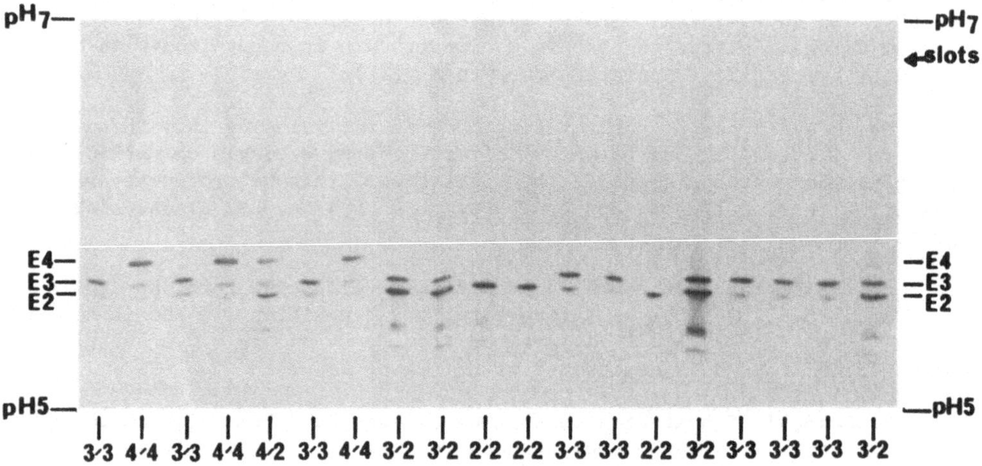

Fig. 1. Immunoblot of an isoelectric focusing slab gel showing the apo E phenotypes of 19 normolipidemic individuals.

Using this rapid micro-method for apo E phenotyping directly in serum, we determined the apo E phenotype distribution and the apo E allele frequencies in 2,018 35-year old males randomly selected from the Dutch population (Table 1).

In Table 2 the apo E allele frequencies obtained in the present study are compared with those observed in some other populations. In this table, only populations with a number of subjects exceeding 300 are considered. Multi-sample χ^2 analysis showed that there are statistically significant differences in the apo E allele frequencies between the different populations (df = 14; χ^2 = 149; $p < 0.001$). Two-sample χ^2 analysis showed that the allele frequencies of the Dutch population differ highly significantly from those of the populations of Japan, Finland, New Zealand ($p < 0.001$) and U.S.A. ($p < 0.005$). No significant differences were found with the German populations and the population of Scotland.

Table 1. Apo E phenotype and allele frequencies in randlomly selected 35-year old males.

phenotype	numbers observed* (relative frequency)	
E4/E4	59	(2.9)
E4/E3	512	(25.4)
E4/E2	45	(2.2)
E3/E3	1128	(55.9)
E3/E2	261	(12.9)
E2/E2	13	(0.7)
total number	2018	(100)

allele	relative frequency
E*4	0.167
E*3	0.750
E*2	0.082

*The phenotype distribution is in Hardy-Weinberg equilibrium (df = 5; χ^2 = 2.83)

Table 2. Apo E gene frequencies in several random population samples.

population sample	number of subjects	apo E allele frequency E*2	E*3	E*4	difference from the Dutch population (p-value)	ref.
The Netherlands	2018	0.082	0.751	0.167	-	this study
Scotland	400	0.083	0.770	0.145	N.S.	13
Germany (Münster)	1000	0.078	0.783	0.139	N.S.	14
Germany (Marburg)	1031	0.077	0.773	0.150	N.S.	5
U.S.A.	1204	0.075	0.786	0.135	< 0.005	8
Finland	615	0.041	0.733	0.227	< 0.001	6
New Zealand	426	0.119	0.739	0.141	< 0.001	15
Japan (Asahikawa)	576	0.037	0.846	0.117	< 0.001	16

The Japanese and New Zealand population differ in apo E allele frequencies from the Dutch population because of their relatively low and high E*2 allele frequencies, respectively.

The Finnish population (6) differs from the Dutch population in apo E allele frequencies both by a decreased E*2 allele and an increased E*4 allele frequency. For the American population (8) the E*4 allele frequency is the major contributor to the difference in apo E allele frequencies.

The frequencies of the E*3 allele appears to be rather similar for all population samples considered. The differences in apo E allele frequencies between the Dutch, the Finnish and Japanese populations may be due to differences in ethnic background and geographical isolation, whereas the differences between the Dutch population, a community in New Zealand (15) and a U.S.A. population (8) might be due to a combination of population admixture and genetic drift.

Compared to the most common E*3 allele, the E*4 allele leads to elevated plasma cholesterol levels, whereas the E*2 allele is associated with a decreased plasma cholesterol concentration (Table 3).

Table 3. Plasma levels of cholesterol, triglyceride, apo B and apo E in relation to apo E phenotype.

phenotype	cholesterol (mmol/l)	triglyceride (mmol/l)	apo B (mg/dl)	apo E (mg/dl)
E4/E4	5.82 ± 1.05	1.72 ± 0.85	127.04 ± 34.64	4.37 ± 1.96
E4/E3	5.70 ± 1.04	1.84 ± 1.20	122.94 ± 38.61	5.00 ± 1.99
E4/E2	5.36 ± 1.06	1.91 ± 1.25	111.91 ± 35.66	6.68 ± 2.14
E3/E3	5.61 ± 1.02	1.70 ± 0.99	117.30 ± 35.15	5.45 ± 1.97
E3/E2	5.31 ± 1.27	1.83 ± 1.31	107.76 ± 42.51	7.29 ± 2.88
E2/E2	4.83 ± 0.98	1.66 ± 0.83	70.33 ± 11.54	13.06 ± 3.59
total population	5.59 ± 1.07	1.76 ± 1.09	117.10 ± 37.42	5.61 ± 2.32

Values represent mean ± standard deviation.

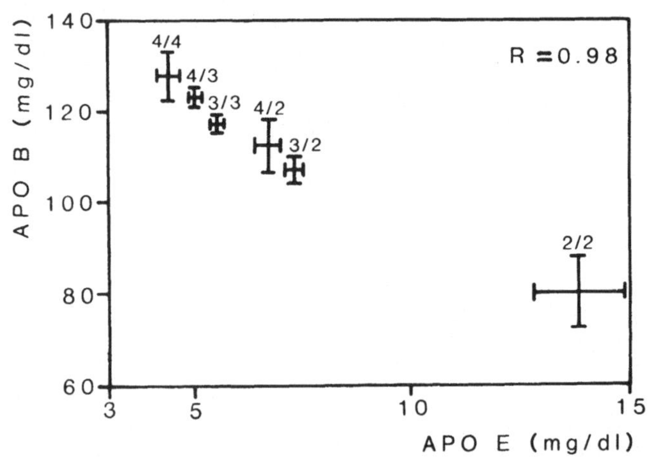

Fig. 2. Inverse relationship between apo B and apo E levels in relation to apo E phenotype.

A statistically significant effect of the allelic variation at the APO E locus on plasma triglycerides was not observed. Parallel to that of plasma cholesterol levels the mean plasma apo B levels also depend significantly on the apo E phenotype, although more pronounced. The effect of allelic substitution at the APO E gene locus on apo E levels is the opposite to that on plasma cholesterol and apo B. The results in Table 3 show that the E*4 allele leads to decreased apo E levels, whereas the E*2 allele is strongly associated with an increased apo E concentration. Considering the mean apo B and apo E levels calculated for each phenotype group separately, a strong inverse relationship was observed between apo B and apo E (Fig. 2).

The mechanisms underlying the relationship between apo E phenotype and plasma lipid and apolipoprotein levels are at present assumed to be the result of: (i) a more efficient catabolism of chlomicron- and VLDL-remnants by the liver in individuals with the E*4 allele and (ii) a less efficient catabolism of these lipoprotein particles in subjects exhibiting the E*2 allele due to a defect in binding of apo E2 to hepatic lipoprotein receptors (7,17).

Table 4. Average effects of the common apo E alleles on the plasma cholesterol, apo B and apo E levels, expressed as percentage of the respective total population means.

population sample (number; reference)	apo E allele	relative frequency	average allelic effect		
			cholesterol	apo B	apo E
The Netherlands (2.018; this study)	E*2	0.082	-4.5	-9.7	37.6
	E*3	0.751	0.3	0.0	-1.8
	E*4	0.167	2.0	4.5	-10.7
Germany (1000; 14)	E*2	0.078	-5.9		
	E*3	0.783	0.2		
	E*4	0.139	2.4		
Canada (102; 12)	E*2	0.078	-7.2	-11.2	
	E*3	0.770	-0.2	0.8	
	E*4	0.152	4.6	1.7	
Finland (207; 6)	E*2	0.063	-8.3	-14.5	
	E*3	0.698	-0.7	-1.3	
	E*4	0.239	4.0	7.8	

We calculated the average effects of the three apo E alleles on the lipid and apolipoprotein levels (Table 4) according to the formula of Sing and Davignon (12). The E*2 allele is accompanied by a reduction in plasma cholesterol (4.5%) and apo B (9.7%) and an increase in apo E levels (37.6%), whereas the E*4 allele induces an opposite effect (2.0, 4.5 and 10.7%, respectively). In comparison to the average effects of E*2 and E*4 alleles, the E*3 allele does not seem to influence these lipoprotein parameters. Similar observations could be drawn from results reported by others (Table 4). The data summarized in Table 4 also demonstrate that the effect of the allelic substitution at the apo E locus on plasma apo B is more pronounced than the effect on total plasma cholesterol levels, whereas the effect on plasma apo E levels is most dramatic and the opposite to that on plasma cholesterol and apo B.

This suggests that the apo E genes primarily affect apo E concentrations and thus that the metabolism of apo E containing lipoproteins thereby regulates the LDL cholesterol and apo B concentrations in plasma.

Although the mean average effects of the allelic substitutions at the APO E gene locus on plasma lipoprotein parameters are statistically significant, a simple relationship between apo E phenotype and atherosclerotic risk has not yet been established. The elevated plasma (LDL-)cholesterol levels in individuals with the E*4 allele is due to a more efficient catabolism of chylomicrons and VLDL in these subjects (18) thereby preventing the accumulation of atherogenic chylomicron- and VLDL-remnants. If the LDL concentrations are only moderately elevated, E*4-bearing individuals will be at lower risk. In individnals with the E*2 allele, the LDL cholesterol levels are low due to an impaired VLDL- and chylomicron-remnant catabolism. These individuals are at low risk as long as the levels of the atherogenic remnant particles remain below the level at which atherosclerotic risk increases. This relationship between the efficiency of chylomicron- and VLDL-remnant catabolism on one side and the level of LDL cholesterol on the other side might be responsible for the lack of a general relationship between apo E phenotype and atherosclerotic risk notwithstanding the contribution of the polymorphic APO E gene locus to the plasma cholesterol, apo B and apo E levels.

These investigations were supported by the Division for Health Research TNO (project no. 900-504-059).

REFERENCES

1. B. C. Sherrill, T. L. Innerarity, and R. W. Mahley, Rapid hepatic clearance of the canine lipoproteins containing only the E apoprotein by a high affinity receptor, J. Biol. Chem. 255:1804 (1980).
2. K. H. Weisgraber, T. L. Innerarity, and R. W. Mahley, Abnormal lipoprotein receptor binding activity of the human E apoprotein due to cysteine-arginine interchange at a single site, J. Biol. Chem. 257:2518 (1982).
3. V. I. Zannis, and J. L. Breslow, Human very low density lipoprotein apolipoprotein E isoprotein polymorphism is explained by genetic variation and posttranslational modification, Biochemistry 20:1033 (1981).
4. G. Utermann, A. Steinmetz, and W. Weber, Genetic control of human apolipoprotein E polymorphism: comparison of one- and two-dimensional techniques of isoprotein analysis, Hum. Genet. 60:344 (1982).
5. G. Utermann, I. Kindermann, H. Kaffarnik, and A. Steinmetz, Apolipoprotein E phenotypes and hyperlipidemia, Hum. Genet. 65:232 (1984).
6. C. Ehnholm, M. Lukka, T. Kuusi, E. Nikkilä, and G. Utermann, Apolipoprotein E polymorphism in the Finnish population: gene frequencies and relation to lipoprotein concentrations, J. Lipid Res. 27:227 (1986).
7. G. Utermann, Apolipoprotein E polymorphism in health and disease, Am. Heart J. 113:433 (1987).
8. J. M. Ordovas, L. Litwack-Klein, P. W. F. Wilson, M. M. Schaefer, and E. J. Schaefer, Apolipoprotein E isoform phenotyping methodology and population frequency with identification of apoE1 and apoE5 isoforms, J. Lipid Res. 28:371 (1987).
9. L. Havekes, P. de Knijff, U. Beisiegel, J. Havinga, M. Smit, and E. Klasen, A rapid micro-method for apolipoprotein E phenotyping directly in serum, J. Lipid Res. 28:455 (1987).

10. M. Rosseneu, N. Vinaimont, R. Vercaemst, W. Dekeersgieter, and F. Belpaire, Standardization of immunoassays for the quantitation of plasma apo B protein, Anal. Biochem. 116:204 (1981).
11. J. Bury, R. Vercaemst, M. Rosseneu, and F. Belpaire, Apolipoprotein E quantified by enzyme-linked immunosorbent assay, Clin. Chem. 32:265 (1986).
12. C. F. Sing, and J. Davignon, Role of the apolipoprotein E polymorphism in determining normal plasma lipid and lipoprotein variation, Am. J. Hum. Genet. 37:268 (1985).
13. A. M. Cumming, and F. W. Robertson, Polymorphism at the apoprotein-E locus in relation to risk of coronary disease, Clin. Genet. 25:310 (1984).
14. H. J. Menzel, R. G. Kladetzky, and G. Assmann, Apolipoprotein E polymorphism and coronary artery disease, Arteriosclerosis 3:310 (1983).
15. M. R. Wardell, P. A. Suckling, and E. D. Janus, Genetic variation in human apolipoprotein E. J. Lipid Res. 23:1174 (1982).
16. M. Eto, K. Watanabe, and K. Ishii, A racial difference in apolipoprotein E allele frequencies between the Japanese and Caucasian populations. Clin. Genet. 30:422 (1986).
17. G. Utermann, Morgagni lecture. Genetic polymorphism of apolipoprotein E - impact on plasma lipoprotein metabolism, in: "Diabetes, Obesity and Hyperlipidemias - III," Crepaldi et al. eds., Elsevier Science Publishers B.V., (1985).
18. R. E. Gregg, L. A. Zech, E. J. Schaefer, D. Stark, D. Wilson, and H. B. Brewer, Abnormal in vivo metabolism of apolipoprotein E4 in humans, J. Clin. Invest. 78:815 (1986).

MOLECULAR GENETICS OF CORONARY HEART DISEASE

David J. Galton

Medical Professorial Unit
St. Bartholomew's Hospital
London EC1

INTRODUCTION

Genes underlying the inheritance of atherosclerosis are implicated by family and twin studies. The aggregation of coronary artery disease in families has been reported by many authors since 1948 (1). For example, Slack and Evans (2) analysed first degree relatives of 121 men and 96 women with coronary artery disease. The increased risks of death from coronary artery disease in such relatives were five and seven-fold greater than in matched controls for males and females respectively. Familial clustering of coronary artery disease was noted especially for female patients. In Southern Finland, 104 out of 296 brothers of patients with coronary artery disease also had arterial disease compared to 8 out of 81 brothers of healthy controls (relative incidence 3.5 for brothers of probands with coronary artery disease (3) A detailed analysis of the Finish data yielded heritability estimates compatible with almost total determination of the disease by additive polygenic factors in the youngest age groups (myocardial infarction prior to the age of 46 years). In another study (4),of 207 patients who had myocardial infarcts before the age of 55 years, the highest "risk-ratios" calculated for nineteen independent variables were found with a positive family history of coronary artery disease (10.5) and lesser "risk ratios" were found with plasma cholesterol levels greater than 270 mg/dl. (4.3) cigarette smoking (4.0) and stroke in a first degree relative (3.5). The "risk ratio" for a family history of coronary artery disease was greater than that for individuals in the highest quintile of cholesterol levels. This observation may suggest that major genetic effects are not necessarily mediated by pathways of cholesterol metabolism. From this study, the heritability of coronary heart disease of early onset was calculated to be 0.63. If the families in which the proband had a monogenic hyperlipidaemia were eliminated, the heritability estimate remained as high as 0.56 (4). Equally, persuasive evidence comes from twin studies (5). Concordance rates for coronary artery disease (diagnosed by angina pectoris or myocardial infarction) in monozygotic twin pairs was found to be 0.65 compared to 0.25 in dizygotic twin pairs in a Norweigan Study.

If twins with coronary artery disease occurring before the age of 60 years were considered alone, the concordance rates were 0.83 and 0.22 for monozygotic and dizygotic pairs respectively (6). Of 17 twin pairs where both members had established coronary artery disease, 12 were monozygotic and 5 dizygotic, a highly significant difference ($p<0.02$). Earlier studies also found and increased concordance rates for coronary artery disease in mono compared to dizygotic twin pairs (7, 8 and 9).

One of the main aims of genetic studies are now to identify and locate major distinct genes that are implicated in excess lipid deposition in arterial walls leading to atherosclerosis. It is possible that two or three major genes could be interacting with several minor genes to account for the polygenic inheritance of this disease. There is widespread variation in the incidence of atherosclerosis throughout world populations and the major genes may also vary amongst world populations. This may arise from differences in environmental factors with which these major genes interact for atherosclerosis to develop (10). For example, a major environmental factor for macrovascular disease in European populations may be high dietary intake of animal fats, and genes involving lipid transport may be of major importance for the inheritance of the disease; whereas in Asian Indian populations, the occurrence of diabetes mellitus may be a major environmental factor and an alternative set of genes may therefore be involved for the occurrence of macrovascular disease. In an attempt to identify these major genes, there is a divergence in approach from the study of monogenetic disease. In the latter group of diseases, a highly successful strategy has been to find linkage markers (as DNA polymorphisms on a particular chromosome) and then to see if the DNA marker tracks with affected individuals in large pedigrees in which the disease segregates. Having found a linkage marker, techniques are available (such as stepwise chromosome walking and chromosome-mediated gene transfer) to identify the disease locus and thus directly study the mutation possible for the inheritance of the disease. Such an approach has already proved successful in the study of the genetics of cystic fibrosis, Duchenne Muscular Dystrophy and Huntington's Chorea. However the approach is not so applicable to the study of polygenic disease. If for example genetic variants at three separate loci are required for transmission of the disease to family members, it is quite possible that unaffected relatives will also possess one or two of these genetic determinants. One needs to know all the disease related alleles to be able to discriminate patients from unaffected siblings in such pedigrees. For example, there is a hypervariable DNA polymorphism at the 5' end of the insulin gene on the short arm of chromosome 11. Homozygosity of a short-insertional allele (genotype 1/1) appears to associate with type 1 diabetes mellitus (11, 12). This may be one of several genetic determinants for the disease; but in pedigree studies, it is not of use to discriminate patients from non-diabetic relatives. For example in 17 diabetic pedigrees, 21 diabetic siblings were found to be homozygous for the short allele, as were 28 non-diabetic siblings (12). If the other genetic determinants were studied at the same time such as DNA variants at the HLA-DR locus on chromosome 6, and used in combination with the insulin gene polymorphism, this may predict more accurately, which individuals will develop the disease (13). But the pedigree approach may not be helpful until all the major genetic determinants for the disease have been established.

A more powerful approach to family studies may be by sib-pair analysis. In this approach affected sib-pairs within a pedigree are genotyped to see if they are concordant for haplotypes at the locus of interest. It may be necessary to study the whole family (either parents or offspring) to find out which of four chromosomes are being transmitted to the affected sibs from their parents. This depends on how polymorphic is the marker, e.g. it depends on the heterozygote frequency. If the site variants are only "plus" or "minus," more complete pedigree data is required than if the site is highly polymorphic because then each of the parental chromosomes can be defined more accurately. This method is specially suited to the study of inherited diseases of variable penetrance and variable age of onset because unaffected siblings (that may go on to develop the disease) are not included in the analysis.

An alternative approach for a polygenic disease is to search for DNA linkage markers that associate in groups of unrelated individuals with atherosclerosis. Having found a disease related marker, it may then be possible to locate the aetiological locus giving rise to the association. The controls are now not first degree relatives (who themselves are very likely to possess the same linkage marker) but carefully defined healthy groups matched for age, race and geographic origins. Since atherosclerosis develops with increasing age, it is preferable that the control group is decades older than the patient group to reduce the number of individuals in the controls who are likely to develop clinical features of atherosclerosis. A useful control group is one demonstrated to be free of macro-vascular disease by angiography. In addition, multiple samples from hospital and non-hospital populations will also help to establish control frequencies of the genetic variants under study. When a disease marker has been identified, it is important to study its frequency - distribution in other world populations to see if the disease association still occurs in other racial groups. The more wide-spread the disease-related marker, the more likely it is to be close to a major aetiological locus.

The ability to detect restriction fragment length polymorphisms (RFLP's) by the techniques of Southern blotting has provided a wealth of new genetic markers. It is possible that approximately 1:200 base pairs may show polymorphic sequence variation in particular DNA segments of the human genome (14). Such RFLP's may occur in exons, introns, flanking sequences or intergenic regions; and at some of these sites (exons and 5' - flanking sequences), DNA polymorphisms could have direct phenotypic effects. DNA polymorphic variants at the apolipoprotein gene clusters (on chromosome 11 and 19) are currently under intensive investigation for possible associations with premature atherosclerosis (15). Studies of the gene cluster for apolipoproteins AI CIII and AIV and the B-apolipoprotein locus on chromosome 2 have so far produced significant results. They will now be reviewed to illustrate the approach.

The Apolipoprotein AI-CIII-AIV Gene Cluster

The above three genes are clustered on the long arm of chromosome 11 on a DNA segment of approximately 14 Kb (16, 17). The organisation of the cluster is presented in Table 1 and shows the following features: 1. The apo CIII gene is transcribed in the opposite direction to the apo A1 and AIV genes, despite being within 3 Kb of the 3' end of the apo Al gene.

2. More than 9 restriction enzyme dimorphisms along the length of this part of the genome have been described, occurring within introns, exons, intergenic sequencies and in flanking sequences (18, 19). Eight studies have been performed in Caucasian populations from the United Kingdom, West Germany and United States, examining the frequencies of allelic variations at these restriction sites to see if any associate with premature atherosclerosis. The results are as follows:

TABLE 1

THE FREQUENCY OF THE S2 ALLELIC VARIANTS OF THE APO CIII GENE IN CONTROL POPULATIONS

Control Groups	n	Allelic Frequencies S1	S2	References
Random medical outpatients	37	0.96	0.04	Rees et al (18)
Health screen clinics:-				
Samples:- 1	52	1.0	0.00	Rees et al (21)
2	74	0.98	0.02	Ferns et al (20)
3	56	0.98	0.02	O'Connor et al (36)
Angio-room:- Normal coronary arteries	68	0.97	0.03	Rees et al (37)
Chest Clinic - random sample	35	0.99	0.01	Trembath et al (38)
Normolipidaemic controls	71	0.94	0.06	Kessling et al (40)
Random normals	101	0.94	0.06	Deeb et al (25)
controls	66	0.98	0.02	Hegele et al (26)

Results are mean allelic frequencies of a mutation involving a C - G transversion in the fourth exon of the apolipoprotein CIII gene on the long arm of chromosome II.

United Kingdom

Table 1 establishes the control frequency of an uncommon allele (the S2 allele) at the Sst 1 restriction site in the fourth exon of the apo CIII gene. Multiple samples since 1983 from Hospital outpatient clinics, health screening clinics and a group free from coronary artery disease as defined by coronary angiography all give a similar range of allele frequencies between 0.01 - 0.04. Patient groups with hyperlipidaemia, coronary and extra-coronary atherosclerosis have been studied for the frequency of this apo C-III polymorphism and the results are summarised in Table II. The S2 allelic frequency is increased between 6 to 10 fold in these groups, compared to controls and suggests that the mutation is acting as a linkage marker for an atherogenic allele in the vicinity (20, 21).

TABLE II

THE APO CIII MUTATION: FREQUENCY OF THE S2 ALLELIC VARIANTS IN DISEASE POPULATIONS

Patient Groups	n	Allelic Frequencies S1	Allelic Frequencies S2	References
Hyperlipidaemic (IV/V)	28	0.80	0.20	Rees et al (18)
Survivors of M I	48	0.88	0.12	Ferns et al (20)
Coronary athero-sclerosis	61	0.89	0.11	Rees et al (21)
Peripheral atherosclerosis	49	0.86	0.12	O'Connor et al (36)
Diabetic survivors of M I	47	0.86	0.14	Trembath et al (38)
Hyperlipidaemia with gout	22	0.88	0.12	Ferns et al (39)
Coronary heart disease	140	0.88	0.12	Deeb et al (25)
Survivors of M I	66	0.96	0.04	Hegele et al (26)

Results are mean allelic frequencies of the mutation described in the Legend of Table 1.

When other restriction site polymorphisms (Mspl and Pst 1) were included in the analysis (22), thereby constructing DNA haplotypes, it was found that one particular haplotype containing the uncommon allele at the Msp 1 and Sst 1 sites was increased from 2% in normolipaemic controls ($n = 48$) to 21% in survivors of myocardial infarction ($n = 47$) giving a relative incidence of 12.7 ($p < 0.01$).

However because of tight linkage disequilibrium between the alleles studied, it was not possible to identify haplotypes associated with any greater risk of premature atherosclerosis than when the Sst-1 polymorphism was considered in isolation. In a study from Edinburgh (23), the frequency of the S2 allele was unaltered between controls and patients with coronary heart disease but the frequency of an uncommon allele (the P2 allele) at the restriction site for Pst 1 was 6% in controls (n = 64) compared to 12% in patients with coronary heart disease. Such results must be interpreted with caution however. A common disease such as premature atherosclerosis may be particularly heterogeneous with different polygenic determinants operating in different geographical localities. The patient groups must be clearly defined and made as homogeneous as possible with regard to racial and geographic origins, and clinical phenotype features. As previously mentioned, atherosclerosis may have a very variable age of onset before the diagnosis can be established with certainty and the control groups may contain individuals who will go on to develop atherosclerosis at a later age. For example the frequency of the S2 allele in healthy subjects from Scotland (23), was reported as 18% (n = 64) compared to 4% in London (n = 47). This may represent a real difference in allelic frequencies or simply be due to differences in exclusion criteria for constituting a control group (i.e. presence or absence of hyperlipidaemia, diabetes or a family history of coronary artery disease etc). It is of interest that the S2 allele frequency in healthy Caucasian groups from Boston and Seattle are: 5% (n = 66) and 6% (n = 101) respectively, (26, 25). This suggests that it is important to establish strict inclusion criteria for healthy control groups. Other studies (21, 27) have used coronary arteriography to define the presence of atherosclerosis. In one report (21), the frequency of the S2 allele was 22% in patients with severe obstructive coronary atherosclerosis (n = 61) compared to 6% in patients with minimal disease (n = 68 $p < 02$).

United States

Studies from Boston, Seattle and New York have been reported. In the first (24) Caucasian patients (n = 88) with severe coronary artery disease were compared to Framingham control population (n = 64) matched for ethnic origin and other clinical criteria were carefully standardised. The frequency of an uncommon allele revealed by the enzyme Pst 1 at a restriction site 314 base pairs 3' to the apo A-1 gene was 32% in patients compared to 4% in matched controls ($p < 0.01$) and 3% in 30 subjects with no angiographic evidence for coronary artery disease giving a relative risk of coronary artery disease in individuals possessing the P2 allele of at least 10. The same rare allele was found at increased frequency in subjects with familial hypoalphalipoproteinaemia and the effect of this mutation may be mediated by lowering the levels of plasma HDL (24). Frequencies of alleles at other polymorphic sites at this locus were not reported. However in a study from Seattle (25) frequencies of alleles revealed by Pst-1 and Sst 1 restriction enzymes were compared in patient groups with coronary artery disease proven by arteriography (n = 140) and random "normals" (inclusion criteria not stated). No differences were observed in the frequency of the P2 allele, however the frequency in the control group was 10% which is three times that of the Boston Study. Clearly this will tend to minimise differences between patients and "normals." With regard to the S2 allele, frequencies were 6% (n = 101) and 12% (n = 140) in "normals" and patients respectively ($p < 0.05$).

In a third Study from New York (26) survivors of myocardial infarcts were examined for allelic frequencies at four polymorphic sites revealed with the enzymes Xmn 1, Msp 1, Pst 1 and Sac 1. The only significant difference was observed with the Xmn 1 polymorphism where the X2 allele frequency was lower in the patient group (n = 57) at 15% compared to 24% in controls (n = 57 $p < 0.05$). When individuals above the age of 60 years were studied, the P2 allele was significantly less frequent in patients (3% versus 21% for controls $p < 0.02$). The authors concluded that DNA polymorphisms near the apo A1 gene may be significantly associated with myocardial infarction.

Federal Republic Of Germany

A study from West Germany (27) examined six polymorphic sites at the apo AI-CIII-AIV gene cluster. These included sites for the enzymes Apa 1, Msp 1, Pst 1, Ban 11 and Pvu II. Only one, the Pvu II site, showed an uncommon allele that was increased in frequency in patients with angiographically proven coronary atherosclerosis (n = 41) giving a relative incidence of 3.59 ($p < 0.02$).

Interpretation

All 7 studies thus far show restriction site polymorphisms that associate in patient groups with atherosclerosis defined either by angiography or myocardial infarction but the involved sites differ markedly within Caucasian populations. This may be expected if the sites are only acting as linkage markers for an atherogenic allele in the vicinity. There is also variability amongst studies regarding the frequencies of allelic variants in control populations and by affecting the comparisons with patient groups may account for some of the inconsistencies. However, some tentative conclusions may be drawn. Firstly the restriction site polymorphisms so far studied probably arise from harmless mutations and are not functioning in any way as aetiological determinants. They are possibly background (or neutral) DNA variants that differ in frequencies amongst Caucasian and other racial populations in the same way that some HLA antigen frequencies are found to vary amongst different Caucasian populations. However the DNA polymorphisms may be acting as linkage markers for a neighbouring atherogenic mutation. There are at least two possible reasons why different neutral polymorphisms may act as linkage markers within a racial group. Firstly the putative atherogenic allele may have mutated more than once on separate chromosomes bearing different background polymorphisms depending on which chromosome the atherogenic mutation initially occurred. The background polymorphisms may have since "hitch-hiked" with the mutated atherogenic allele. An example of this is the sickle cell mutation in the B-globulin gene in West Africans that occurred on a chromsome carrying an Hpal polymorphism 13 Kb downstream from the globin gene (28). This Hpa 1 polymorphism acts as a linkage marker for the sickle-cell mutation but only in West African populations. Another example of this is in the field of lipid metabolism is familial hypertriglyceridaemia due to mutations within the apo C-II gene. In a pedigree study (29) from North Italy and Holland, it was observed that different apo-C-II alleles revealed by a Taq 1 restriction site polymorphism tracked with affected members of each pedigree. In the North Italian family, the affected members were associated with a 3.8 Kb allele; in the Dutch family by a 3.5 Kb allele.

The simplest explanation for these results is the apo C-II mutation has occurred at least twice on different chromosomes carrying different background polymorphisms at the Taq restriction site. Subsequent studies of the CII apolipoproteins have shown the occurrence of different animo-acid replacements amongst pedigrees with this form of familial hypertriglyceridaemia (30).

Acknowledgements

Financial support for the work presented comes from the Medical Research Council, British Diabetic Association and Wellcome Trust to which the author is very grateful.

References

1. Yater WM, Traum AH, Brown WG, Fitzgerlald RP, Geisler MA, Wilcox BB. Coronary artery disease in men eighteen to thirty-nine years of age. Am Heart J 36 : 334 - 372 (1948)

2. Slack J, Evans KA: The increased risk of death from ischaemic heart disease in first-degree relatives of 121 men and 96 women with ischaemic heart disease. J Med Genet 3 : 239 - 259 (1966)

3. Rissanen AM. Familial aggregation of coronary heart disease in a high incidence area (North Karelia, Finland) Br Heart J. 42 : 294 - 303 (1979).

4. Nora JJ, Lortscher RH, Spangler RD, Nora AH, Kimberling WJ : Genetic-epedemiologic study of early onset ischaemic heart disease. Circulation 61 : 503 - 508 (1980)

5. Berk K. Twin studies of coronary heart disease and its risk factors Acta Genet. Med Gemellol. 33 : 349 - 361 (1984)

6. Berg K : Genetics of coronary heart disease in Progress in Medical Genetics volume 35 ed. Steinberg AG, Bearn AG, Motulsky AR, Childs B. WB Saunders, Philadelphia 1983 p 35 - 90

7. Kahler OH, Weber R : Zur erbpathologise von Herz und Kreislaufer Krankungen. Z Klin Med 137 : 507 - 575 (1940)

8. Liljefors I : Coronary heart disease in male twins. Acta Med Scand Suppl 511 (1970)

9. De Faire U : Ischaemic heart disease death in discordant twins. Acta Med Scand. Suppl 568 (1974)

10. Galton DJ : Molecular Genetics of Common Metabolic Disorders publ Ed Arnold , London (1985)

11. Bell GI, Horita S, Karam JH. A highly polymorphic locus near the human insulin gene is associated with insulin-dependent diabetes. Diabetes 33 : 176 - 183 (1984)

12. Hitman GA, Tarn AC, Winter RM, Williams LG, Bottazzo GF, Galton DJ. Type I (insulin-dependent) diabetes and a highly variable locus close to the insulin gene on chromosome II. Diabetologia 28 : 218 - 222. (1985)

13. Ferns GAA, Hitman GA, Trembath R, Williams LG, Gale EA, Galton DJ. DNA polymorphic haplotypes on the short arm of chromosome II and the inheritance of Type I Diabetes Mellitus. Journal of Medical Genetics 23 : 210 - 216 (1986).

14. Jeffrey AJ (1979). DNA sequence variants in the Gg Ag, Delta and beta globulin genes of Man : Cell 18 : 1 - 12 (1979)

15. Galton DJ : DNA polymorphisms for the Genetic Analysis of Atherosclerosis. G Schlierf, H Morl (Eds) Expanding Horizons in Atherosclerosis Research pp 187 - 197. Springer-Verlag Berlin 1987

16. Karathanasis SK, Zannis VI, Breslow JL : Linkage of human apolipoprotein AI and CIII genes. Nature 304 : 371 - 374 (1983)

17. Karathanasis SK, Apolipoprotein multigene family : tandem organisation of human apolipoprotein AI, CIII and AIV genes. Proc. Natl. Acad. Sci USA 82 : 6374 6378 (1985)

18. Rees A, Stocks J, Shoulders CC, Galton DJ, Baralle FE : DNA polymorphism adjacent to the human apoprotein AI gene in relation to hypertriglyceridaemia. Lancet i : 444 - 447 (1983)

19. Seilhamer JJ, Protter AA, Frossard P, Levy-Wilson B. Isolation and DNA sequence of full length cDNA of the entire gene for human apolipoprotein AI. Discovery of a new polymorphism. DNA 3 : 309 - 317 (1984)

20. Ferns GAA, Stocks J, Ritchie C, Galton DJ : Genetic polymorphisms of apolipoprotein CIII and insulin in survivors of myocardial infarction. Lancet i : 300 - 304 (1985)

21. Rees A, Jowett NI, Williams LG, Stocks J, Vella MA, Camm J, Galton DJ : DNA polymorphisms flanking the insulin and apolipoprotein CIII genes and atherosclerosis. Atherosclerosis 58 : 269 - 275 (1985)

22. Ferns GAA, Galton DJ : Haplotypes of the human apoprotein AI-CIII-AIV gene cluster in coronary atherosclerosis. Human Genetics 73 : 245 - 249 (1986)

23. Price WH, Morris SW, Kitchin AH : Allele frequencies at five polymorphic DNA restriction enzyme sites in the apolipoprotein AI-CIII-AIV gene cluster and coronary heart disease in a Scottish Population. Clinical Sci 75 : suppl 16 pp 46P (1987)

24. Ordovas JM, Schaeffer EJ, Salem D, Ward RH, Glueck CJ, Vergani C, Wilson PWF, Karathanasis SK : Apolipoprotein AI gene polymorphism associated with premature coronary artery disease and familial hypoalphalipoproteinaemia. New Engl. J Med. 314 : 671 - 677 (1986)

25. Deeb S, Failor A, Brown BG, Brunzell JD, Albers JJ, Motulsky AG : Molecular Genetics of Apolipoproteins and Coronary Heart Disease : Cold Spring Harbor Symp. Quantit. Biol. LI : 403 - 409 (1987)

26. Hegele RA, Herbert PN, Blum CB, Buring JE, Hennekens CH, Breslow JL : Apolipoprotein AI and AII gene DNA polymorphisms and myocardial infarction (in press)

27. Frossard PM. Coleman R, Funke H, Assman G : Molecular genetics of the human apo AI-CIII-AIV gene complex. Application to Detection of Susceptibility to Atherosclerosis. ed W Hauss, R Wissler, J Grunwald. Recent Advances in Atherosclerosis Research, Westdeutscher Verlag, Dusseldorf 1987

28. Kan YW, Dozy AM : Polymorphisms of DNA sequence adjacent to human beta globin structural gene : relation to sickle cell mutation. Proc. Natl. Acad. Sci USA 75 : 5631 - 5636 (1978)

29. Humphries SE, Williams LG, Stahenhoef AF, Baggio G, Crepaldi C, Galton DJ, Williamson R : Familial apolipoprotein CII deficiency: a preliminary analysis of the gene defect in 2 pedigrees. Human Genet. 65 : 151 - 156 (1984)

30. Baggio G, Gabelli C, Manzato E. Martini S, Previato L, Verlato S, Brewer HB, Crepaldi C. (eds) Sirtori, Franceschini. Proceedings of NATO Advanced Research Workshop on Apolipoprotein Mutants. Apo - CII Padova: a new apoprotein variant in 2 patients with apo CII deficiency syndrome pp 203 - 210. (1986)

31. Knott TJ, Rall SC Jr. Innerarity TL, Jacobson SF, Urdea MS, Levy-Wilson B, Powell LM, Pease RJ, Eddy R, Nakai H, Byers M, Priestly LM, Robertson E, Rall LB, Betsholtz C, Shows TB, Mahley RW, Scott J. Human apolipoprotein B : Structure of carboxy-terminal domains, sites of gene expression and chromosomal localisation. Science 230 : 37 - 43 (1985)

32. Shoulders CC, Myant N. Sidoli A, Rodriguez JC, Cortese C, Baralle FE. Molecular cloning of human LDL apolipoprotein B cDNA. Atherosclerosis 58 : 277 - 2 (1985)

33. Knott TJ, Pease RJ, Powell LM, Wallis SC, Rall SC. Innerarity TL, Blackhart B, Taylor WR, Lusis AJ, McCarthy BJ, Mahley RW, Levy-Wilson B, Scott J. Human apolipoprotein B : Complete cDNA sequence and identification of domains of the protein. Nature 323 : 734 - 738 (1986)

34. Ferns GAA, Galton DJ. Frequency of Xba I polymorphism of the apolipoprotein B Gene in Myocardial Infarct Survivors. Lancet IIi : 572 (1986)

35. Hegele RA, Huang LS, Herbert PN, Blum CB, Buring JE, Hennekens CH, Breslow JL : Apolipoprotein B-gene DNA polymorphisms associated with myocardial infarction. New Engl J Med. 315 : 1509 - 1515

36. O'Connor G, Stocks J, Lumley J, Galton DJ : A DNA polymorphism of the apolipoprotein CIII gene in extracoronary atherosclerosis Clin Sci 74 : 289 - 292 (1988)

37. Rees A, Stocks J, Sharpe CR, Shoulders CC, Jowett NI, Baralle FE, Galton DJ. DNA polymorphisms in the apo AI-CIII gene cluster : association with hypertriglyceridaemia. Journal of Clinical Investigation 76 : 1090 - 1095 (1985)

38. Trembath RC, Thomas DJ, Hendra T, Yudkin J, Galton DJ
A DNA polymorphism of the apo AI-CII-AIV gene cluster associates with coronary heart disease in non-insulin dependent diabetes. Brit Med Journ. 294 : 1577 - 1579 (1987)

39. Ferns GAA, Lanham J, Galton DJ. Polymorphisms of the apolipoprotein AI-CIII gene cluster in subjects with hypertriglyceridaemia associated with primary gout. Human Genetics 75 : 121 - 129 (1987)

40. Kessling AM, Horsthenke B, Humphries SE. A study of DNA polymorphisms around the human apolipoprotein AI gene in hypolipidaemic and normal subjects. Clin. Genet. : 296 - 306 (1985)

MOLECULAR BIOLOGY OF HUMAN APOLIPOPROTEIN B AND RELATED DISEASES

Vassilis I. Zannis, M. Mahmood Hussain, Margarita Hadzopoulou-Cladaras, Anastasia Kouvatsi, Dimitris Kardassis, and Christos Cladaras

Section of Molecular Genetics
Cardiovascular Institute
Departments of Medicine and Biochemistry
Boston University Medical Center
80 East Concord Street
Boston, MA 02118

INTRODUCTION

Apolipoprotein B is the main protein component of LDL and comprises 23.8% of the LDL particle (1,2). ApoB is the ligand for the cellular recognition and catabolism of LDL by the LDL receptor (1,3). The LDL receptor-apoB interaction and subsequent catabolism mediates the clearance of LDL from plasma and regulates cellular cholesterol biosynthesis (1,3). Early studies had shown that the human and rat apoB exist in two primary forms designated apoB-100 and apoB-48 (4-12). Apolipoprotein B seems to be important for lipoprotein assembly and secretion (14,15). Thus, absence of apoB in abetalipoproteinemia (16) or inhibition of its synthesis by cyclohexamide treatment of cells (15) abolishes secretion of chylomicrons and VLDL.

Primary Structure of ApoB-100

The primary sequence of apoB-100 has been derived recently by four independent research groups (17-20) from the corresponding sequence of overlapping apoB-100 cDNA clones. The main features of the complementary mRNA and protein sequence of apoB-100 are shown in Table I.

Distribution of Cysteines and N-glycosylation Sites in the ApoB-100 Sequence

ApoB-100 sequence contains 25 cysteines and 20 N-glycosylation sites (Fig. 1) (17-20). It is interesting that 18 of these cysteine residues are located between residues 1 and 1635, i.e., approximately the N-terminal one-third of the molecule. Computer analysis of beta turn potential shows that this region contains a greater frequency of strongly predicted

Table I. Features of ApoB-100 mRNA and Protein Sequence

The total length of apoB-100 mRNA is 14,112 nucleotides long.
The 5' untranslated region is 128 nucleotides long.
The 3' untranslated region is 301 nucleotides long.
The coding region is 13,680 nucleotides long.
The length of apoB-100 is 4,560 amino acids.
The signal peptide is 24 amino acids long.
The mature protein is 4,536 amino acids long.
The calculated molecular weight of apoB is 513,000 daltons.
Based on the molecular weight of apoB, it can be calculated that there is one apoB molecule per LDL particle.

beta turn residues than the center third of the molecule which contains only two cysteine residues. The extremely strongly predicted turn locations in the C-terminal third of the molecule again coincide with the location of 5 additional cysteine residues. This polarized location of both cysteine residues and potential beta turns may suggest domain structure for the tertiary folding of the protein (17).

Earlier studies showed that apoB-100 contains N- (21) and possibly O-linked carbohydrate chains (22,23). Examination of the derived amino acid sequence showed 20 potential N-linked glycosylation sites (17-20). Direct protein sequence showed that 13 of these (residues 956, 1341, 1350, 1496, 2955, 3074, 3197, 3309, 3331, 3384, 3438, 3868, and 4210) were glycosylated and four (residues 7, 429, 2212, and 2533) were not (19).

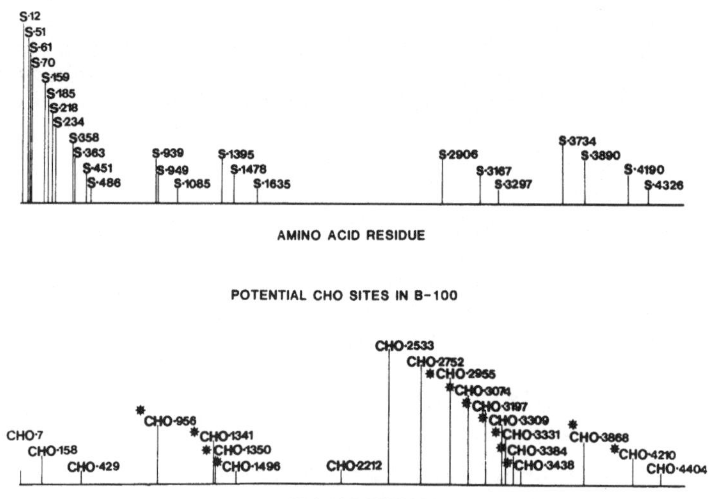

Fig. 1. Distribution of cysteines and potential N-glycosylation sites in the apoB-100 sequence. The symbol * indicates actual glycosylation sites

Practically nothing is known about the location and the physiological significance of O-linked carbohydrate chains of apoB-100.

The Receptor and Heparin Binding Domains of ApoB-100

A well characterized function of the human apoB is the binding to the LDL receptor which leads to the cellular catabolism of LDL (1,3). Domains of this protein may be important for binding to heparin (24-26), to the immunoregulatory receptors of lymphocytes (27), and for the receptor-independent catabolism of LDL by arterial wall cells (28). Chemical modification studies of Arg or Lys residues have shown that these positively charged residues are important for binding of apoB or apoE to the LDL receptor (29,30). Following the elucidation of the structure of the LDL receptor (31), it has been proposed that these positively charged amino acids of apoB or apoE interact with negatively charged residues that are present in the putatively identified ligand binding site of the LDL receptor (31). Similarly positively charged residues may be involved in the binding of apoB to heparin (24-26). Utilization of hydrophobicity and hydrophobic moment calculations for analysis of the structure of apoB identified several potential helical domains with strong polar character with a strong net positive charge and high hydrophobic moment. Four regions defined by residues 3144-3162, 3352-3369, 3595-3613, and 3666-3681 have five, five, four and three overall positive charges, respectively (Fig. 2 and Table IIA). One of these very polar regions with the highest hydrophobic moment occurs between residues 3352 and 3369. The sequence between residues 3359 and 3367 has 67% homology to the apoE sequence between residues 142 and 150 (32) that has been implicated previously in receptor binding (33,34) (Table IIB). Thus, although the region between residues 3352 and 3369 may play an important role in receptor binding, it is also possible that one or more of the adjacent regions participate in the formation of the receptor binding domain of apoB-100 (17). It is also interesting that three putative heparin binding domains of apoB-100 are defined by residues 3150-3157, 3361-3368 and 3670-3677 and overlap with peptides 1, 2, and 4 of Table IIA, that were defined by hydrophobicity and hydrophobic moment calculations. Four of the heparin binding peptides are defined by residues 5-99, 205-279, 875-932 and 2016-2151 (25,26). The implication that the region between residues 3350 and 3370 in receptor binding is supported further by the inhibition of binding of LDL to LDL receptor by antibodies that recognize the carboxy terminal T2 thrombolytic fragment of apoB-100 that originates at amino acid residue 3249 (35). It is also supported by the ability of a synthetic peptide corresponding to residues 3345-3381 of apoB to restore uptake of trypsin treated hypertriglyceridemic VLDL by cultured skin fibroblasts (19).

Lipid Binding Domains of ApoB-100

Computer aided analysis suggests that apoB-100 contains helical and beta structures. Some of the alpha helical and beta sheet regions can assume amphipathic structure. The assignment of a precise region of the structure responsible for lipid binding cannot be made clearly with this level of structural analysis. Amphipathic helical structure has been

Table IIA. Potential Receptor Binding Domains of ApoB-100

Region	Residue Range	Maximum Hydrophobic Moment	Minimum Mean Hydrophobicity	Total Charge Difference
1*	3352-3369	0.48	-0.74	+5
2*	3144-3162	0.39	-0.92	+5
3	3595-3613	0.45	-0.75	+4
4*	3666-3681	0.45	-0.56	+3

* The asterisk indicates that peptides 1, 2, and 4 have also been proposed as the heparin binding domains of apoB-100 (25).

Table IIB. Putative Receptor and Ligand Binding Domains of ApoE and ApoB-100

ApoE aa 141-150	Leu-Arg-Lys-Leu-Arg-Lys-Arg-Leu-Leu-Arg
ApoB aa 3358-3367 (domain 1)	Thr-Arg-Leu-Thr-Arg-Lys-Arg-Gly-Leu-Lys
LDL receptor binding domain (consensus sequence)	Asp-X-X-X-Asp-Cys-X-Asp-Gly-Ser-Asp-Glu

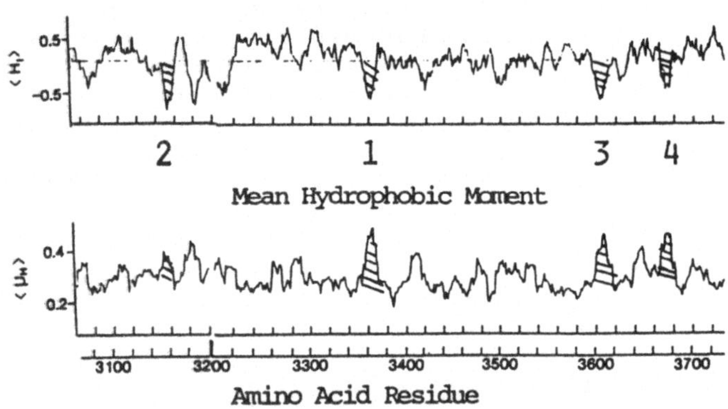

Fig. 2. Predicted mean hydrophobicity and hydrophobic moment values of the apoB-100 sequence. These values were derived by computer analysis of the apoB-100 structure as described in reference 17. Numbers 1, 2, 3, and 4 indicate the potential receptor binding domains 1, 2, 3, and 4 of apoB-100 shown in Table IIA.

strongly implicated in the lipid binding of other water soluble, exchangeable apolipoproteins (36,37). However, amphipathic helices are not unique to the plasma apolipoproteins. Many water soluble proteins with well defined tertiary structure contain segments of amphipathic helices which are frequently associated with each other (38).

The unique physicochemical properties of apoB-100, its insolubility and non-exchangeability, perhaps suggest that a different structural mechanism is involved in binding to the lipoprotein surface. Strongly apolar segments of secondary structure may associate to form a hydrophobic anchor domain, perhaps similar to that found in membrane proteins, which are then associated with the lipid interface. The predicted apolar

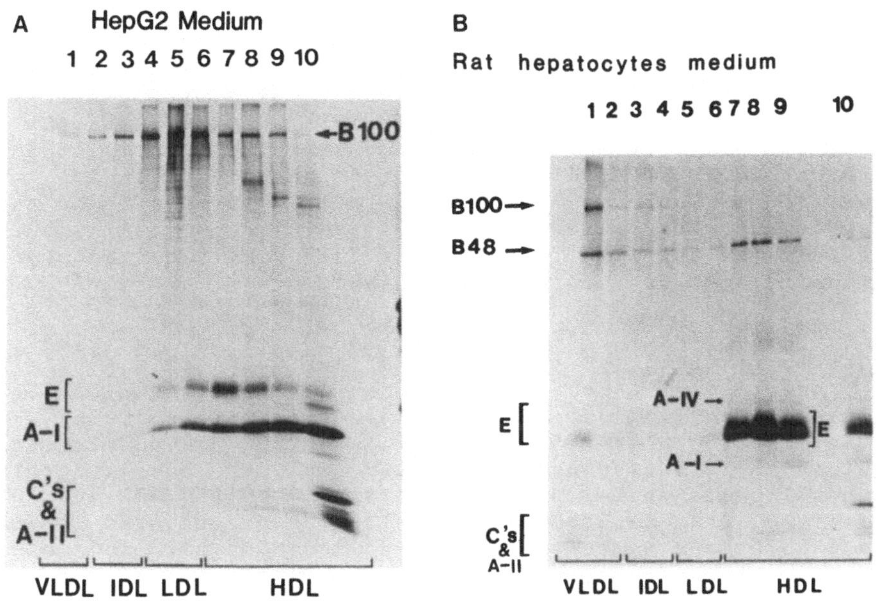

Fig. 3. One-dimensional gradient polyacrylamide gel electrophoretic analysis of nascent lipoproteins secreted by HepG2 cells and primary cultures of adult rat hepatocytes following purification by density gradient ultracentrifugation. HepG2 cells and rat hepatocytes in Panel A and B were labeled with ^{35}S-methionine for five hours and the medium was analyzed by density gradient ultracentrifugation (77). Twelve fractions 0.95ml each were collected. Aliquots of 0.7ml from the top ten fractions were analyzed by one dimensional gradient polyacrylamide gel electrophoresis and autoradiography. This figure shows the autoradiogram obtained from this analysis. The average density in each fraction (g/ml) was as follows: 1 = 1.01; 2 = 1.018; 3 = 1.024; 4 = 1.038; 5 = 1.056; 6 = 1.08; 7 = 1.11; 8 = 1.14; 9 = 1.17; 10 = 1.206.

helical and beta sheet regions may play a role in generating such domains. Representative alpha helices with an overall neutral to apolar character have been localized in the vicinity of residues 820-860, 2760-2791, and 3930-3960. Similarly representative apolar beta sheet regions have been localized in the vicinity of residues 380-400 and 1300-1320 (17). Systematic protein sequencing of surface and lipid bound apoB peptides indicated that these two kinds of peptides were alternating in the apoB-100 sequence. However, the amino terminal region (amino acids 1-1000) was greatly enriched in surface peptides and the middle region (amino acids 1360-3160) was enriched in lipid bound peptides (19).

Synthesis and Flotation Properties of Nascent ApoB-100 and ApoB-48 Forms

A quantitative study in the rat has indicated that the two major sites of apoB synthesis are the liver and the intestine and they contribute 84 and 16% respectively to total plasma apoB pool (39). Numerous studies have indicated that rat liver produces two forms of apoB designated apoB-48 and apoB-100, respectively, whereas small intestine synthesizes only the apoB-48 form (4-10). ApoB synthesis by cells of intestinal and hepatic origin has been reported (9-13). We have used the human hepatoma (HepG2) cell line and primary cultures of adult rat hepatocytes to study the synthesis and lipid binding properties of apoB-100 and apoB-48 forms. We have found that in agreement with previous findings (9-13) HepG2 cells synthesize only the apoB-100 form whereas the primary rat hepatocytes synthesize both apoB forms (Fig. 3A&B).

To study the flotation properties of secreted nascent apoB forms, we fractionated the medium of cultured cells grown in the presence of ^{35}S-methionine by density gradient ultracentrifugation. Ten lipoprotein fractions of densities 1.006-1.206g/ml were collected and analyzed by SDS polyacrylamide gradient gel electrophoresis. These analyses showed that in the HepG2 cells apoB-100 was found in VLDL, IDL, LDL and HDL regions. In rat hepatocytes apoB-100 was distributed in VLDL plus IDL regions whereas apoB-48 was found in all lipoprotein fractions (Fig. 3A&B).

More detailed information on the sites of apoB synthesis has been obtained by blotting analysis of RNA isolated from different tissues and cell lines of human origin (40). This analysis showed the presence of apoB mRNA in fetal human and adult monkey liver and intestine, as well as in HepG2 and Caco-2 cells. ApoB mRNA was absent from a variety of fetal human and adult monkey tissues and cells including adrenal gland, brain, gonads, spleen, lung, kidney, heart, stomach, muscle, lymph nodes, thyroid gland, artery, aorta, monocyte macrophages and skin fibroblasts (Fig. 4A).

Relationship Between ApoB-100 and ApoB-48 Forms

The relationship of the two apoB forms and their specific expression by hepatic and intestinal cells has been the subject of extensive investigation (4-12,41-44). Genetic and biochemical evidence is consistent with the hypothesis that the two forms of apoB are products of the same gene (16,41-44). These data include (i) the absence of both apoB forms in patients with abetalipoproteinemia (16); (ii) the description of monoclonal antibodies that recognize both apoB-100 and apoB-48 forms (41); (iii) the presence of the same genetic polymorphism in both apoB-100 and apoB-48 forms in human subjects (42); and (iv) the recognition of both apoB-100 and apoB-48 by antisera raised against a synthetic peptide corresponding to the amino terminal region of apoB-100 (44). Finally, the selective deficiency of apoB-100 reported by Malloy et al. (43) could be caused by a nonsense mutation at a codon near the middle of the apoB mRNA transcript.

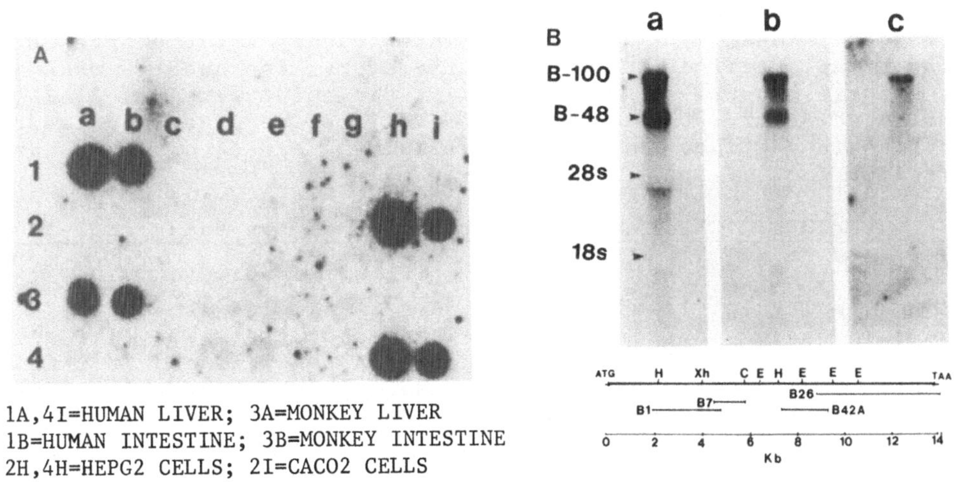

Fig. 4. Panel A. Dot blot analysis of RNA isolated from various human and monkey tissues as well as of various cell lines of human origin. Rows 1 and 2 contain 10ug of RNA obtained from fetal human tissues and human cells. Row 1: (a) liver, (b) intestine, (c) adrenal, (d) brain, (e) spleen, (f) gonads, (g) lung, (h) kidney and (i) heart. Row 2: (a) stomach, (b) thymus, (c) muscle, (d) pancreas, (e) peripheral blood human monocyte macrophage culture, (f) SV40 transformed human fibroblasts, (g) U937 cells, (h) HepG2 cells, and (i) Caco-2 cells. Rows 3 and 4 contain 10ug of RNA isolated from monkey tissues. Row 3: same order of tissues as in row 1. Row 4: (a) stomach, (b) muscle, (c) lymph nodes, (d) thyroid, (e) artery, (f) aorta, (g) pancreas, (h) HepG2 cells, and (i) human liver. The monkey liver RNA was obtained from cebus, and all other RNAs were obtained from cynomolgus monkeys.
Panel B. Blotting analysis of RNA isolated from rabbit intestine. The blots shown in lanes a and b were hybridized with cDNA nick translated probes corresponding to clones lambda B1 and lambda B7, respectively, shown at the bottom of the figure. The combined length of clones used covers the area between nucleotides 1900 and 5800. The blot shown in lane c was hybridized with a mixture of nick translated probes lambda B42 and lambda B26, shown at the bottom of the figure. The combined length of the clones used covers the area between nucleotides 7300 and 14112. Restriction sites were abbreviated as follows: H, HindIII; E, EcoRI; C, ClaI; Xh, XhoI.

Blotting analysis of RNA and by hybridization of the blots with amino and carboxy terminal cDNA probes has shown that carboxy terminal apoB cDNA probes detect a single (15kb) intestinal mRNA species whereas amino terminal probes detect two mRNA species (15 and 8kb) (Fig. 4B). This observation suggested that the apoB-48 mRNA was identical or had extensive sequence homology with the amino terminal region of apoB-100 (17). The length of the hybridization probes used and the expected amino acid composition of apoB-48 (2) were consistent with the hypothesis that the common sequences of the two apoB forms extend from the amino terminus of apoB-100 to the vicinity of amino acid residue 2100 (17). Recent findings suggest that the apoB-48 mRNA is generated by a tissue specific post transcriptional mRNA processing which causes a C-U change of codon CAA specifying Gln 2153. Therefore, the apoB-48 mRNA and the corresponding protein represent the amino terminal 2152 amino acids of apoB-100 (45,46). Utilization of antibodies to synthetic peptides has also localized the end of apoB-48 near residue 2152 of apoB-100 (47).

Structural and Chromosomal Localization of the Human ApoB-100 Gene

Overlapping lambda phage clones containing the entire human apoB-100 have been isolated and characterized. The human apoB-100 gene is 43kb long and contains 29 exons and 28 introns (48). The apoB-100 gene has been mapped on the distal end of the short arm of chromosome 2 (49,50).

Lipoprotein Receptors Recognizing ApoB-100

a) LDL (B/E) Receptor. Biochemical and genetic studies have demonstrated that the cell surface of cultured human fibroblasts and a variety of other cells contain high-affinity receptors for LDL, the major cholesterol transport protein in human plasma (51-53). The binding of LDL is Ca^{2+} dependent and pronase-sensitive (1,3,54). The LDL receptor binds LDL by interacting with apoB (1,3,29,30,54). Modification of apoB lysine (30) or arginine (29) residues destroys its ability to bind to the receptor. The LDL receptor has been purified (55,56) recently. In addition, the gene and a full length cDNA clone corresponding to the human LDL receptor have been isolated and characterized (31,57). The mature receptor is a glycoprotein of apparent molecular weight of 160,000, and isoelectric point 4.3 (55,56), and contains both N-linked and O-linked oligosaccharides (58). LDL binds to the LDL receptor at $4^{o}C$ but is not internalized (1,3). The dissociation constant of the human LDL-LDL receptor complex is 2.8 nM (54). When cell cultures are incubated at $37^{o}C$, the coated pits with the LDL receptor complex invaginate into the cell and pinch off to form endocytic vesicles called endosomes that carry LDL to lysosomes (59). Fusion of endosomes with primary lysosomes results in the hydrolytic degradation of apoB to amino acids and hydrolysis of cholesteryl esters by the lysosomal enzyme, acid lipase (51,60). Liberated cholesterol is used by cells for membrane synthesis (61), and also plays an important role in the regulation of cellular cholesterol biosynthesis (1,3).

b) Modified LDL receptor? Mouse peritoneal macrophages contain a receptor that binds specifically and with high affinity to modified LDL, but not to native LDL (62,63). The human monocyte-macrophage system contains specific, high-affinity receptors for both LDL and modified LDL (64). The bound, modified LDL is internalized and the protein moiety degraded to amino acids (62,63). The regulation of cellular cholesterol homeostasis following modified LDL degradation is fundamentally different from that following LDL degradation by reticuloendothelial cells (3). The cholesteryl ester moiety of modified LDL is hydrolyzed presumably by a nonlysosomal cholesteryl esterase (65). When the cells are grown in serum-containing medium, half of the free cholesterol is secreted and the remainder is reesterified by acyl CoA:cholesterol acyl transferase and stored in the cytoplasm as cholesteryl ester droplets (63). In the continuous presence of modified LDL in the culture medium, the macrophages apparently fail to down regulate their receptor activity and this results in a dramatic increase in cellular cholesteryl ester content (63). Such cholesteryl ester accumulation in the monocyte-macrophages may lead to the formation of foam cells that are found in atherosclerotic lesions (66-69). This hypothesis is further supported by the observation that aortic foam cells have receptors for both LDL as well as modified LDL (69). The modified LDL receptor has been purified from tumors induced by injection of the murine macrophage cell line p388D into syngeneic mice (70). The dissociation constant of the modified LDL-modified LDL receptor complex is 3×10^{-8} (70). The purified receptor has a molecular weight of 260,000 daltons and an isoelectric point of 6.0, and maintains its specificity for modified LDL. A molecular weight of 200,000 has been reported for the modified LDL receptor isolated from rabbit alveolar macrophages (71). Recent data indicate solubilized fragments of modified apoB obtained by incubation of LDL with endothelial cells or by chemical oxidation with Cu^{2+} are taken up and degraded by mouse peritoneal macrophages (72). The uptake of the modified apoB fragments was inhibited competitively by modified LDL but not by native LDL. These findings suggest but do not prove that domain(s) of modified apoB may be involved in the cellular recognition and catabolism of modified LDL by the modified LDL receptor.

Genetic Variation in Human ApoB-100

Abetalipoproteinemia. Abetalipoproteinemia is a rare autosomal recessive disorder characterized by fat malabsorption, failure to thrive, ataxic neuropathy, retinitis pigmentosa, and acanthocytosis (16). Approximately 50 cases of abetalipoproteinemia have been reported to date (16). These patients are characterized by low plasma cholesterol and triglycerides (73) and complete absence of plasma apoB, VLDL, LDL, and chylomicrons (74). Severe deficiency of apoCIII-1 has also been observed in this condition (75,76). In abetalipoproteinemia, all the other apolipoproteins are found in the HDL fraction (77). Immunofluorescence techniques have shown absence of apoB in intestinal biopsies obtained from patients with abetalipoproteinemia (78). However, recent studies have suggested normal apoB-100 mRNA in liver biopsies obtained from different abetalipoproteinemic patients (79). It is possible that these cases of abetalipoproteinemia result from apoB gene mutations that prevent synthesis of apoB or produce an unstable aberrant apoB mRNA or protein molecule (78,79). Whatever the nature of the defect, these diseases demonstrate the requirement of apoB synthesis for the secretion of chylomicrons and VLDL. Similar conclusions can be drawn from tissue culture experiments in which blockage of apoB synthesis by cyclohexamide prevented the secretion of VLDL (15).

Genetically Altered Betalipoprotein Levels. Individuals with half normal LDL and betalipoprotein levels have been described. These people are said to have hypobetalipoproteinemia and are usually phenotypically normal. Patients homozygous for hypobetalipoproteinemia have been described who lack plasma apoB, VLDL, LDL and chylomicrons and are phenotypically indistinguishable from patients with abetalipoproteinemia. The only difference appears to be that parents of these patients have half normal LDL levels (80-84), whereas parents of abetalipoproteinemic patients have normal LDL levels (85-88). A patient has been described who lacks plasma VLDL and LDL but has chylomicrons. This individual has a selective deficiency of apoB-100 (43) but produces the apoB-48 form (43). Another distinct case of hypobetalipoproteinemia has been reported which was characterized by very low levels, but not total absence, of LDL cholesterol (4-8mg/dl) and complete absence of apoCIII-1 (88). Analysis of the chylomicron and VLDL apoB of this patient and members of his kindred showed that they contain a distinct apoB form designated apoB-37 in addition to the normal apoB-100 and apoB-48 forms (89-91). Genetic analysis showed that the apoB-100 and apoB-37 forms of these patients are products of two alleles (89-91). The molecular basis of these diseases is not known. As suggested earlier it is possible that the selective apoB-100 deficiency which is associated with synthesis of only the apoB-48 form as well as synthesis of the apoB-37 form could be caused by nonsense mutations which introduce termination codons at positions near the 48% and 37% points, respectively, of the apoB-100 transcript. Recently some patients with moderate forms of familial hypercholesterolemia have been described who displayed decreased fractional catabolic rate for autologous LDL (92). LDL obtained from these patients had reduced affinity for the LDL receptor of normal cultured human skin fibroblasts (93). Genetic analysis indicated that the affected members were heterozygous for this disorder. The finding suggests that these conditions are caused by structural apoB-100 gene mutations which affect the LDL-LDL receptor interactions and are reminiscent of familial type III hyperlipoproteinemia which is caused by structural mutations in the apoE gene (94).

Numerous population studies indicate that elevated LDL cholesterol levels are associated with increased risk of coronary heart disease (95,96). In contrast, it has been suggested that moderately decreased LDL cholesterol levels are associated with longevity (97). Finally, subjects have been described having a condition characterized as hyperapobetalipoproteinemia (98). These people have elevated LDL apoB but normal LDL cholesterol levels and are at increased risk for developing coronary atherosclerosis (98). Some of the variations in LDL cholesterol and apoB levels observed in these subjects may be the result of as yet unidentified structural apoB gene abnormalities.

CONCLUSIONS

1. Apolipoprotein B-100 is the main protein component of LDL and comprises 23.8% of the LDL particle. ApoB-100 and apoB-48 are synthesized by the liver and the intestine and associate intracellularly with lipids to form lipoprotein particles. Following synthesis one molecule of nascent apoB associates intracellularly with lipids to form lipoprotein particles floating in the VLDL, IDL and LDL regions. ApoB mediates the catabolism of LDL by the LDL (B/E) receptor.

2. The structure of apoB mRNA and protein has been derived from the sequences of overlapping cDNA clones and has the following features:

a. The apoB mRNA consists of 14,112 nucleotides including the 5' and 3' untranslated regions which are 128 and 301 nucleotides, respectively.

b. The DNA-derived protein sequence shows that the unmodified apoB-100 is 513,000 daltons and contains 4560 amino acids including a 24-amino acid-long signal peptide. The MW of apoB-100 implies that there is one apoB molecule per LDL particle.

c. ApoB-100 contains 25 cysteines and 20 potential N-glycosylation sites. The majority of cysteines are distributed in the amino terminal portion of the protein. Direct protein sequence showed that 13 of the potential sites (residues 956, 1341, 1350, 1496, 2955, 3074, 3197, 3309, 3331, 3384, 3438, 3868, and 4210) are glycosylated and four (residues 7, 429, 2212, and 2533) are not.

d. A region of apoB-100 between residues 3359 and 3367 has 67% homology with the presumed receptor binding domain of apoE between residues 142 and 150. Adjacent domains between residues 3144 and 3681 may also play a role in receptor binding. Finally, the region between residues 3144 and 3681 contains three heparin binding domains of apoB-100.

e. Computer analysis of the predicted secondary structure of the protein showed that some of the potential alpha helical and beta sheet structures are amphipathic, whereas others have non-amphipathic neutral to apolar character. These latter regions may contribute to the formation of the lipid binding domains of apoB-100.

3. The human apoB gene is 43kb long and contains 29 exons and 28 introns. It maps on the distal end of the short arm of chromosome 2.

4. ApoB exists in two forms designated apoB-100 and apoB-48. Blotting analysis of intestinal RNA and hybridization of the blots with carboxy apoB cDNA probes produce a single 15kb hybridization band whereas hybridization with amino terminal probes produced two hybridization bands 15 and 8kb. Recent findings suggest that the apoB-48 mRNA is generated by a tissue specific post transcriptional change of codon CAA specifying Gln-2153 to a termination codon, UAA.

Acknowledgements

Dr. Vassilis I. Zannis is an Established Investigator of the American Heart Association. This research was performed at Housman Medical Research Center of Boston University Medical Center. This work was supported by grants from the National Institutes of Health (HL33952 and 26335).

REFERENCES

1. Goldstein, J.L., and Brown, M.S., Ann. Rev. Biochem. 46:897-930 (1977).
2. Hardman, D.A., and Kane, J.P., in: Methods Enzymol. Vol. 128 (J.P. Segrest and J.J. Albers, eds.), pp. 262-272 (1986).
3. Goldstein, J.L., and Brown, M.S., in: The Metabolic Basis of Inherited Disease, 5th ed. (J.B. Stanbury, J.B. Wyngaarden, D.S. Fredrickson, J.L. Goldstein, and M.S. Brown, eds.), pp. 672-712, McGraw-Hill, New York (1982).
4. Kane, J.P., Hardman, D.A., and Paulus, H.E. Proc. Natl. Acad. Sci. USA 77:2465-2469 (1980).
5. Krishnaiah, K.V., Walker, L.F., Borensztajn, J., Schonfeld, G., and Getz, G.S. Proc. Natl. Acad. Sci. USA 77:3806-3810 (1980).
6. Sparks, C.E., and Marsh, J.B. J. Lipid Res. 22:514-518 (1981).
7. Elovson, J., Huang, Y.O., Baker, N., and Kannan, R. Proc. Natl. Acad. Sci. USA 78:157-161 (1981).
8. Wu, A.L., and Windmueller, H.G. J. Biol. Chem. 256:2615-3618 (1981).
9. Bell-Quint, J., Forte, T., and Graham, P. Biochem. Biophys. Res. Commun. 99:700-706 (1981).
10. Vance, D.E., Weinstein, D.R., and Steinberg, D. Biochim. Biophys. Acta 792:39-47 (1984).
11. Thrift, R.N., Forte, T.M., Cahoom, B.E., and Shore, V.G. J. Lipid Res. 27:236-250 (1986).
12. Hughes, T.E., Sasak, W.V., Ordovas, J.M., Forte, T.M., Lamon-Fava, S., and Schaefer, E.J. J. Biol. Chem. 262: 3762-3767 (1987).
13. Bostrom, K., Wettesten, M.,Boren, J., Bondjers, H., Siklund, O., and Olofsson, S.O. J. Biol. Chem. 261:13800-13806 (1986).
14. Davis, R.A., and Boogaerts, J.R. J. Biol. Chem. 257:10908-10913 (1982).
15. Siuta-Mangano, P., Janero, D.R., and Lane, M.D. J. Biol. Chem. 257:11463-11467 (1982).
16. Herbert, P.N., Assmann, G., Gotto, A.M., Jr., and Fredrickson, D.S., in: The Metabolic Basis of Inherited Disease, 5th ed. (J.B. Stanbury, J.B. Wyngaarden, D.S. Fredrickson, J.L. Goldstein, and M.S. Brown, eds.), pp. 589-651, McGraw-Hill, New York (1982).
17. Cladaras, C., Hadzopoulou-Cladaras, M., Nolte, R.T., Atkinson, D., and Zannis, V.I. The EMBO J. 5:3495-3507 (1986).
18. Law, S.W., Grant, S.M., Higuchi, K., Hospattankar, A., Lackner, K., Lee, N., and Brewer, H.B., Jr., Proc. Natl. Acad. Sci. USA 83:8142-8146 (1986).
19. Yang, C-Y., Chen, S-H., Gianturco, S.H., Bradley, W.A., Sparrow, J.T., Tanimura, M., Li, W-L., Sparrow, D.A., DeLoof, H., Rosseneu, M., Lee, F-S., Gu, Z-W., Gotto, A.M., Jr., and Chan, L., Nature 323:738-742 (1986).
20. Knott, T.J., Pease, R.J., Powell, L.M., Walils, S.C., Rall, S.C., Jr., Innerarity, T.L., Blackhart, B., Taylor, W.H.,

Marcel, Y., Milne, R., Johnson, D., Fuller, M., Lusis, A.J., McCarthy, B.J., Mahley, R.W., Levy-Wilson, B., and Scott, J. Nature 323:734-839 (1986).
21. Suita-Mangano, P., Howard, S.C., Lennarz, W.J., and Lane, M.D. J. Biol. Chem. 257:42929-43000 (1982).
22. Lee, P., and Breckenridge, W.C. Can. J. Biochem. 54:829-833 (1976).
23. Swaminathan, N., and Aladjem, F. Biochemistry 15:1516-1520 (1976).
24. Mahley, R.W., Weisgraber, K.H., Innerarity, T.L., Biochim. Biophys. Acta 575:81-91 (1979).
25. Hirose, N., Blankenship, D.T., Krivanek, M.A., Jackson, R.L., and Cardin, A.D. Biochemistry 26:5505-5512 (1987).
26. Weisgraber, K.H., and Rall, S.C., Jr. J. Biol. Chem. 262:11097-11103 (1987).
27. Hui, D.Y., Harmony, J.A.K., Innerarity, T.L., and Mahley, R.W. J. Biol. Chem. 255:11775-11781 (1980).
28. Carew, T.E., Pittman, R.C., Marchand, E.R., and Steinberg, D. Arteriosclerosis 4:214-224 (1984).
29. Mahley, R.W., Innerarity, T.L., Pitas, R.E. Weisgraber, K.H., Brown, J.H., and Gross, E. J. Biol. Chem. 252:7279-7287 (1977).
30. Weisgraber, K.H., Innerarity, T.L., and Mahley, R.W. J. Biol. Chem. 253:9053-9062 (1978).
31. Yamamoto, T., Davis, C.G., Brown, M.S., Schneider, W.J., Casey, M.L., Goldstein, J.J., and Russell, D.W. Cell 39:27-38 (1984).
32. Rall, S.C., Weisgraber, K.H., and Mahley, R.W. J. Biol. Chem. 257:4171-4178 (1981).
33. Innerarity, T.L., Friedlander, E.J., Rall, S.C., Jr., Weisgraber, K.H., and Mahley, R.W., J. Biol. Chem. 258: 12341-12347 (1983).
34. Weisgraber, K.H., Innerarity, T.L., Harder, K.J., Mahley, R.W., Milne, R.W., Marcel, Y.L., and Sparrow, J.T., J. Biol. Chem. 258:12348-12354 (1983).
35. Knott, T.J., Rall, S.C., Jr., Innerarity, T.L., Jacobson, S.F., Urdea, M.S., Levy-Wilson, B., Powell, I.M., Pease, R.J., Eddy, R., Nakai, H., Byers, M., Priestley, L.M., Robertson, E., Rall, L.B., Betsholtz, C., Shows, T.B., Mahley, R.W., and Scott, J. Science 230:37-43 (1985).
36. Segrest, J.P., Jackson, R.L., Mormsell, J.D., and Gotto, A.M., Jr. FEBS Lett. 38:247-258 (1976).
37. Atkinson, D., and Small, D.M. Annu. Rev. Biophys. Biochem. 15:403-450 (1986).
38. Schiffer, M., and Edmundson, H.R. Biophys. J. 7:121-135 (1967).
39. Wu, A.L., and Windmueller, H.G. J. Biol. Chem. 254:7316-7322 (1979).
40. Cladaras, C., Hadzopoulou-Cladaras, M., Avila, R., Nussbaum, A.L., Nicolosi, R., and Zannis, V.I. Biochemistry 25:5351-5357 (1986).
41. Marcel, Y.L., Hogue, m., Theolis, R., Jr., and Milne, R.W. J. Biol. Chem. 257:13165-13168 (1982).
42. Young, S.G., Bertics, S.J., Scott, T.M., Dubois, B.W., Curtiss, L.K., and Witztum, J.L. J. Biol. Chem. 261:2995-2998 (1986).
43. Malloy, M.J., Kane, J.P., Hardman, D.A., and Hamilton, R.L. J. Clin. Invest. 67:1441-1450 (1981).
44. Protter, A.A., Hardman, D.A., Schilling, J.W., Miller, J., Appleby, V., Chen, G.C., Kirsher, S.W., McEnroe, G., and Kane, J.P. Proc. Natl. Acad. Sci. USA 83:1467-1471 (1986).

45. Powell, L.M., Wallis, S.C., Pease, R.J., Edwards, Y.H., Knott, T.J., and Scott, J. Cell 50:831-840 (1987).
46. Chen, S.H., Habib, G., Yang, C.Y., Gu, Z.W., Lee, B.R., Weng, S.A., Silberman, S.R., Cai, S.J., Deslypere, J.P., Rosseneu, M., Gott, A.M., Jr., Li, W.H., and Chan, L. Science 238:363-366 (1987).
47. Innerarity, T.L., Young, S.G., Poksay, K.S., Mahley, R.W., Smith, R.S., Milne, R.W., Marcel, Y.L., and Weisgraber, K.H. J. Clin. Invest. 80:1794-1798 (1987).
48. Blackhart, B.D., Ludwig, E.M., Pierotti, V.R., Caiati, L., Onasch, M.A., Wallis, S.C., Powell, L., Pease, R., Knott, T.J., Chu, M-L., Mahley, R.W., Scott, J., McCarthy, B.J., and Levy-Wilson, B. J. Biol. Chem. 261:15364-15367 (1986).
49. Law, S.W., Lackner, K.J., Hospattankar, A.V., Anchors, J.M., Sakaguchi, A.Y., Naylro, S.L., and Brewer, H.B., Jr. Proc. Natl. Acad. Sci. USA 82:8340-8344 (1985).
50. Deeb, S.S., Disteche, C., Motulsky, A.G., Lebo, R.B., and Kan, Y.W. Proc. Natl. Acad. Sci. USA 83:419-422 (1986)
51. Goldstein, J.L., and Brown, M.S. J. Biol. Chem. 249:5153-5162 (1974).
52. Brown, M.S., and Goldstein, J.L. Cell 6:307-316 (1975).
53. Brown, M.S., Kovanen, P.T., and Goldstein, J.L. Science 212:628-635 (1981).
54. Pitas, E., Innerarity, T.L., Arnold, K.S., and Mahley, R.W. Proc. Natl. Acad. Sci. USA 76:2311-2315 (1981).
55. Tolleshaug, H., Goldstein, J.L., Schneider, W.J., and Brown, M.S. Cell 30:715-724 (1982).
56. Beisiegel, U., Schneider, W.J., Brown, M.S., and Goldstein, J.L. J. Biol. Chem. 257:13150-13156 (1982).
57. Sudhof, T.C., Goldstein, J.L., Brown, M.S., and Russell, D.W. Science 228:815-822 (1985).
58. Brown, M.S., Anderson, R.G.W., and Goldstein, J.L. Cell 32:663-667 (1983).
59. Goldstein, J.L., Anderson, R.G.W., and Brown, M.S. Nature 279:679-685 (1979).
60. Goldstein, J.L., Dana, S.E., Faust, J.R., Beaudet, A.L., and Brown, M.S. J. Biol. Chem. 250:8487-8495 (1975).
61. Brown, M.S., Faust, J.R., and Golstein, J.L. J. Clin. Invest. 55:783-793 (1975).
62. Goldstein, J.L., Ho, Y.K., Basu, S.K., and Brown, M.S. Proc. Natl. Acad. Sci. USA 76:333-337 (1979).
63. Brown, M.S., Goldstein, J.L., Krieger, M., Ho, Y.K., and Anderson, R.G.W. J. Cell Biol. 82:597-613 (1979).
64. Brown, M.S. and Goldstein, J.L. Annu. Rev. Biochem. 52:223 (1983).
65. Brown, M.S., Ho, Y.K., and Goldstein, J.L. J. Biol. Chem. 255:9344-9352 (1980).
66. Mahley, R.W. Atheroscler. Rev. 5:1 (1979).
67. Faggiotto, A., Ross, R., and Harker, L. Circulation 66(II):225 (1982).
68. Mahley, R.W. in: Medical Clinics of North American: Lipid Disorders, Vol. 66 (R.J. Havel, ed.), pp. 375-402, W.B. Saunders, Philadephia (1982).
69. Pitas, R.E., Innerarity, T.L., and Mahley, R.W. Arteriosclerosis 3:2-12 (1983).
70. Via, D.P., Dresel, H.A., Cheng, S.L., and Gotto, A.M., Jr. J. Biol. Chem. 260:7379-7386 (1985).
71. Wong, H., Fogelman, A.M., Haberland, M.E., and Edwards, P.A. Circulation 68III:50 (1983).
72. Parthasarathy, S., Fong, L.G., Otero, D., and Steinberg, D. Proc. Natl. Acad. Sci. USA 84:537-540 (1987).

73. Herbert, P.N., adn Fredrickson, D.S. in: Handbuch der Inneren Medizin, VII/4: Fettstoffwechsel (G. Schettler, H. Greten, G. Schlierf, and D. Seidel, eds.), p. 485, Springer-Verlag, Heidelberg (1976).
74. Cooper, R.A., and Gulbrandsen, C.L. J. Lab. Clin. Med. 78:323-335 (1971).
75. Gotto, A.M., Levy, R.I., John, K., and Fredrickson, D.S. N. Engl. J. Med. 284:813-818 (1971).
76. Scanu, A.M., Aggerbeck, L.P., Kruski, A.W., Lim, C.T., and Kayden, H.J. J. Clin. Invest. 53:440-453 (1974).
77. Zannis, V.I., and Breslow, J.L. Mol. Cell. Biochem. 42:3-20 (1982).
78. Glickman, R.M., Green, P.H., Lees, R.S., Lux, S.E., and Kilgore, A. Gastroenterology 76:288-292 (1979).
79. Hospattankar, A.V., Law, S.W., Lackner, K.J., Brewer, H.B., Jr. Circulation 72:10 (1985).
80. Mars, H., Lewis, L.A., Robertson, A.L., Jr., Butkus, A., and Williams, C.H., Jr. Am. J. Med. 46:886-900 (1969).
81. Ricket, G., Durepaire, H., Hartmann, L., Ollier, M.P., Polonovski, J., and Maitrot, B. Presse Med. 77:2045-2048 (1969).
82. Levy, R.I., Langer, T., Gotto, A.M., and Fredrickson, D.S. Clin. Res. 18:539 (1970).
83. Fosbrooke, A., Choksey, S., and Wharton, B. Arch. Dis. Chil. 40:729-732 (1978).
84. Tamir, I., Levtow, O., Lotan, D., Lequin, C., Heldenberg, D., and Werbin, B. Clin. Genet. 9:149 (1976).
85. Forsyth, C.C., Lloyd, J.K., and Fosbrooke, A.S. Arch. Dis. Child. 40:47-52 (1965).
86. Khachadurian, A.K., Freyha, R., Shamma's, M.M., and Baghdassarian, S.A. Arch. Dis. Child. 46:871-873 (1971).
87. Kostner, G., Holasek, A., Bohlmann, H.G., and Thiede, H. Clin. Sci. Mol. Med. 46:457-468 (1974).
88. Steinberg, D., Grundy, S.M., Mok, H.Y.I., Turner, J.D., Weinstein, J.J., Brown, W.V., and Albers, J.J. J. Clin. Invest. 64:292-301 (1979).
89. Young, S.G., Peralta, F.P., Dubois, B.W., Curtiss, L.K., Boyles, J.K., and Witztum, J.L. J. Biol. Chem. 262:16604-16611 (1987).
90. Young, S.G., Bertics, S.J., Curtiss, L.K., and Witztum, J.L. J. Clin. Invest. 79:1831-1841 (1987).
91. Young, S.G., Bertics, S.J., Curtiss, L.K., Dubois, B.W., and Witstum, J.L. J. Clin. Invest. 79:1842-1851 (1987).
92. Vega, G.L., and Grundy, S.M. J. Clin. Invest. 78:1410-1414 (1986).
93. Innerarity, T.L., Weisgraber, K.H., Arnold, K.S., Mahley, R.W., Krauss, R.M., Vega, G.L., and Grundy, S.M. Proc. Natl. Acad. Sci. USA 84:6919-6923 (1987).
94. Breslow, J.L., Zannis, V.I., SanGiacomo, T.R., Third, J.L.H.C., Travy, T., and Glueck, C.J. J. Lipid Res. 23:1224-1235 (1982).
95. Kannel, W.B., Castelli, W.P., Gordon, T., and McNamara, P.M. Ann. Intern. Med. 74:1-12 (971).
96. Kannel, W.B., Castelli, W.P., and Gordon, T. Ann. Intern. Med. 90:85-91 (1979).
97. Siervogel, R.M., Morrison, J.A., Kelly, K., Meliles, M., Gartside, P., and Glueck, C.J. Clin. Genet. 17:13 (1980).
98. Sniderman, A., Shapiro, S., Marpole, D., Skinner, B., Teng, B., and Kwiterovich, P.O., Jr. Proc. Natl. Acad. Sci. USA 77:601-603 (1980).

SECONDARY AND TERTIARY STRUCTURE OF APOLIPOPROTEINS

Mary T. Walsh, James A. Hamilton, David Atkinson and Donald M. Small

Biophysics Institute, Housman Medical Research Center, Departments of Biochemistry and Medicine, Boston University School of Medicine, 80 East Concord Street, Boston MA 02118

The current knowledge of the primary structures of apolipoproteins (ApoLP's) is impressive. Traditional protein sequencing technology has provided the primary amino acid sequences of the major exchangeable apo-LP's - ApoAI, AII, the C's and E (1-4). Recently, simultaneously in several laboratories, the complete amino acid sequence of ApoB100 has been deduced from its cDNA. Due to its extremely large size (550,000 daltons) and insolubility in water in the absence of lipids or denaturants, direct sequencing methods for ApoB100 have been hindered.

The secondary structure of apolipoproteins (apoLP's) in native lipoproteins, in detergents or solvents or in reconstituted lipid systems has been under intense investigation in many laboratories by a number of chemical and physical chemical methods for some years. Little is yet known, however, about the tertiary structure of apoLP's. Recently, exciting advances have been made in the development of new techniques for the study of secondary and tertiary structure of apoLP's. These include (1) conformational and epitope mapping of apoprotein topology using monoclonal antibodies; (2) 2-dimensional NMR methods; (3) X-ray crystallography; (4) secondary structure prediction and molecular modelling; (5) thermodynamic and spectroscopic analysis; (6) optical diffraction-image reconstruction.

In this brief paper, we will present data from ongoing studies on the secondary and tertiary structure of human ApoAI, E3, B100 and an interesting synthetic peptide which comprises a lipid-binding and receptor binding region of ApoE. The physical methods which will be discussed are high resolution microcalorimetry, circular dichroic spectroscopy, ^{1}H-NMR (1- and 2-d) and computer-aided secondary structure prediction.

HIGH RESOLUTION CALORIMETRY OF APOLIPOPROTEINS

Calorimetry is a very important technique utilized in the study of changes in state of proteins and apoLP's. It represents the only method available for the direct measurement of the enthalpy associated with temperature-induced changes in state of a protein. DSC has been used to reveal important information about the structural and functional domains in fibrinogen (8), prothrombin (9), and plasminogen (10). The folding-unfolding process of proteins or domains of proteins is generally a highly cooperative process, but, the physical reasons governing the process are not completely clear. Highly cooperative protein melting may proceed by cooperative expansion of the polypeptide chain, cooperative 'melting' of a secondary structure or 'melting' of a specific tertiary structure. At present, the phenomena of "protein folding" and the principles which govern these processes is one of the most widely debated topics in the field of structural biology as well as molecular biology.

In an effort to further characterize the thermodynamics of unfolding of apoLP's (11,12), high resolution calorimetric measurements were performed on two of the water-soluble apoLP's, human ApoAI and ApoE3 and ApoB100, a non-water soluble apoLP, which is solubilized in sodium deoxycholate mixed micelles.

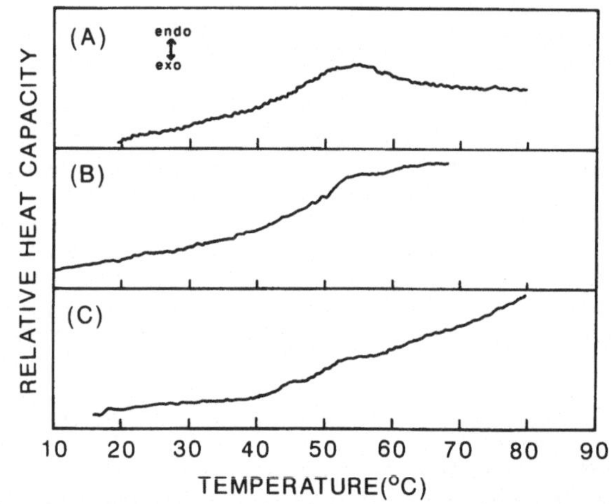

FIGURE 1 **HIGH RESOLUTION MICROCALORIMETRY OF HUMAN APOLIPOPROTEINS.** (A) ApoAI, 3.4 mg/ml; (B) ApoE3, 1.2 mg/ml; (C) ApoB100, 1.8 mg/ml. Sample volume = 1.202 ml. Calorimetric data was collected and analysed by computer using a Microcal-2 Calorimeter (Amherst, MA) at heating rates of 90°/hour. The reference cell contained dialysis buffer. For ApoB100, the dialysis buffer contained 10mM NaDC.

Figure 1 shows high resolution calorimetric scans of these apoLP's. Human ApoAI, as reported previously by Tall et al (13), undergoes a reversible folding-unfolding transition with a T_{max} of 55°C (Fig. 1A). Based on the calorimetric enthalpy and the van't Hoff enthalpy (11,12), calculation of the size of the cooperative unit involved in this process shows its molecular weight to be ~28,000 daltons, the molecular weight of ApoAI. Thus, the ApoAI molecule is unfolding in a "simple" completely reversible two-state process.

Preliminary calorimetric analyses of the thermal behavior of ApoE3 and ApoB100 are shown in Figure 1 B and C. [More extensive presentations of the thermodynamic properties of these apoLP's will be presented elsewhere.] Unlike ApoAI, the folding-unfolding of ApoE3 and ApoB100 are more complex. The calorimetric scans for both are characterized by an overall change in heat capacity and are essentially irreversible after heating to high temperatures. The thermal transition of ApoE3 is characterized by a broad onset starting at 37°C and a low enthalpy thermal transition with T_{max} of 54°C. ApoE3 is irreversibly denatured by 61°C. ApoB100 solubilized in NaDC-micelles also undergoes a multi-stage unfolding process. Similar to ApoE3, the unfolding of ApoB100 exhibits a broad onset starting at 36°C followed by two reversible calorimetric events (T_{max} = 45° and 54°). Heating above 65°C results in irreversible loss of secondary structure of ApoB100 and disruption of apoB-NaDC micelles (14). This is accompanied by a low enthalpy calorimetric transition with T_{max} of 65-70°C. The multiple thermal events and thermal behavior of these apolipoproteins may suggest the presence of "domains" which are undergoing secondary or tertiary structural changes in response to temperature. This hypothesis is currently under investigation in this laboratory.

CIRCULAR DICHROISM OF APOLIPOPROTEINS

CD spectroscopy is a physical method which has been widely used for many years to investigate the overall secondary structure of apoLP's in their native lipid environments (15-17), in aqueous solvents or detergents (18-21) or in reassembled particles with lipids (14,20-23). The great utility of CD studies applied to the determination of the conformation of proteins is the fact that the observed optical activity of peptide transitions depends critically upon the overall conformation of the protein.

We have shown previously that the secondary structure of human ApoB100 varies in response to alterations in either its lipid or its solvent environment (14,20,22,23). As shown in Figure 2 (center spectrum) ApoB100 in native LDL possesses a secondary structure (24,25) which is predominantly α-helical (45%) with a moderate amount of β-sheet (15%) and the remainder as random coil (40%). Delipidation and solubilization of ApoB100 in NaDC micelles maintains this "native" conformation. As with other apoproteins, reconstitution of ApoB100 by detergent dialysis into small phospholipid vesicles, induces a conformational change. In this lipid bilayer environment, ApoB100 is characterized to have ~60% α-helix, essentially no β-sheet and the remainder as random coil. In the absence of lipids, detergents or high concentrations of denaturants, ApoB100 aggregates to form insoluble complexes (18,19).

Figure 3 - Panel A shows CD spectra of human ApoAI in aqueous solution at "high" and "low" concentrations. Similar to observations made by other investigators (16,26,27), CD spectra of ApoAI exhibits concentration dependent molar ellipticities indicative of a protein which undergoes non-covalent self-association. Figure 3 - Panel B shows CD spectra of ApoAI which has been "refolded" from guanidine-HCl (gu-HCl). Secondary structural analysis of ApoAI in a monomeric form shows AI to have 45% α-helix, 22% β-sheet, 5% β-turn and 28% random coil. ApoAI in 4M gu-HCl contains residual structural elements as shown in the top spectrum in Figure 3-Panel B. At ~2M gu-HCl, the α-helical content begins to increase and at ~1M is 45% (not shown).

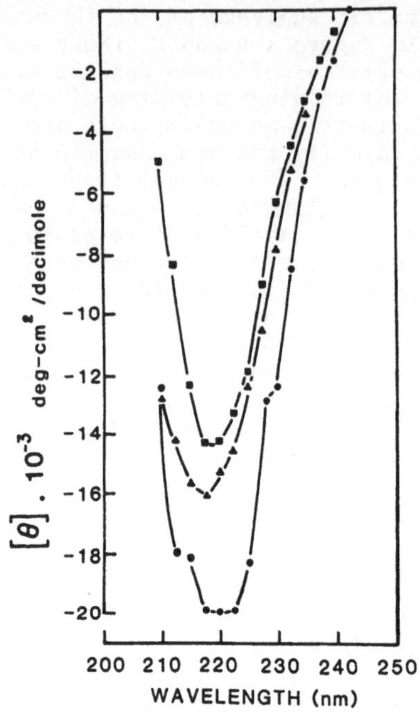

FIGURE 2 CIRCULAR DICHROIC SPECTRA OF HUMAN ApoB100
(●-●) ApoB in DMPC Vesicles; (▲-▲) Native LDL; (■-■) ApoB in NaDC Micelles. (For experimental details, see Ref. 20.)

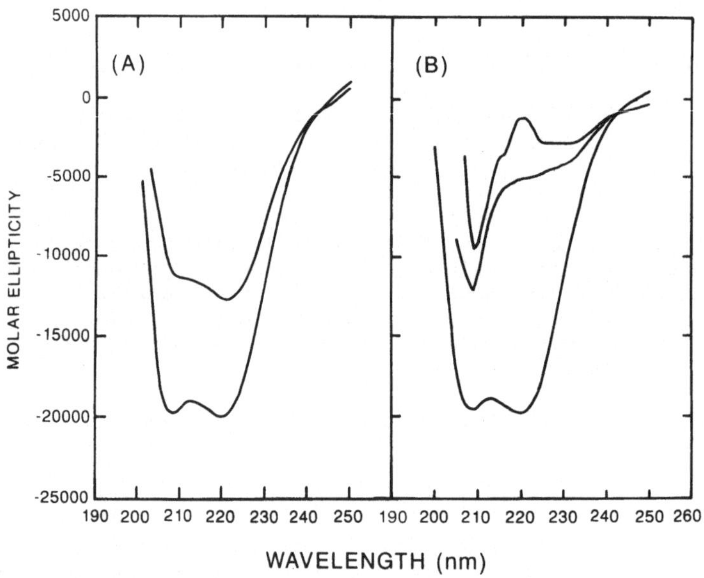

FIGURE 3 CIRCULAR DICHROIC SPECTRA OF HUMAN ApoAI
All CD spectra were recorded on a Cary 61 CD Spectropolarimeter at ambient temperature (unless otherwise indicated) using either 0.02 or 0.1 cm pathlength cells. Protein or peptide concentrations were 0.03-0.10 mg/ml.
Panel A: Top-Dilute, 0.15 mg/ml; Bottom-Concentrated, 2.0 mg/ml
Panel B: Top-4M Guanidine-HCl; Middle-2M Guanidine-HCl; Bottom-Aqueous Buffer, pH 7.4; 2.0 mg/ml ApoAI.

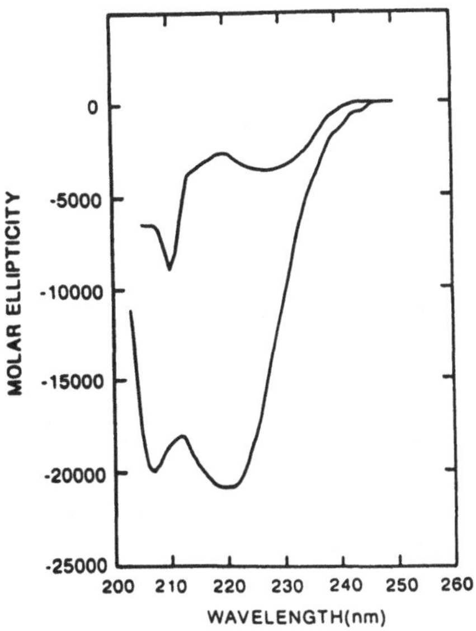

FIGURE 4 CIRCULAR DICHROIC SPECTRA OF ApoE3
Top-7M Guanidine-HCl; Bottom-Aqueous Buffer, pH 7.4.

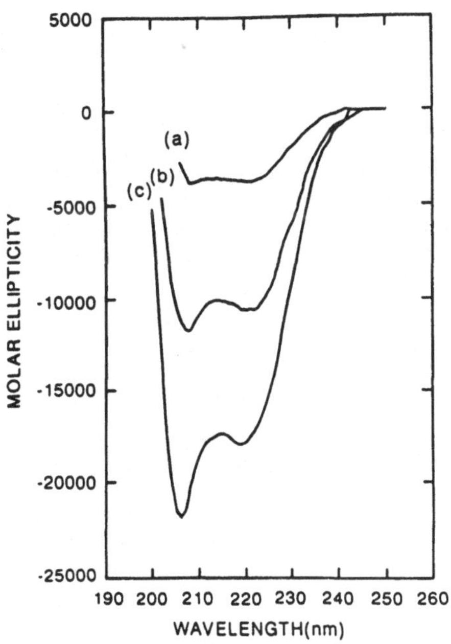

FIGURE 5 CIRCULAR DICHROIC SPECTRA OF EFRAG
(a) 0.1M KBr, 10 mM Potassium phosphate, pH 7.4; (b) 10 mM Potassium phosphate, pH 7.4; (c) 50% Trifluoroethanol/D_2O.

Figure 4 shows CD spectra of ApoE3 in 7M gu-HCl and in aqueous buffer after complete denaturant removal. It is interesting that even in such a high concentration of gu-HCl, ApoE3 retains a great deal of residual secondary structure - 18% β-sheet, 18% β-turn and the remainder as random coil. Reduction of the gu-HCl concentration by dialysis to 4M results in the increase of the α-helical content of ApoE3 to 30%. Complete removal of denaturant results in the secondary structure which has been previously reported (28,29) - 50% α-helix, 10% β-sheet, 3% β-turn and 37% random coil.

STRUCTURAL STUDIES OF A SYNTHETIC PEPTIDE: APOE FRAGMENT (129-169)

A synthetic peptide comprising residues 129-169 of human apoE was synthesized by Dr. James T. Sparrow (30). This peptide (EFRAG) contains both a putative lipid-binding region (residues 130-150) and the receptor binding domain of human apoE. EFRAG is an interesting model system for the investigation of the physical properties of apoproteins by CD to examine overall secondary structure and by ^{1}H NMR (one- and two-dimensional) to examine the secondary and tertiary structure in solution. It is also an excellent candidate for developing protocols for the growth of good single crystals for x-ray crystallographic analysis of apoproteins since it is chemically pure.

CD spectra of apoE fragment recorded in several solvents are shown in Figure 5. In aqueous buffers of low ionic strength EFRAG exhibits two negative minima (Fig. 5b) at 222 and 208 nm, similar to CD spectra of ApoAI and ApoE3 presented in a previous section of this paper and characteristic of a protein containing a moderate amount of α-helix and β-sheet. Secondary structural analysis (24,25) shows the fragment to contain 22% α-helix, 30% β-sheet, 7% β-turn and 41% random coil. Computer aided structural analysis of its primary amino acid sequence (31,32) predicts EFRAG to have the structure modelled in Fig. 6, with a single α-helical region of 10 amino acids, one β-turn and two sections of β-sheet. At high ionic strength (0.1M KBr) in the same aqueous buffer, the CD spectrum is characterized by very low molar ellipticity over all wavelengths examined (Fig. 5a). SDS-polyacrylamide gels performed on the peptide show that in low ionic strength buffer all of the peptide comigrates at M_w ~4000 daltons, whereas at high ionic strength, higher molecular weight aggregates are observed. Thus, a high salt environment promotes the formation of aggregates which are not disrupted by SDS. The CD spectrum of EFRAG recorded in trifluoroethanol/D_2O (Fig. 5C) is also characterized by two negative minima at 222 and 208 nm; however, the magnitude of these minima are greater than those of the peptide in low salt buffer. Structural analysis shows the peptide to be more highly α-helical in this solvent system - 72% α-helix, 17% β-sheet and 11% random - than in aqueous systems.

1H NMR studies (33) have been carried out at 200 MHz on EFRAG. Initial spectra showed that the peptide amide protons exchanged with solvent deuterons completely within 10 minutes of dissolving the fragment in 2H20 buffer solution (pH 7.4, 0.1M KBr, 0.01M phosphate). Spectra were obtained as a function of pH (2.0-9.0) and temperature (25-50°C) at concentrations of ~10 mg/ml (2.5mM). The one-dimensional (1-D) ^{1}H NMR spectrum at pH 7.4 and 25°C is illustrated in Figure 7 (main spectrum) and shows a poorly resolved aliphatic region (0.8-4.5 ppm) but a relatively simple aromatic region (~6.6-8.0 ppm and inset). EFRAG contains a single His(140) and Tyr(162), and the aromatic region of the spectrum reveals all expected aromatic peaks. At pH 7.4 the Tyr C3,5H and C2,6H resonate at 6.75 and 7.05 ppm, respectively. The His C2H appears at 7.70 ppm, while the His C4H overlaps the Tyr C2,6H at 7.05 ppm. Titration to acid pH shifts the His C4H peaks to 7.25 ppm, clearly separating it from the

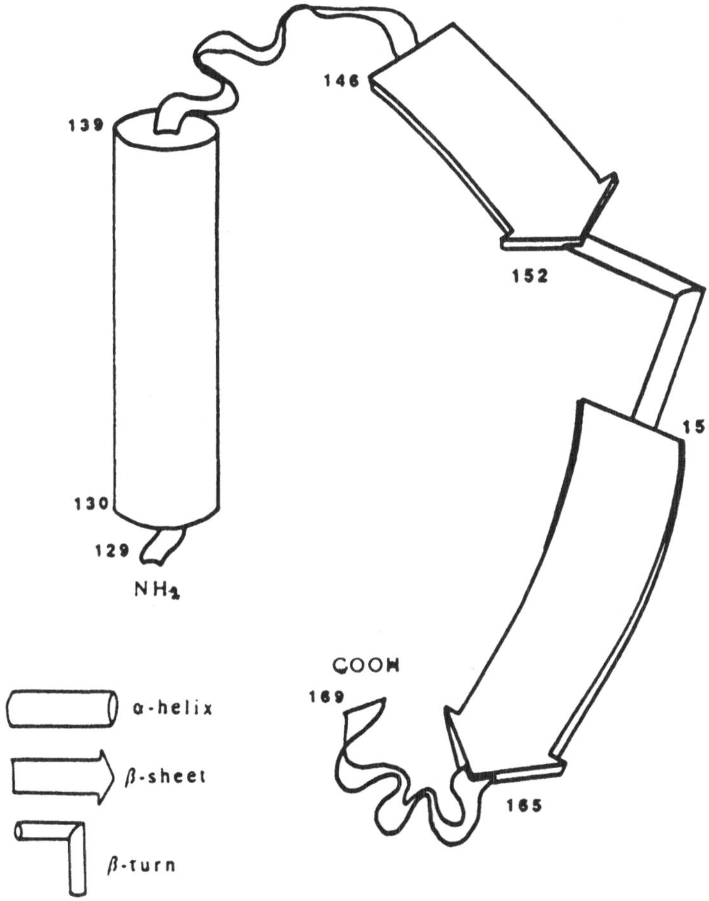

FIGURE 6 SECONDARY STRUCTURAL MODEL OF EFRAG

Tyr peak C4H (top inset) and also results in a larger shift for the C2H than the C4H proton resonance, as expected. The Tyr resonances begin to shift at pH 8.0, signaling onset of the phenolic side chain ionization (precipitation of the protein at pH >8.0 precluded titration of this basic group). A titration curve for the single His was fitted to a curve with pK′ of 5.8 and a Hill coefficient of 1.0. Thus, the His pK_a is depressed from the value in a small peptide by ~1.0 pH unit, indicating the possibility of a nearby positively charged group; the unity Hill coefficient shows that there are no nearby titrating groups in the titration range of the His.

Decreasing the sample pH from 7.4 to 4.8 resulted in considerable narrowing of most ^1H peaks. The extent of narrowing was significantly greater than that obtained by increasing the sample temperature from 25 to 50°C (at pH 7.4). The pH 4.8 spectrum is approximated by a "random coil spectrum" calculated from chemical shifts and relative amounts of constituent amino acids. The tyrosine peaks (6.75 and 7.05 ppm) illustrate the generalized narrowing of ^1H peaks at acid pH and a splitting of 7-8Hz from coupling of the Tyr C3,5H with the C2,6H protons is now observed (Figure 7, Inset 2, aromatic region). At both pH 7.4 and pH 4.8 all the aromatic peaks are narrow, indicating that the His and Tyr are

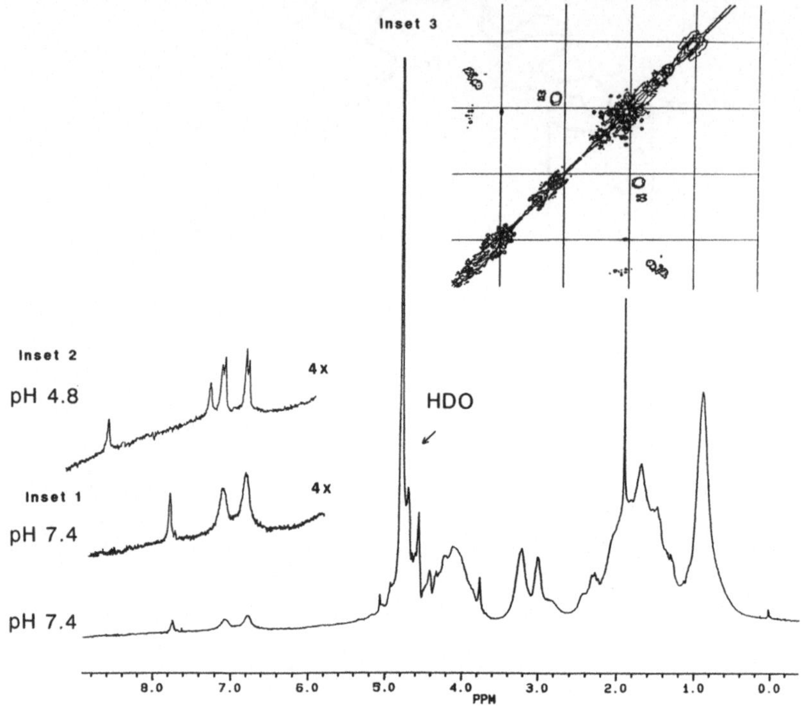

FIGURE 7	**NMR SPECTRA OF EFRAG** One-dimensional NMR spectrum at 200 MHz of EFRAG at 25°C and pH 7.4 at ~10 mg/ml (2.5mM) in 0.10 M KBr.
Inset 1:	4x vertical expansion of His-Tyr region of spectrum at pH 7.4.
Inset 2:	4x vertical expansion of His-Tyr region at pH 4.8.
Inset 3:	1H NMR 2-D COSY spectrum at 200 MHz and 25°C. With a contour representation. The 1-D spectrum is related to the diagonal elements of the 2-D spectrum (by direct projection in this Figure, except that a small scaling difference exists between the two spectra). Off-diagonal cross peaks reveal couplings between different protons. For example, the coupling of Lys CδH (3.0 ppm) and Lys CγH (1.8 ppm) is indicated. This coupling is clearly separated from the coupling of Arg CδH (3.2 ppm) and Arg CγH (at slightly >1.8 ppm).

not motionally restrained. Thus, the NMR studies show a significant unfolding of protein structure between pH 7.4 and pH 4.8 without a large effect on His and Tyr mobilities and environments.

The poorly resolved aliphatic region of the 1-D spectrum at pH 7.4 reveals no proton couplings (Figure 7). The method of 2-dimensional correlated spin spectroscopy (2-D COSY) is suitable for deciphering spin couplings between covalently linked protons in complex spectra, such as those of small proteins. The aliphatic region of 2-D COSY spectrum at pH 7.4 and 25°C is shown above the corresponding region of the 1-D spectrum. The 2-D spectrum shows strong pairs of cross peaks (off diagonal peaks) which reflect couplings between adjacent protons. These peaks have been tentatively assigned to protons in Leu, Ala , Arg and Lys. The aromatic region of the 2-D COSY spectrum (not shown) revealed coupling of the two Tyr peaks.

SUMMARY

The advent of these and other high-powered techniques for the detailed study of apoLP organization will allow us to obtain a high resolution picture of apoLP conformation both in solution and on native lipoprotein particles.

ACKNOWLEDGEMENTS

The authors wish to thank Dr. James T. Sparrow of Baylor College of Medicine for the synthetic ApoE fragment, Dr. Karl H. Weisgraber of Gladstone Foundation Laboratories for the ApoE3, Dr. David H. Croll for performing the NMR experiments and John W. Steiner for calorimetry of ApoE3. The work presented in this paper was supported by USPHS Program Project HL-26335.

REFERENCES

1. Brewer, H.B., Jr., Fairwell, T., LaRue, A., Ronan, R, Houser, A. and Bronzert, T. (1978) Biochim. Biophys. Res. Commun., 80, 623- .
2. Brewer, H.B., Jr., Lux, S.E., Ronan, R. and John, K.M. (1972) Proc. Natl. Acad. Sci. (USA), 69, 1304- .
3. Morrissett, J.D., Jackson, R.L. and Gotto, A.M., Jr. (1975) Ann. Rev. Biochem., 44, 183-207.
4. Rall, S.C., Jr., Weisgraber, K.H. and Mahley, R.W. (1982) J. Biol. Chem., 257, 4171-4178.
5. Knott, T.J., Pease, R.J., Powell, L.M., Wallis, S.C., Rall, S.C., Jr., Innerarity, T.L., Blackhart, B., Taylor, W.H., Marcel, Y., Milne, R., Johnson, D., Fuller, M., Lusis, A.J., McCarthy, B.J., Mahley, R.W., Levy-Wilson, B. and Scott, J. (1986) Nature, 323, 734-738.
6. Yang, C.Y., Chen, S.H., Gianturco, S.H., Bradley, W.A., Sparrow, J.T., Tanimura, M., Li, W.H., Sparrow, D.A., DeLoof, H., Rosseneu, M., Lee, F.S., Gu, Z.W., Gotto, A.M., Jr. and Chan, L. (1986) Nature, 323, 738-742.
7. Cladaras, C., Hadzopoulou-Cladaras, M., Nolte, R.T., Atkinson, D. and Zannis, V.I., (1986) The EMBO J., 5, 3495-3507.
8. Donovan, J.W. and Mihalyi, E. (1974) Proc. Natl. Acad. Sci. (USA), 71, 4125-4128.
9. Ploplis, V.A., Strickland, D.K. and Castellino, F.J. (1981) Biochemistry, 20, 15-21.
10. Castellino, F.J., Ploplis, V.A., Powell, J.R. and Strickland, D.K., (1981) J. Biol. Chem., 256, 4778-4782.
11. Privalov, P.L. and Khechinashvili, N.N. (1974) J. Mol. Biol., 86, 665-684.
12. Sturtevant, J.M. (1974) Annu. Rev. Biophys. Bioeng., 3, 35-51.
13. Tall, A.R., Shipley, G.G. and Small, D. M. (1976) J. Biol. Chem., 251, 3749-3755.
14. Walsh, M.T. and Atkinson, D. (1986) J. Lipid Res., 27, 316-325.
15. Scanu, A. and Hirz, R. (1968) Nature, 218, 200-201.
16. Scanu, A., Pollard, H., Hirz, R. and Kothary, K., (1969) Proc. Natl. Acad. Sci. (USA), 62, 171-178.
17. Dearborn, D.G. and Wetlaufer, D.B., (1969) Proc. Natl. Acad. Sci. (USA), 62, 179-185.
18. Steele, J.C.H., Jr. and Reynolds, J.A., (1979) J. Biol. Chem., 254, 1633-1638.

19. Cardin, A.D., Witt, K.R., Barnhart, C.L. and Jackson, R.L. (1982) Biochemistry, 21, 4503-4511.
20. Walsh, M.T. and Atkinson, D. (1983) Biochemistry, 22, 3170-3178.
21. Watt, R.M. and Reynolds, J.A. (1981) Biochemistry, 20, 3897-3901.
22. Ginsburg, G.S., Walsh, M.T., Small, D.M. and Atkinson, D., (1984) J. Biol. Chem., 259, 6667-6673.
23. Walsh, M.T. and Atkinson, D. (1986) Meth. Enzymol, 128, 582-607.
24. Greenfield, N. and Fasman, G.D. (1969) Biochemistry, 8, 4108-4116.
25. Mao, D. and Wallace, B.A. (1984) Biochemistry, 23, 2667-2673.
26. Osborne, J.C., Jr., Schaefer, E.C., Powell, G.M., Lee, N.S. and Zech, L.A., (1984) J. Biol. Chem., 259, 347-353.
27. Patterson, B.W. and Lee, A.M. (1986) Biochemistry, 25, 4953-4957.
28. Roth, R.I., Jackson, R.L., Pownall, H.J. and Gotto, A.M., Jr. (1977) Biochemistry 16, 5030-5036.
29. Chen, G.C., Guo, L.S.S., Hamilton, R.L., Gordon, V., Richards, E. G. and Kane, J.P. (1984) Biochemistry 23, 6530-6538.
30. Sparrow, J.T., Sparrow, D.A., Culwell, A.R. and Gotto, A.M., Jr. (1985) Biochemistry, 24, 6984-6988.
31. Chou, P.Y. and Fasman, G.D. (1978) Meth. Enzymol., 47, 45-148.
32. Whitlow, M. (1986) Doctoral Thesis, Chemistry Dept., Boston Univ.
33. Wuthrich, K. (1986) NMR of Proteins and Nucleic Acids, John Wiley and Sons, New York, 292 p.

PREDICTION OF THE TERTIARY STRUCTURE OF APOLIPOPROTEIN A-II BY COMPUTER MODELING

J-L. De Coen, C. Delcroix, J-F. Lontie, and C.L. Malmendier

Laboratoire de Chimie Générale 1, Université Libre de Bruxelles, 50, avenue F. Roosevelt, B-1050 Brussels, and Fondation de Recherche sur l'Athérosclérose, 2, rue Evers, B-1000 Brussels

The lack of information about the tertiary structure of apolipoproteins as well as the absence of precise description at the molecular level of specific lipid-protein binding sites have prevented until now a deep understanding of the different events taking place during the metabolism of lipoproteins. The determination of the tertiary structure of apolipoproteins using X-Ray diffraction analysis was not yet possible due to the failure in obtaining crystalline material. In this paper, we want to point out the possibility to obtain detailed structural informations about apolipoproteins using computer modeling, including conformational energy calculations and search for sequence analogies with proteins of known tertiary structure.

The observation of sequence homologies between Apo A-I, A-II, C-I, C-II, C-III, E (1), and the suggestion that all these proteins could have derived from a common ancestor (2) by a serie of gene duplications, fusions and evolutions of a single small gene unit containing eleven amino acid residues led us, quite obviously, to look at the conformational properties of such repetitive units. Applying different methods of prediction of the secondary structure (Chou and Fasman algorithm, Edmunson helical wheel representation, calculation of the probability distribution of conformers), we reached the same conclusion, that is: many repetitive units of eleven amino acids encountered in the sequence of the apolipoproteins have a high propensity to adopt the α-helical conformation. Since this prediction was corroborated by the observation of a high α-helical content for all the above mentioned apolipoproteins in solution (3), we decided to go one step further in trying to build an approximated tertiary structure for these proteins by computer modeling.

Looking in the protein data bank for sequence homology between apoliproteins and proteins of known tertiary structure, we select uteroglobin (4), a small steroid binding protein, 70 amino acids long, as a possible initial template for apolipo-

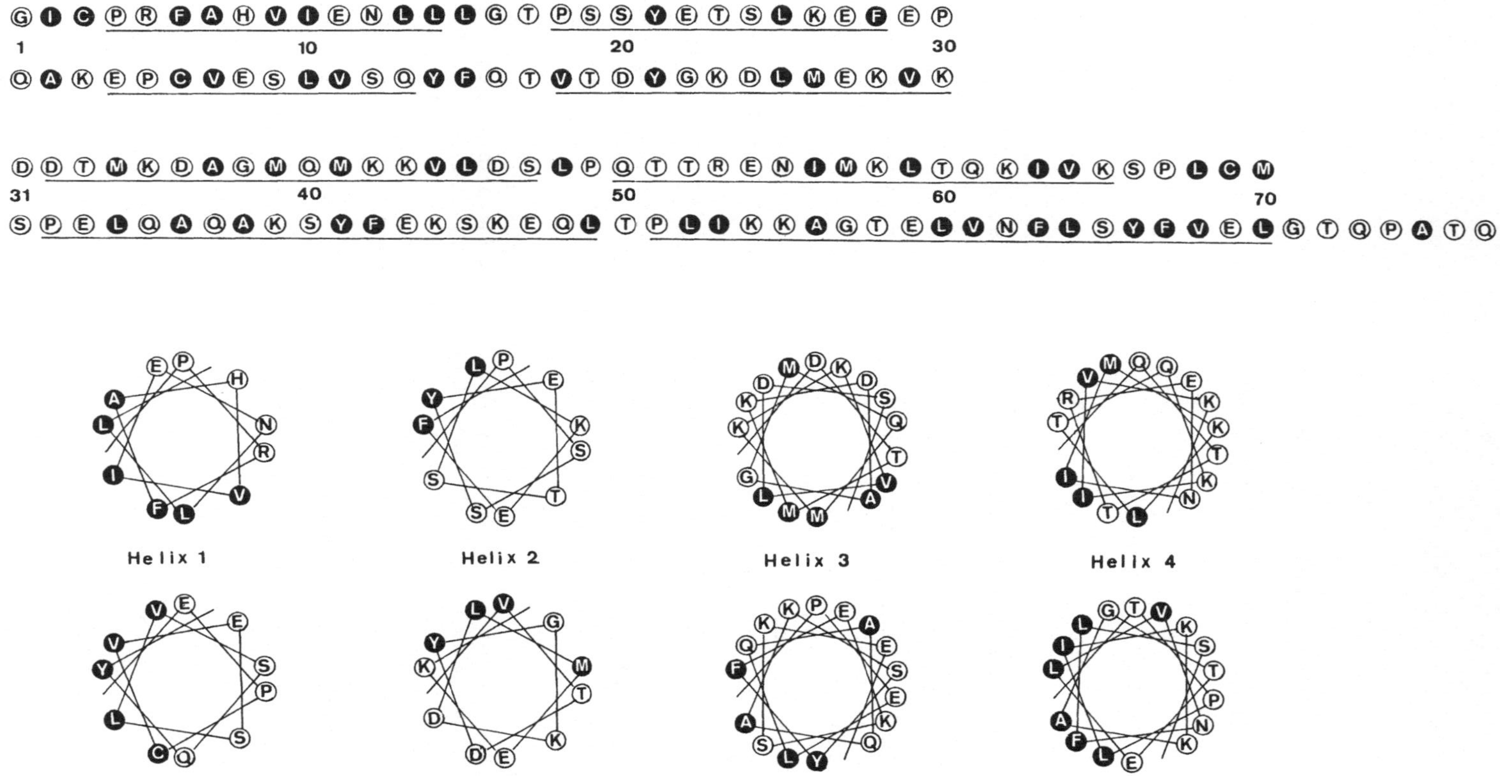

Fig. 1. Comparison between the amino acid sequences of uteroglobin and apolipoprotein A-II. Helical regions observed in uteroglobin and predicted for apo-II are underlined and presented as Edmunson helical wheels. Hydrophobic residues are shown as black dots.

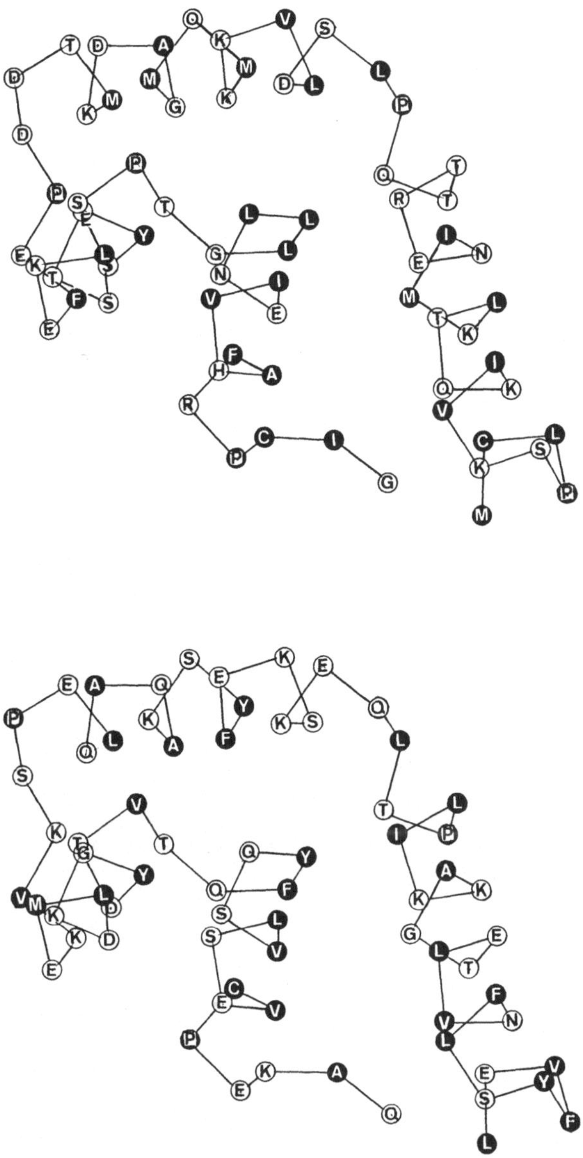

Fig. 2. Folding pattern of uteroglobin observed in the crystal (above) used as an hypothetical template for the sequence of apo A-II (under).

protein A-II (5). The comparison between the primary structure of the two molecules, shown on Fig 1, indicates two very interesting common structural features: the presence of four proline residues located at similar position in the sequence and a distribution of hydrophobic residues giving rise to hydrophobic clusters when the residues are arranged on Edmunson α-helical wheels.

The calculation of the van der Waals energy Evdw (I,J) between all residues I and J for the crystalline conformation of uteroglobin and the tentative model of apo A-II (Fig 2) gave results which are illustrated on Fig 3.

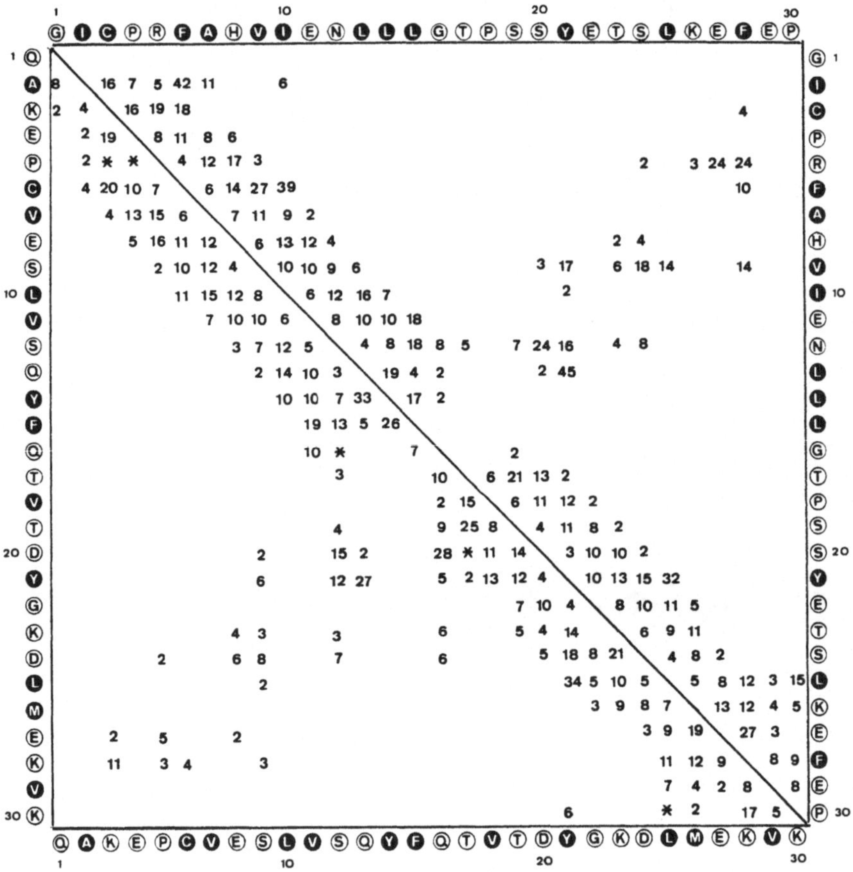

Fig 3. van der Waals interaction energy map calculated for the 30 first residues of the crystalline conformation of uteroglobin (above the diagonal) compared to that obtained for the same fragment of the apo A-II tentative model. The energy values are given in -0.1 kcal/mole. The symbols * indicate pair of residues whose side chains are in short van de Waals contact.

The figures reported above the diagonal correspond to the van der Waals interaction energies between all residues of the uteroglobin fragment (1-30) which folds in the crystal as a pair of antiparallel α-helices (See helices 1 and 2 in Fig 1 and Fig 2).

The figures reported under the diagonal correspond to the same interaction energy terms calculated between the side-chains of the amino acid residues occurring at the same position in the sequence of apo A-II when fitted on the tertiary structure observed for uteroglobin. It is noteworthy that beside a few pair of side chains which appear to be in short van der Waals contact (see the * on the map of Fig 3) all the other residues provide van der Waals interactions of the same magnitude as the one calculated for uteroglobin.

The similarity observed between the two energy plots, supports strongly the hypothesis that the two molecules could assume the same folding profile.

The existence of many common structural features between apo A-II and other apolipoproteins (apo A-I, apo C-I, apo C-II, apo C-III, apo E,...) prompts the possibility that the same scheme of folding and interactions might be useful for these latter molecules also. If such is the case, a major step towards the elucidation of the factors governing the stability of lipoproteins as well as the exchange of particles between them might be achieved.

REFERENCES

1. J.C. Osborne, and H.B. Brewer,Jr., The plasma lipoproteins, Adv.Protein Chem., 31:253 (1977).
2. C-C. Luo, W-H. Li, M.N. Moore, and L. Chan, Structure and evolution of the apolipoprotein multigene family, J.Mol. Biol., 187:325 (1986).
3. H.J. Pownall, Q. Pao, D. Hickson, J.T. Sparrow, and A.M. Gotto,Jr., Thermodynamics of lipid-protein association in human plasma lipoproteins, Biophys. J., 37:175 (1982).
4. I. Morize, E. Surcouf, M.C. Vaney, Y. Epelboin, M. Buehner, F. Fridlansky, E. Milcrom, and J-P. Mornon, Refinement of the C 222_a crystal form of oxidized uteroglobin at 1.34 Å resolution, J. Mol. Biol., 194:725 (1987).
5. J-L. De Coen, M. De Boeck, C. Delcroix, J-F. Lontie, C.L. Malmendier, A proposed folding pattern for apolipoprotein A-II based on a structural analogy with uteroglobin, Proc. Natl. Acad. Sci. USA, 1988 (in the press).

RELATIONSHIP BETWEEN STRUCTURE AND METABOLISM OF HDL APOLIPOPROTEINS: STUDY WITH SYNTHETIC PEPTIDES

Gabriel Ponsin

INSERM U. 197, Laboratoire de Métabolisme des Lipides
Hôpital de l'Antiquaille
Lyon, France

INTRODUCTION

In blood, lipids are transported in complex structures referred to as lipoproteins. These are formed of a monomolecular surface of phospholipids surrounding a core of neutral lipids (cholesterol esters and triglycerides)[1]. The polar phospholipid surface contains unesterified cholesterol and apolipoproteins. Apolipoproteins play several roles which may be summarized as follows :

- They participate to the structural stability and to the solubility of lipoprotein particles.
- Certain apolipoproteins act as effectors of lipolytic enzymes[2,3].
- Apolipoproteins may be recognized by specific receptors[4-7].

Because of their critical roles, the mechanism of interaction of apolipoproteins with lipoproteins has emerged as an essential field of investigation. A general model, the amphipathic helical theory, has been proposed for the lipid binding of exchangeable apolipoproteins[8]. Hypothetically, an apolipoprotein has helical regions in which polar and non polar residues lie on opposite faces of the helix. The non polar face penetrates the lipid matrix and the polar face interacts with the aqueous phase. According to this theory, an exchangeable apolipoprotein is predicted to have : a) a high helical potential ; b) a high degree of hydrophobicity ; c) a minimal amphipathic length and d) a specific arrangement of charged residues. The polar face is supposed to contain negatively charged aminoacids while positively charged aminoacids are believed to occur at the polar - non polar interface.

Several experimental evidences involving native apolipoproteins have supported this theory[9,10]. However, detailed analysis has come from studies using synthetic peptides specifically designed for investigating the main features of the theory. The properties of these peptides in aqueous phase or as components of lipid-peptide complexes have been analyzed using various methodologies. The intrinsic fluorescence of tryptophan has been very useful to monitor lipid-peptide interactions while circular dichroism has been largely used for the determination of the peptide secondary structures. Thermodynamical characteristics of lipid-peptide interaction have been obtained following experiments involving either differential scanning calorimetry or binding studies. Gel permeation chromatography, electron microscopy and non denaturing gradient gel electrophoresis have been largely used for the determination of the size and homogeneity of the particles resulting from lipid-peptide interaction.

LIPID-PEPTIDE INTERACTION

Peptides designed to form optimal amphipathic helices were shown to be capable of forming complexes with phospholipids[11]. Using 3 Lipid Associating Peptides of 16, 20 and 24 aminoacids (LAP-16, LAP-20 and LAP-24, respectively)[12], it was established that the ability of a peptide to interact with a lipid matrix was dependent upon its amphipathic length. However, varying the amphipathic length of a peptide may alter both its potential to form an α-helix and its hydrophobic content. Therefore, the relative importance of these two criteria has not been easily determined since they are not independent. To study the quantitative effect of hydrophobicity on the lipid-peptide interaction independent of the α-helix potential, a family of acylated lipid associating peptides was synthesized[13]. A saturated fatty acyl chain of various number of carbons ($0 < n < 18$) was coupled to the N-terminus of a peptide the sequence of which was that of the C-terminal fragment of LAP-20. The Cn-LAPs had a hydrophobic content that increased with the acyl chain length (Figure 1). The ability of Cn-LAPs to bind a phospholipid matrix or model HDL was dramatically dependent upon their hydrophobicity ; the equilibrium constants (Keq) measured by equilibrium dialysis increased by 3 orders of magnitude as the length of the acyl chain increased from 0 to 16 carbon units. Addition of each carbon unit decreased the free energy of association (ΔGa) by a constant value demonstrating a clear hydrophobic effect (Figure 2). It was concluded from these data that for a given helical potential, the binding of a lipid-associating peptide to lipoprotein is governed by its hydrophobicity.

More recently the influence of helical potential of peptides on lipid-peptide interaction was studied using another family of peptides derived from LAP-20[14]. These were obtained by a simple proline substitution at residues 1, 5, 8, 11, 15, 19. The peptide hydrophobicity was maintained constant while the position of proline substitution was predicted to have a major effect on their putative helicities. Experimental determination of ellipticity based on circular dichroism spectra confirmed this prediction and showed that the helical potential decreased dramatically as the proline position was moved closer to the midpoint of the peptide (table 1). When added to single bilayer vesicles of POPC, all Pro-LAP-20 peptides exhibited a blue shift of their tryptophane fluorescence, reflecting their interaction

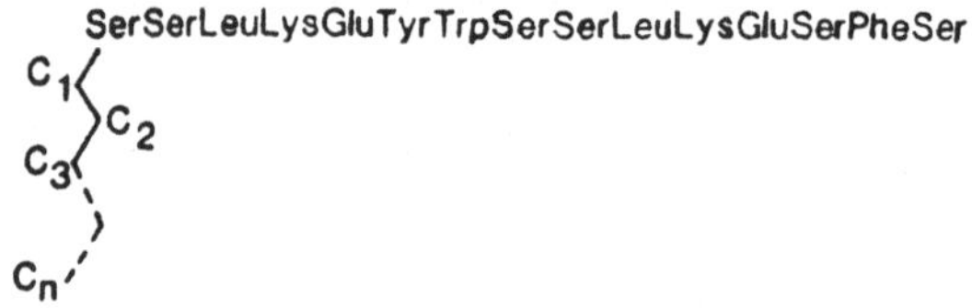

Fig. 1. Structure of the acylated lipid-associating peptide. Saturated fatty acids of various lengths (0-18 carbons) were covalently bound to the N-terminal serine of the peptide through a peptide bond.

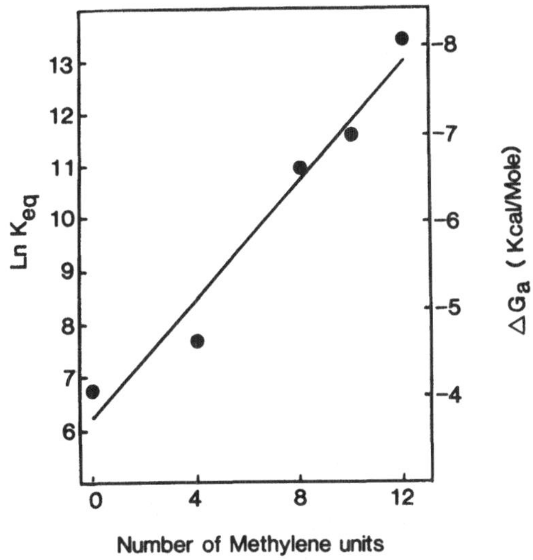

Fig. 2. Effect of the acyl chain length on the binding of acylated LAPs to POPC/apo A-I. The slope of the straight line corresponded to a variation of the free energy of association of -0.35 kcal mol^{-1} per methylene unit.[13]

with the lipid. However, this effect varied markedly from one peptide to another since a shift of 10 nm required the addition of 0.04, 1.2, 4.0 and 7.5 mM POPC to LAP-20, 5-Pro,15-Pro and 8-Pro-LAP-20, respectively. Determination by equilibrium dialysis of the equilibrium constants (Keq) for the binding of these peptides to model HDL revealed the same decreasing affinity as that mentioned above (Table I). A noticeable finding of this work was the observation that the affinity of the Pro-LAP-20 peptides for lipid or for model HDL was closely related to their α -helical potential. These data, together with those obtained with the Cn-LAP series led to two main conclusions. Firstly, α -helicity and hydrophobicity are independent determinants of the affinities of model peptides to lipid surfaces and secondly strong lipid-peptide interaction requires both a high hydrophobicity and α -helical potential.

Beside the major roles of peptide hydrophobicity and helicity predicted by the amphipathic helical theory, it was proposed that a specific distribution of charged residues might be important for the stability of lipid-protein interaction. Since Lys or Arg residues are amphiphilic in nature, their occurence at the polar-non polar interface would permit their alkyl side chains to contribute to the hydrophobicity of the hydrophobic face while their positive charges might form ionic interactions with the negatively charged phosphate group of the phospholipid. This question of charge topography has been addressed in a series of works involving specifically designed peptides[15-18]. Peptide 18 A was synthesized to have polar and non polar faces of approximately equal surfaces areas. It contains Lys residues at the polar-non polar interface and negatively charged Asp and Glu at the center of the polar face. The lipid-binding interaction

Table I. Lipid binding properties of Pro-LAP-20 peptides[14]

Peptide	Ellipticity ($-\Theta_{222} \times 10^{-2}$) Buffer	DMPC	Keq ($\times 10^{-2}$)
1-Pro	67	170	350
5-Pro	24	130	57
8-Pro	23	39	7
11-Pro	18	20	ND
15-Pro	10	120	25
19-Pro	86	130	ND

The equilibrium constants (Keq) for the binding of apopeptides to POPC : Apo A-I complex (100:1, molar ratio) were determined after equilibrium dialysis. They were calculated from Keq = (B/F) (W/P) where B and F represent the concentration of bound and free peptide, respectively ; W and P are the respective molar concentrations of water and POPC. ND : Not determined.

of Peptide 18A was compared to that of peptide reverse 18A. This is the same peptide as 18A except that the positively and negatively charged residue positions were reversed. Results indicated that this reversion resulted in a dramatic decrease of the ability of the peptide to associate with DMPC. Moreover, in contrast to Peptide 18A, reverse 18A was unable to displace significant amount of apo A-I from HDL. Detailed analysis of the properties of this peptide family appeared in a recent review[19].

LECITHIN : CHOLESTEROL ACYL TRANSFERASE ACTIVATION

One of the most interesting findings resulting from the studies of synthetic apolipopeptideshas been the observation that they were able to activate LCAT[12-14,19-21]. As an example the ability of Pro-LAP-20 peptides to activate LCAT is depicted in Table 2 . The apparent Vmax of the enzyme was strongly dependent upon the peptide tested while the apparent Km values were approximately the same for all peptides and for Apo A-I. This and other studies have clearly shown that the LCAT stimulating activity of peptides was correlated to their ability to interact with lipid. Determination of LCAT activation by the Peptide 18A series revealed that 18A peptide was a good LCAT stimulator while reverse 18A had no LCAT stimulating activity. An other interesting information has come from the observation that Peptide 18A-pro-18A, designed to test the effect of multiple amphipathic helical domains, had a LCAT activation ability better than that of Apo A-I. Taken together, these data support the concept that the model apopeptides likely stimulate LCAT through the activation of a phospholipid surface by their bound helical segments.

IN VIVO METABOLISM OF SYNTHETIC APOPEPTIDES

The peptides of the Cn-LAP series were injected to rats to determine whether or not the peptide hydrophobicity was an important determinant of their metabolic behaviour[22]. It was observed that Cn-LAPs were able to specifically interact with HDL. The HDL-peptide interaction was almost nil

Table 2. LCAT activation of Pro-LAP-20 peptides[14]

Activator	App. Km mmol/l	App. Vmax nmol/h	Relative Vmax %
Apo A-I	0.21	2.4	100
1-Pro-LAP-20	0.17	0.6	25
5-Pro-LAP-20	0.23	0.07	2.9
15-Pro-LAP-20	0.08	0.02	0.8
8-Pro-LAP-20	-	0	0

LCAT activity was assayed at 37°C by measuring the formation of [^{3}H] cholesterol esters from substrates which consisted of DMPC-[^{3}H] cholesterol (100:2, molar ratio) containing 0.46 mol % apo A-I or equivalent mass of peptides.

with Co-LAP whereas it was close to 100 % in the case of C16-LAP(Figure 3). The serum clearances of acyl peptides were consistent with the correlation between the acyl chain length of the LAPs and their affinities for HDL. The plasma residence times increased with the acyl chain length (Figure 4). The apparent half-life of C16-LAP (≃ 8h 30mn) was comparable to that of Apo A-I in rat, which was consistent with the expected high binding of C16-LAP to HDL. The analysis of tissue distribution of CnLAPs showed that the peptides unbound to HDL distributed in many tissue compartments and were degraded mainly in the kidneys. As the acyl chain length increased the tissue distribution of Cn-LAPs shifted to tissues known to metabolize HDL in rat (i.e. liver, adrenals and ovaries). These metabolic data were fitted to a theoretical model[23] from which it was possible to estimate for each LAP the partition coefficient between HDL and the aqueous phase. A log-linear relationship was found between the calculated constants of equilibrium and the acyl

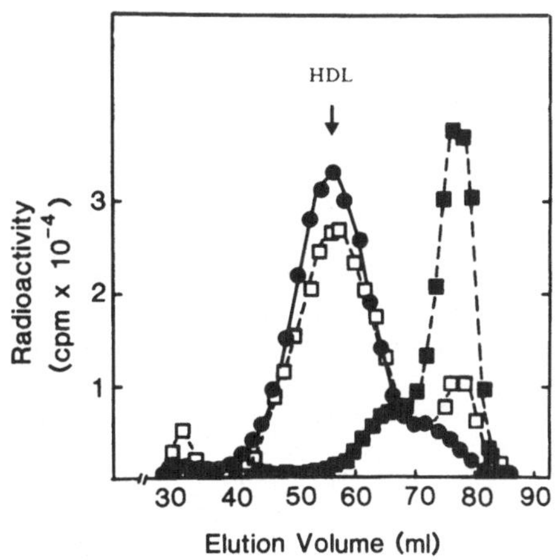

Fig. 3. Elution profiles of serum on a Sepharose CL-4B column, 30 min after injection of Co-(■), C8-(□) or C16-LAP (●).[22]

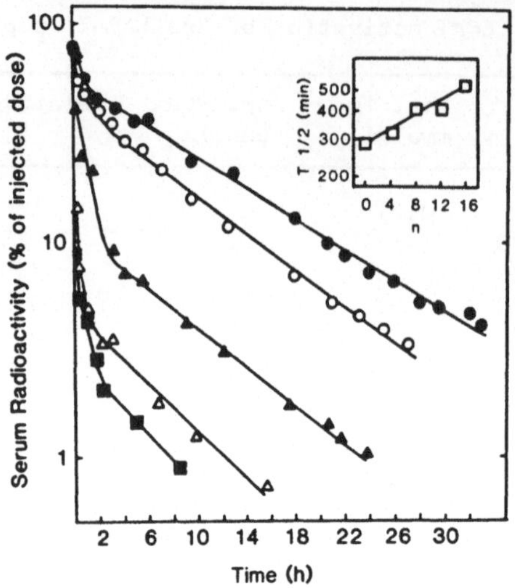

Fig. 4. Serum decay curves of ^{125}I-Co-(■) ; C4-(△) ; C8-(▲) ; C12-(○) ; and C16-LAP(●). After 2.5 h the curves reached an apparent linearity. The inset shows the corresponding apparent half-lives plotted vs the number of carbon units of the acyl chain of the LAP. Each point corresponds to the mean of at least 6 rats.[22]

chain length of Cn LAPs. Moreover there was a strict correlation between these in vivo Keq's and those previously calculated in vitro (Figure 5). This clearly demonstrated that the hydrophobic properties of the LAPs were fully expressed in vivo. This finding strongly supports the in vivo validity of the amphipathic helical theory.

DISCUSSION

The impressive mass of data now available from synthetic apopeptide studies confirms the theoretical predictions. However, synthetic peptides are only models which have structural features necessarily less complex than that of naturally occuring apolipoproteins. The studies summarized above have emphasized the importance of secondary structures of peptides. It remains to determine to which extent the structural properties resulting from the superior levels of structure of native apolipoproteins (i.e. tertiary structure and protein-protein interaction at the lipoprotein surface) might overcome those predicted from their secondary structures. Although this question is far from being answered at least two groups of experimental evidences have indicated that the secondary structure of native apolipoproteins was a major determinant of their lipid-binding properties. Firstly, in vitro studies have shown that the lipid-binding properties of native apolipoproteins were consistent with those predicted by the amphipathic helical theory[9,10,24,25]. The second group of evidences has come from studies involving genetic variants of apolipoproteins. This may be illustrated by the following example. According to the theory, the deletion of an aminc-acid in an amphipathic domain of the apoprotein may lead to the rotation of the polar and/or non polar face of the helix ;

thereby decreasing the lipid-protein interaction. To control this point, an analog of Peptide 18A was synthesized, in which Valine 10 was deleted [17-19,21]. This resulted in a major decrease of the affinity of des-Val^{10}-18A for lipid and in the total loss of its LCAT-stimulating activity. The same observations were made following the discovery of a deletion mutant of Apo A-I (Apo A-I[$Lys_{107} \rightarrow 0$] previously referred to as Apo A-I Marburg or Apo A-I Munster-2. This deletion was predicted to result in the forma-

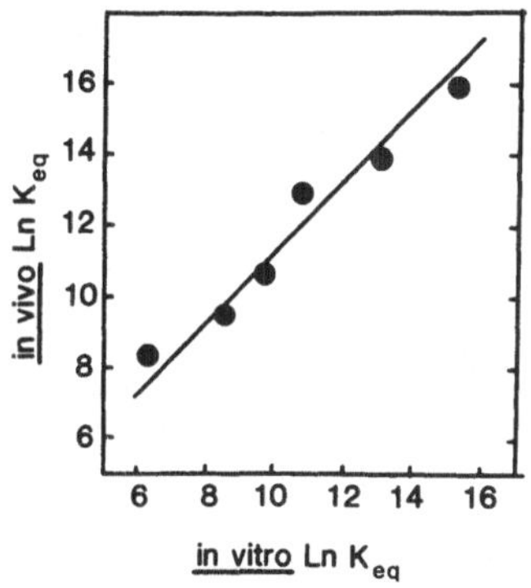

Fig. 5. Correlation between the constants of equilibrium for the binding of the LAPs to HDL in vivo, with those previously observed in vitro.[22] (see fig. 2).

tion of an angle of approximately 90° between the polar and the non polar faces. In agreement with these structural alterations it was observed that both the LCAT-stimulating activity[26] of Apo A-I [$Lys_{107} \rightarrow 0$] and its affinity for lipids[27] were partially deficient. Taken together, these data emphasize the major importance of secondary structures of apolipoproteins, although they do not exclude the possibility that tertiary structure may contribute to their various functions.

REFERENCES

1. L. C. Smith, H. J. Pownall, and A. M. Gotto Jr., The plasma lipoproteins. Structure and metabolism, Ann. Rev. Biochem. 47:751 (1978).
2. C. J. Fielding, V. G. Shore, and P. E. Fielding, A protein co-factor of lecithin : cholesterol Acyltransferase, Biochem. Biophys. Res.Commun. 47:1493 (1972).
3. J. C. Larosa, R. I. Levy, P. N. Herbert, S. E. Lux, and D. S. Fredrickson, A specific apoprotein activator for lipoprotein lipase, Biochem. Biophys. Res. Commun. 41:57 (1970).
4. J. L. Goldstein, M. S. Brown, The low density lipoprotein pathway and its relation to atherosclerosis, Ann. Rev. Biochem. 46:897(1977).

5. R. W. Mahley, D. Y. Hui, T. L. Innerarity, and K. H. Weisgraber, Two independent lipoprotein receptors on hepatic membranes of dog, swine and man, Apo-B, E and Apo : E receptors, J. Clin. Invest. 68:1197(1981).
6. R. Biesbroeck, J. R. Oram, J. J. Albers, and E. L. Bierman, Specific high affinity binding of high density lipoproteins to cultured human skin fibroblasts andarterial smooth muscle cells, J. Clin. Invest.71: 525 (1983).
7. N. M. Fidge, and P. J. Nestel, Identification of apolipoproteins involved in the interaction of human high density lipoprotein 3 with receptors on cultured cells, J. Biol. Chem. 260:3570 (1985).
8. J. P. Segrest, R. L. Jackson, J. D. Morrisett, and A.M. Gotto, Jr., A molecular theory of lipid-protein interactions in the plasma lipoproteins, FEBS Lett. 38:247 (1974).
9. H. J. Pownall, J. B. Massey, S. K. Kusserow, and A. M. Gotto Jr., Kinetics of lipid-protein interactions. Interaction of apolipoprotein A-I from human plasma high density lipoproteins with phosphatidylcholines, Biochemistry 17:1183 (1978).
10. H. J. Pownall, D. Hickson, and A.M. Gotto, Jr., The free energy of association of lecithin with reduced and carboxymethylated apolipoprotein A-II from human plasma high density lipoprotein, J. Biol. Chem. 256:9849 (1981).
11. S. Yokoyama, D. Fukushima, J. P. Kupferberg, F. J. Kezdy, and E. T. Kaiser, The mechanism of activation of lecithin:cholesterol acyltransferase by apolipoprotein A-I and an amphiphilic peptide, J. Biol. Chem. 255:7333 (1980).
12. H. J. Pownall, A. M. Gotto Jr., and J. T. Sparrow, Thermodynamics of lipid-protein association and the activation of lecithin:cholesterol acyltransferase by synthetic model apolipopeptides, Biochim. Biophys. Acta 763:149 (1984).
13. G. Ponsin, K. Strong, A. M. Gotto, Jr., J. T. Sparrow , H.J. Pownall, In vitro binding of synthetic acylated lipid associating peptides to high density lipoproteins. Effects of hydrophobicity, Biochemistry 23: 5337 (1984).
14. G. Ponsin, L. Hester, A. M. Gotto, Jr., H. J. Pownall, and J. T. Sparrow, Lipid-peptide association and activation of lecithin:cholesterol acyltransferase : effect of helicity, J. Biol. Chem. 261:9202 (1986).
15. P. Kanellis, A. Y. Romans, B. J. Johnson, H. Kercret, R. Chiovetti Jr., T. M. Allen, and J. P. Segrest, Studies of synthetic peptide analogs of the amphipathic helix. Effect of charged amino-acid residue topography on lipid affinity, J. Biol. Chem. 255:11464 (1980).
16. J. P. Segrest, B. H. Chung, C.G. Brouillette, P. Kanellis, and R. Mc Gahan, Studies of synthetic peptide analogs of the amphipathic helix. Competitive displacement of exchangeable apolipoproteins from native lipoproteins, J. Biol. Chem. 258:2290 (1983).
17. G. M. Anantharamaiah, J. L. Jones, C. G. Brouillette, C. F. Schmidt, B. H. Chung, T.A. Hughes, A.S. Bhown, and J. P. Segrest, Studies of synthetic peptide analogs of the amphipathic helix, J. Biol. Chem. 260:10248 (1985).
18. B. H. Chung, G. M. Anantharamaiah, C. G. Brouillette, T. Nishida, and J. P. Segrest, Studies of synthetic peptide analogs of the amphipathic helix. Correlation of structure with function, J. Biol. Chem. 260:10256 (1985).
19. G. M. Anantharamaiah, Synthetic peptide analogs of apolipoproteins,in: Methods in Enzymology, Plasma lipoproteins, part A, vol. 128, J. P. Segrest and J. J. Albers, eds., Academic Press, Inc. (1986).
20. H. J. Pownall, A. Hu, A. M. Gotto, Jr., J. J. Albers, and J. T. Sparrow, Activation of lecithin:cholesterol acyltransferase by a synthetic model lipid-associating peptides, Proc. Natl. Acad. Sci. USA 77:3154 (1980).

21. R. M. Epand, A. Gawish, M. Iqbal, K. B. Gupta, C. H. Chen, J. P. Segrest, and G. M. Anantharamaiah, Studies of synthetic peptide analogs of the amphipathic helix, Effect of charge distribution, hydrophobicity, and secondary structure on lipid association and lecithin: cholesterol acyltransferase activation, J. Biol. Chem. 262:9389 (1987).
22. G. Ponsin, J. T. Sparrow, A. M. Gotto Jr., and H. J. Pownall, In vivo interaction of synthetic acylated apopeptides with high density lipoproteins in rat, J. Clin. Invest. 77:559 (1986).
23. G. Ponsin, and H. J. Pownall, Equilibrium of apoproteins between high density lipoprotein and the aqueous phase : modelling of in vivo metabolism, J. Theor. Biol. 112:183 (1985).
24. S. J. T. Mao, J. T. Sparrow, E. B. Gilliam, A. M. Gotto Jr., and R. L. Jackson, Mechanism of lipid-protein interaction in the plasma lipoproteins : lipid-binding properties of synthetic fragments of apolipoprotein A-II, Biochemistry 16:4150 (1977).
25. A. Fukushima, S. Yokoyama, A.J. Kroon, F. J. Kezdy, and E. T. Kaiser, Chain length function correlation of amphiphilic peptides. Synthesis and surface properties of a tetratetracontrapeptide segment of apolipoprotein A-I, J. Biol. Chem. 255:10651 (1980).
26. S. C. Rall, Jr., K. H. Weisgraber, R.W. Mahley, Y. Ogama, C.J. Fielding, G. Utermann, J. Haas, A. Steinmetz, H. J. Menzel, and G. Assmann, Abnormal lecithin : cholesterol acyltransferase activation by a human apolipoprotein A-I variant in which a single lysine residue is deleted, J. Biol. Chem. 259:10063 (1984).
27. G. Ponsin, A. M. Gotto, Jr., G. Utermann, and H. J. Pownall, Abnormal interaction of the human apolipoprotein A-I variant (Lys 107→0) with high density lipoprotein, Biochem. Biophys. Res. Commun. 133:856 (1986).

ACKNOWLEDGEMENTS

The author is grateful to Drs Henry J. Pownall, James T. Sparrow and Antonio M. Gotto Jr. from Baylor College of Medicine, Houston, TX, where the most part of the work described here was carried out.

HETEROGENEITY IN THE CONFORMATION OF APO A-I ON THE SURFACE OF HDL PARTICLES

M. Ayrault-Jarrier*, E. Bekaert, E. Petit, D. Pastier,
J. Polonovski, B. Pau**, F. Paolucci, E. Hervaud and M. Laprade

*C.N.R.S. UA 524, UFR Saint-Antoine, 27 rue Chaligny
75571 - Paris Cédex 12, France
**Centre de Recherche SANOFI, 34000 - Montpellier, France

The plasma apolipoproteins are distributed on several sets of discrete particles whose structures are constantly changing because of their interactions with each other, with enzymes in plasma and with cells. Subtle alterations in the conformation of fonctionnally important domains of apolipoproteins on the surface of particles occur which can be detected by sensitive immunological technics. Monoclonal antibodies, because of their high degree of specificity and of the unlimited quantities which may be produced, can be used as probes of the antigenic structure and of the protein conformation of lipoproteins.

Here, we have developed some strategies allowing localization of two epitopes recognized to monoclonal antibody and demonstration of the heterogeneity of apo A-I conformation on the surface of HDL particles.

Among thirty monoclonal antibodies obtained from two fusions of spleen cells of mice immunized with apo A-I or HDL, three monoclonal antibodies were selected (after two successive clonings) and identified as F28 2G11 1F3, F59 4A12 2F4 and F59 4B11 1B4. These antibodies reveled with both lipid-free apo A-I and HDL-bound apo A-I.

Assignment of epitopes in apo A-I

After cyanogen bromide fragmentation of apo A-I, the epitope defined by Mab 4A12 was localized to the COOH-terminal fragment (CNBr 1), while the epitope defined by Mab 2G11 was localized to the NH_2-terminal fragment (CNBr 4), (1). Mab 4B11 did not react with any fragments suggesting that the structure of this epitope was dependent on protein conformation whereas the epitope to Mab 4A12 and to Mab 2G11 may depend either on amino acid sequence or on fragment conformation, this being preserved after cleavage.

Mabs REACTIVITY ON CNBr FRAGMENTS

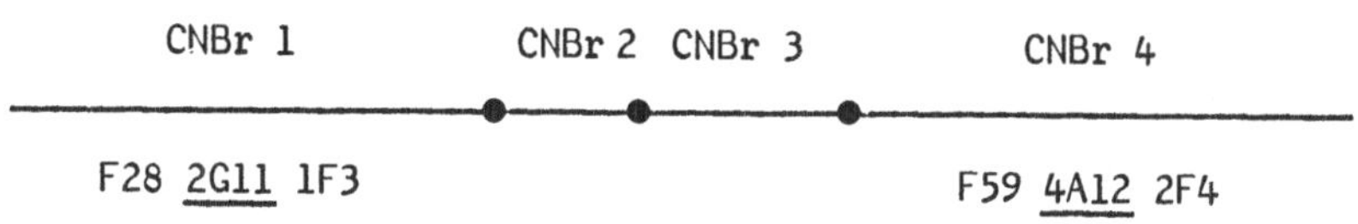

The chemical modifications of certain amino acid residues was performed and the reactivity of monoclonal antibodies to modified apo A-I determined. The results are presented in Table 1.

TABLE I SELECTIVE CHEMICAL MODIFICATIONS OF APO A-I AMINO ACIDS

	Modified Residues			Reactivity of Mabs 2G11	4A12	4B11
[a]CHD-apo A-I	Arg	R	●	+	+	+
[b]Acetyl-apo A-I	Lys	K	□	+	-	+
[c]NO_2-apo A-I	Tyr	Y	▽	-	-	-
[d]I-apo A-I (1 I per mol)	Tyr	Y	▽	+	-	+
[e]Cationized-apo A-I	Glu Asp	E D	○ ○	-	-	-

a - treatment with 1.2 cyclohexanedione according to Mahley et al. (2)
b - treatment with acetic anhydride according to Basu et al. (3)
c - treatment with tetranitromethane according to Chacko et al. (4)
d - treatment with chloramine T according to Greenwood (5)
e - treatment with N,N dimethyl 1-3 propane diamine according to Basu et al. (6).

All modifications were controlled by differencies in mobility upon agarose gel electrophoresis and by the extent of modification as estimated by measurement optical density (NO_2-apo A-I) or with the trinitrobenzene-sulfonic acid assay (acetyl-LDL), (7) or by amino acid analysis. The change in reactivity was studied by immunoblot after electrotransfer from SDS-PAGE and by competitive binding assays. No change in the reactivity of CHD-apo A-I was obtained suggesting that arginyl residues were not implicated in the epitopes. The negative reactivity of cationized-apo A-I and of nitrated-apo A-I, favours a role of tyrosine and glutamic or aspartic acid residues in the epitopes. In addition, only the reactivity of Mab 4A12 was abolished after acetylation of apo A-I, thereby suggesting that a lysyl residue contributed to the epitope recognized by Mab 4A12.

Considering the sequence of the CNBr 1 and CNBr 4 fragments (Fig. 1), two regions of each fragment can be implicated in the recognition of Mabs, i.e. Asp_{12} to Leu_{22} or Asp_{28} to Gly_{40} with Mab 2G11 and Leu_{189} to Glu_{198} or Phe_{229} to Glu_{243}, including Glu_{234} to Lys_{239} with Mab 4A12.

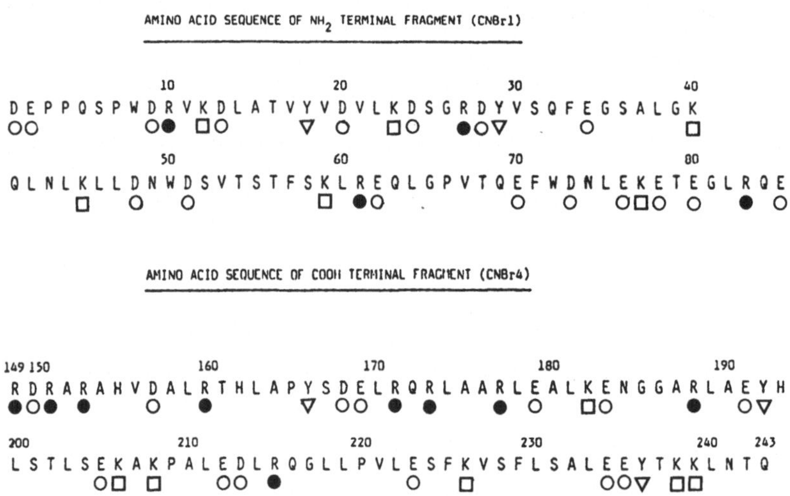

FIGURE 1

A cross-reactivity was shown between human apo A-I and canine (8) and chicken (9) apo A-I, the binding of Mabs were assayed with the apo A-I's of several animal species whose amino acid sequences were partially or totally known. Canine and badger apo A-I were recognized by Mab 2G11 but no reactivity was found with baboon, rat and chicken apo A-I (Fig. 2).

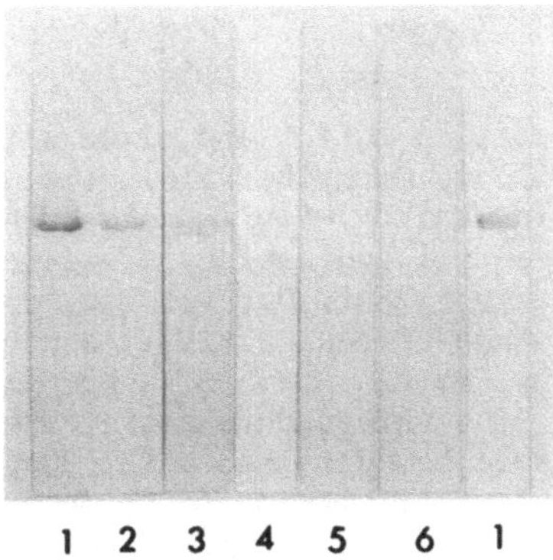

FIGURE 2

Immunotransfer revealed with Mab 2G11
1. Human HDL, 2. Canine HDL, 3. Badger HDL, 4. Baboon HDL, 5. Rat HDL, 6. Chicken HDL.

The NH_2-terminal sequences of these apo A-I are compared in Fig. 3. In considering the two domains previously determined for Mab 2G11 binding, it appears that rat and chicken apo A-I, as well as canine and badger apo A-I, possess several mutations in the region of Asp_{28} to Gly_{40} although the antibody recognized only canine and badger apo A-I. Conversely, if it is accepted that residues 21 and 22, which are substituted in all animal apo A-I's, and which do not form part of the antigenic site then only the domain Asp_{13} to Asp_{20} is identical for human, canine and badger apo A-I while rat, baboon and chicken apo A-I show certain substitutions at residues 14, 15 and 14 to 16 respectively. These results are in accordance with the binding of Mab 2G11 at this domain.

APO A-I NH2 TERMINAL

```
                           10                  20                  30                  40
HUMAN[1]   D E P P Q S P W D R V K D L A T V Y V D V L K D S G R D Y V S Q F E G S A L G K
DOG[2]     - - -   - - - - - - - - - - - - - - - - A V - - - - - - - - A - - - - - - - - -
BADGER[3]  - - -   - - - - - - - - - - - - - - - - A V - - G - - - - - A - - - - - - - - -
BABOON[3]  - - - P - - - - - - - - - - - V - - - - - A L Q - S - - - - - A
RAT[2]     - - -   - - Q - - - - - - F A - - - - - A V K - - - - - - - S G - - S - T - - -
CHICKEN[2] - - -   - T P L - - I R - M V D - - L E T V - A - - K - A I A - - - S - A V - -
```

FIGURE 3

[1]The human sequence is taken from Brewer et al. (10)
[2]The dog, rat and chicken sequences from Chung et al. (11), Haddad et al. (12), Rajavashisth et al. (13) respectively
[3]The badger and baboon sequences from Beaubatie et al. (14) and Blaton et al. (15) respectively.

No recognition of Mab 4A12 was obtained with the animal apo A-I's except with baboon apo A-I. The importance of tyrosyl and lysyl residues in expression of this epitope was confirmed using iodinated apo A-I (1 mol I per mol apo A-I) and glucosylated apo A-I isolated from diabetic patients. It is quite possible that lysyl residues 238 and 239, situated at the extremity, may be easily altered after Amadori formation of glucosyl-lysine adduction, modifying the accessibility of Mab 4A12. If this assumption is correct, it could be concluded that the 234 to 239 domain may be important in the epitope directed to Mab 4A12 rather than the 191 to 192 domain which is implicated in an α-helix according to the predicted secondary structure of apo A-I (16).

Immunoprecipitation of ^{125}I-labeled HDL

Antibody dilutions were prepared in reaction buffer and mixed with ^{125}I-labeled HDL (17) and incubated as previously described (18). The antigen-Mab complexes were precipitated, after incubation with IgG sheep anti-mouse IgG, by Staphylococcus Protein A. A control-containing ^{125}I-labeled HDL without antibodies was included in each experiment. Our results, presented in Fig. 4, were expressed as a percentage of radioactivity bound, i.e. $B/B_o \times 100$ where B = cpm precipitated by Protein A and B_o = cpm precipitated in control by trichloracetic acid. The values obtained at the plateau show that 95 % of ^{125}I-labeled HDL were bound by Mab 4A12, 80 % by Mab 2G11 and 40 % by Mab 4B11. No significant additivity could be observed with pairs of Mabs (4A12 + 2G11 and 4A12 + 4B11); in contrast, with the mixture 2G11 + 4B11, 95 % of ^{125}I-labeled HDL was bound, thereby demonstrating the complementarity of these antibodies. It appears that all apo A-I containing particles could be precipitated with Mab 4A12 and therefore the 5 % of particles which did not precipitate must correspond to particles without apo A-I, i.e. LPA-II, LPD, LPF, LPG (19-21).

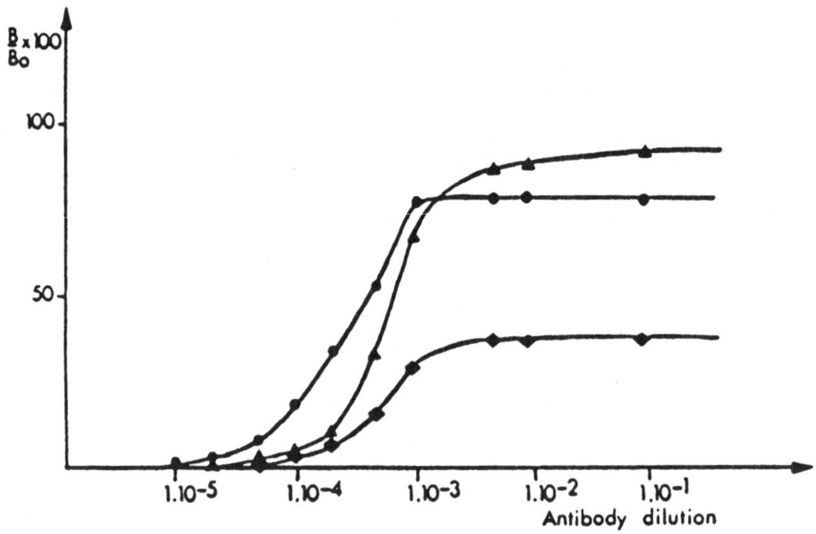

FIGURE 4 Immunoprecipitation of ^{125}I labelled HDL (1.085 - 1.21 g/ml)

Similar results were also observed when ^{125}I-labeled HDL_2 and HDL_3 were used in immunoprecipitation (Table II). These experiments confirm that all apo A-I-containing particles express at least one epitope directed to Mab 4A12. In contrast, the other apo A-I epitopes are not homogeneously expressed on all lipoproteins, since 46 to 86 % of particles were precipitated with Mab 2G11, and 39 to 56 % of particles with Mab 4B11 when HDL_2 and HDL_3 preparations were obtained from the blood of different donors. These results may reflect a variability in particles expressing each epitope. This variability depends on particle distribution in the HDL of these individuals as previously demonstrated by Curtiss and Edgington (22) and Marcel et al. (23).

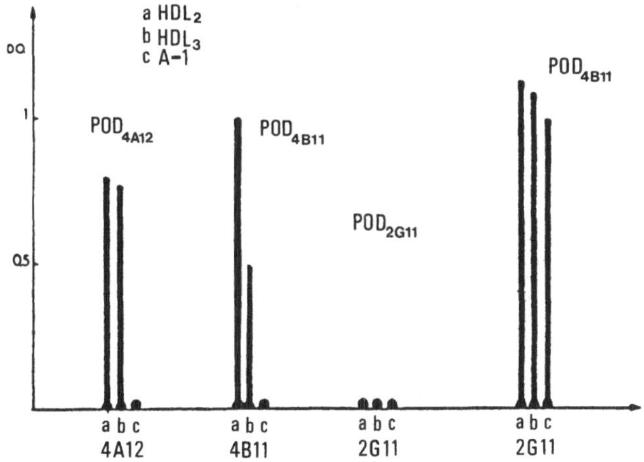

FIGURE 5 DDIA: Double determinant immunoassay

Heterogeneity of apo A-I conformation

As we had seen before all apo A-I-containing particles expressed at least one epitope to Mab 4A12. We therefore undertook to verify whether all apo A-I molecules contained in each HDL particle were able to express this epitope, and used the double determinant immunoassay method.

For this assay, the surface of the microtiter well was coated with monoclonal antibody and residual binding sites blocked with an irrelevant protein such as albumin. Then after washing, the plate was incubated with HDL_2 or HDL_3 samples or apo A-I control to allow particles to bind the coated monoclonal antibody. After further washing, the same monoclonal antibody labeled with peroxidase enzyme was added and incubated in the wells. The antibody remaining in solution was washed out and substrate (o-phenylenediamine dihydrochloride containing hydrogen peroxide) added. The amount of enzyme bound to the plate was determined by optical absorbance at 490 nm.

When HDL_2 and HDL_3 were bound, by one epitope, to monoclonal antibodies 4A12 immobilized on wells, the homologous peroxidase-labeled Mab 4A12 still reacted with the particles (Fig. 5). As it exist only one epitope to Mab 4A12 by apo A-I molecule, the absorbance values, identical with both HDL samples, suggest that, on average, the number of apo A-I molecules expressing the epitope to Mab 4A12 is similar on HDL_2 and HDL_3 particles. When the same samples were bound, by one epitope, to monoclonal antibody 4B11 immo-

TABLE II IMMUNOPRECIPITATION OF ^{125}I LABELLED-HDL

	Mab 4A12		Mab 2G11		Mab 4B11	
	Mean	Range	Mean	Range	Mean	Range
^{125}I-HDL_2 [a] 1.085-1.125 g/ml	93 [b]	90-97	60 [b]	46-86	46 [b]	39-56
^{125}I-HDL_3 1.125-1.21 g/ml	93	90-95	59	45-85	45	35-52

[a] Lipoproteins were iodinated using the iodine monochloride method (2 mC I/mg protein of HDL).

[b] Values are the mean of percentage of radioactivity bound to Protein A at the plateau.

bilized on wells. The homologous peroxidase-labeled 4B11 still reacted with HDL_2 and HDL_3 particles. However this epitope is not homogeneously expressed since the absorbance values with HDL_2 samples were significantly higher than those with HDL_3 samples suggesting that certain apo A-I molecule did not possess this epitope on the surface of HDL_3. The large hydrophobic core of particles and the higher molar ratio of the apo A-I : apo A-II content of HDL_2 samples could regulate the expression of this epitope. When the wells were coated with Mab 2G11, the HDL_2 and HDL_3 particles expressing at least one epitope to Mab 2G11 were bound. The homologous peroxidase-labeled Mab 2G11 did not react with the particles. The binding of heterologous peroxidase-labeled Mab 4B11 confirmed that Mab 2G11 had been able to bind to both HDL_2 and HDL_3 particles. The absence of reactivity of homologous Mab 2G11 suggested that only one apo A-I molecule per particle expressed this epitope, thereby reflecting a heterogeneous conformation of apo A-I itself or steric hindrance by lipids or other apolipoproteins.

Conclusions

The localization of epitopes to Mab 2G11 and to Mab 4A12 in the segment corresponding to residues 13 to 20 and 233 to 239, respectively, demonstrates that the epitope situated at the end of the COOH-terminal fragment is more hydrophilic than the epitope situated at the onset of the NH_2-terminal fragment.

The privileged position of the epitope to Mab 4A12 in the secondary structure of apo A-I can account for its accessibility on all apo A-I-containing particles and on each apo A-I molecule of such particles. In contrast, the epitope to Mab 2G11 situated on a β sheet configuration according to the predicted secondary structure, is not expressed on all apo A-I containing particles. Futhermore this epitope is expressed only once per particle irrespective of the number of apo A-I molecules contained in the HDL particles. The epitope to Mab 4B11 was expressed on about 40 % of ^{125}I-labeled HDL and is more highly represented on the particles of large size (HDL_2) than on those of the small size particles (HDL_3). It is possible thus that the accessibility of this epitope is influenced by the dimensions of the hydrophobic core and/or by protein-protein and lipid-protein interactions. The variations in the reactivities of Mab 4B11 and Mab 2G11 indicate that certain domains of apo A-I are sterically cryptic in lipoproteins and that the accessibility of these domains may vary in individual lipoproteins.

The observed immunological heterogeneity with both lipid-free apo A-I as well as lipid-associated apo A-I is the reflection of an organizational expression of this apolipoprotein. The conformation of apo A-I on the surface of HDL may interfere with the recognition of enzymatic sites or the binding to cellular sites.

Further studies must be undertaken to determine the influence of protein-protein and protein-lipid interactions on the expression of these epitopes.

Acknowledgments

Doctor G. Ponsin is gratefully acknoledged for the preparation of glucosylated-apo A-I and iodinated apo A-I, Doctor M. Laplaud for the gift of badger HDL and Doctor L. Frémont for the providing of chicken plasma. We wish to thank D. Sérébrénik and M. Wertenschlag for their excellent secretariat.

References

1. E. Petit, M. Ayrault-Jarrier, D. Pastier, H. Robin, J. Polonovski, I. Aragon, E. Hervaud and B. Pau (1987) Biochim. Biophys. Acta, 919, 287-296.
2. R.W. Mahley, T.L. Innerarity, R.E. Pitas, K.H. Weisgraber, J.H. Brown and E. Gross (1977) J. Biol. Chem., 252, 7279-7287.
3. S.K. Basu, J.L. Goldstein, R.G.W. Anderson and M.S. Brown (1976) Proc. Natl. Acad. Sci. USA, 73, 3178-3182.
4. G.K. Chacko (1985) J. Lipid Res., 26, 745-754.
5. F.C. Greenwood, W.M. Hunter and J.S. Glover (1963) Biochemical J., 89, 114-123.
6. S.K. Basu, J.L. Goldstein, R.G.W. Anderson and M.S. Brown (1976) Proc. Natl. Acad. Sci. USA, 73, 3178-3182.
7. A.F.S.A. Habeeb (1966) Anal. Biochem., 14, 328-336.
8. J.B. Swaney (1980) Biochim. Biophys. Acta, 617, 489-502.
9. R.L. Jackson, Y.Hu Lin, L. Chan and A.R. Means (1976) Biochim. Biophys. Acta, 420, 342-349.
10. H.B. Brewer, T. Fairwell, A. Larue, R. Ronan, A. Hanser and T. Bronzert (1978) Biochem. Biophys. Res. Commun., 80, 623-630.
11. H. Chung, A. Randolph, I. Reardon and R.L. Heinrikson (1982) J. Biol. Chem., 257, 2961-2967.
12. I.A. Haddad, J.M. Ordovas, T. Fitzpatrick and S.K. Karathanasis (1986) J. Biol. Chem., 261, 13268-13277.
13. T.B. Rajavashisth, P.A. Dawson, D.L. Williams, J.E. Shackelford, H. Lebherz and A.J. Lusis (1987) J. Biol. Chem., 262, 7058-7065.
14. L. Beaubatie, M. Laplaud, S.C. Rall and D. Maurel (1986) J. Lipid Res., 27, 140-149.
15. V. Blaton, R. Vercaemst, M. Rosseneu, J. Mortelmans, R.L. Jackson, A.M. Gotto and H. Peeters (1977) Biochemistry, 16, 2157-2163.
16. A.L. Andrews, D. Atkinson, M.D. Barratt, E.G. Finer, H. Hauser, R. Henry, A.B. Leslie, N.L. Owens, M.C. Phillips and R.N. Robertson (1976) Eur. J. Biochem., 64, 549-563.
17. D.W. Bilheimer, S. Eisenberg and R.I. Levy (1972) Biochim. Biophys. Acta, 260, 212-221.
18. E. Bekaert, M. Ayrault-Jarrier, E. Petit, C. Bétourné, H. Robin and J. Polonovski (1988) Clin. Chem., in press.
19. W.J. Mc Conathy and P. Alaupovic (1976) Biochemistry, 15, 515-520.
20. E. Koren, W.J. Mc Conathy and P. Alaupovic (1982) Biochemistry, 21, 5347-5351.
21. M. Ayrault-Jarrier, J.F. Alix and J. Polonovski (1978) Biochimie, 60, 65-70.
22. L.K. Curtiss and T.S. Edgington (1985) J. Biol. Chem., 260, 2982-2983.
23. Y. Marcel, D. Jewer, C. Vezina, P. Milthorps and P.K. Weech (1987) J. Lipid Res., 28, 768-777.

STRUCTURAL PROPERTIES OF THE HEPARIN-BINDING DOMAINS OF HUMAN APOLIPOPROTEIN E

Alan D. Cardin and Richard L. Jackson

Merrell Dow Research Institute
2110 E. Galbraith Road
Cincinnati, OH 45215

INTRODUCTION

Apolipoprotein B-100 (apoB-100) and apolipoprotein E (apoE) are the major protein constituents of human plasma low density lipoproteins (LDL), very low density lipoproteins (VLDL) and chylomicrons (1-4). ApoB-100 and apoE mediate the cellular uptake of these lipoproteins by binding to specific membrane receptors on hepatic and extrahepatic tissues. In addition, apoB-100 and apoE interact with sulfated mucopolysaccharides, e.g. glycosaminoglycans (GAG), of arterial tissue (5-10) and with heparin (11). The interaction of apoB-100 and apoE containing lipoproteins with GAG of the extracellular matrix may contribute to the accumulation of cholesterol in the arterial wall and, hence, promote atherosclerosis (12).

Recently, we identified the amino acid sequence domains in apoB-100 and apoE that bind heparin (13,14). Figure 1 shows the location of these domains with respect to the thrombin and kallikrein cleavage sites (15). Heparin-binding domains II, III and IV of apoB-100 are clustered at the carboxyl-terminal end of apoB-100 and flank the T_2-T_3 thrombin cleavage site. A thrombin cleavage site in apoE occurs between residues Arg_{191} and Ala_{192} (16) to yield amino- and carboxy-terminal heparin-binding fragments of ≈ 22,000 and 12,000 daltons, respectively (14). The amino acid sequences of the heparin-binding domains of apoB-100 and apoE are shown in Figure 2. The sequence domains are characterized by clusters of 2-3 basic residues interspersed with 1-3 hydropathic residues terminated by one or more basic residues. Clustering of the lysine and arginine residues could generate a high positive charge density capable of interacting electrostatically with regions of high negative charge on heparin and other glycosaminoglycans. A clustering of basic and hydropathic residues in the heparin-binding regions of human vitronectin and platelet factor-4 have also been noted (13). Moreover, sequence organizations of basic and hydropathic residues similar to those in apoE, apoB-100, vitronectin and platelet factor-4 are present in other heparin-binding proteins such as human hepatic triglyceride lipase (17), endothelial cell growth factor (18), glia-derived nexin (19), antithrombin-III (20) and purpurin (21).

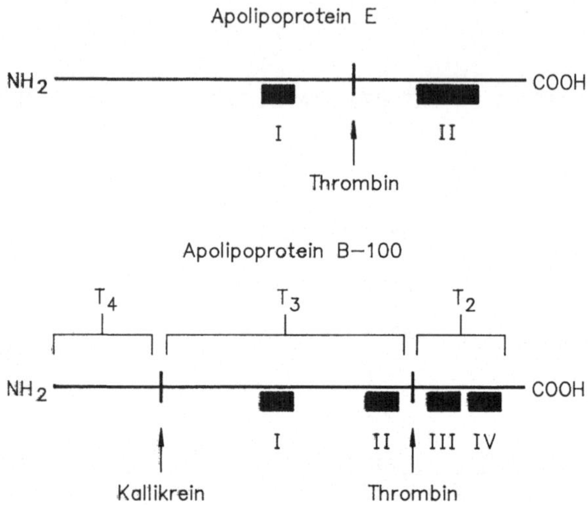

Fig. 1. The major thrombolytic fragments of apoE and apoB-100. Thrombin treatment of apoE yields two major fragments each containing a domain for heparin-binding (14). These domains are designated I and II and their locations are indicated by the filled boxes. Thrombin treatment of apoB-100 in LDL generates two major fragments designated T_1 and T_2; T_3 and T_4 are complementary thrombolytic fragments of T_1 (15). The major heparin-binding domains of apoB-100, designated I-IV (see filled boxes), are in fragments T_2 and T_3.

PEPTIDE DOMAIN		SEQUENCE
B-I_a	2081–2088	Val Arg Lys Tyr Arg Ala Ala Leu
B-I_b	2119–2126	Thr Lys Lys Tyr Arg Ile Thr Glu
B-II	3150–3157	Tyr Lys Lys Asn Lys His Arg His
B-III	3361–3368	Thr Arg Lys Arg Gly Leu Lys Leu
B-IV	3670–3677	Gly Arg Arg Gln His Leu Arg Val
E-I	144–151	Leu Arg Lys Arg Leu Leu Arg Asp
E-II_a	212–219	Glu Arg Leu Arg Ala Arg Met Glu
E-II_b	223–230	Ser Arg Thr Arg Asp Arg Leu Asp

Fig. 2. Amino acid sequences of the putative heparin-binding regions of apoB-100 (B-I-IV) and apoE (E-I and E-II).

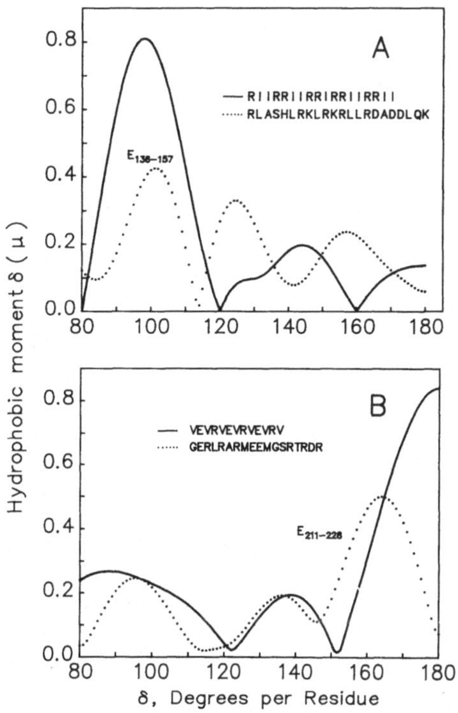

Fig. 3. Eisenberg plots (23) of the E-I and E-II heparin-binding domains of apoE. Peptide R I I R R I I R R I R R I I R R I I is a model amphipathic α-helix. Peptide V E V R V E V R V E V R V is a model amphipathic β-strand (24,26).

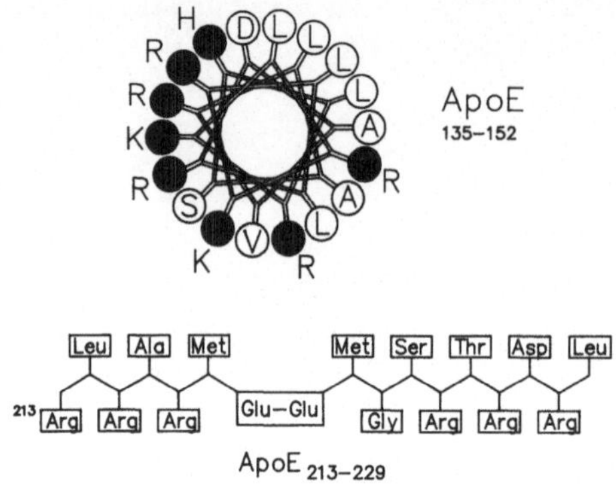

Fig. 4. Models for the secondary structures of the E-I (Panel A) and E-II (Panel B) heparin-binding domains of apoE. The E-I domain (residues 135-152) conforms to an amphipathic α-helix; the E-II domain (residues 213-229) conforms to an amphipathic β-strand.

STRUCTURE OF THE HEPARIN-BINDING DOMAINS OF APOLIPOPROTEIN E

The heparin-binding sequences for apoE shown in Figure 2 have been delineated by structure-activity studies on isolated thrombolytic fragments, synthesis of peptides corresponding to these regions and antibody mapping studies (14,22). However, the secondary structural features of these apoE peptides in the presence of heparin have not been described. Secondary structure analysis by the method of Eisenberg (23) predicts different conformations for E-I and E-II. As is shown in Figure 3A, residues 136-157 of E-I give a major periodicity at 100° and a minor periodicity at 120° characteristic of an amphipathic helical structure. Like E-I, the model amphipathic α-helix (R I I R R I I R R I R R I I R R I I) of Eisenberg (23) gives a major periodicity at 100°. In contrast, residues 211-228 in Figure 3B (E-II) show a major periodicity at 160°. This periodicity is very similar to the model amphipathic β-strand (MABS) peptide (V E V R V E V R V E V R V) reported by Osterman et al. (24).

The structures of E-I and E-II are modeled in Figure 4A and B, respectively. As shown by a helical wheel diagram (25), E-I (residues 135-152) forms an amphipathic α-helix with the basic residues segregating to one side of the helix forming a positively charged face. The opposite face is predominantly hydrophobic. Modeling of E-II (residues 213-229) by the method of Kaiser and Kezdy (26) shows that this region conforms to an amphipathic β-strand structure in which the alternation of basic and nonbasic residues segregate to opposite sides of the β-strand forming a positively charged face and a hydrophobic face. The binding interaction with the negative charges on heparin would be expected to stabilize this configuration and minimize charge repulsion between neighboring arginines (27).

To investigate further the structure of the heparin-binding domains of apoE, two fragments were synthesized corresponding to apoE$_{129-169}$ and apoE$_{211-243}$ and their structural conformations were determined by circular dichroism (CD) in the absence and presence of heparin. Figure 5A shows the CD spectrum of apoE$_{129-169}$. In the absence of heparin, apoE$_{129-169}$ gives a $[\theta]_{208}$ = -4590 deg.cm^2/dmol and a $[\theta]_{222}$ = -2448 deg.-cm^2/dmol corresponding to 22% α-helix and 5% β-strand. The addition of heparin to apoE$_{129-169}$ results in an increase in the negative ellipticity between 195-240 nm with pronounced minima at 208 and 222 nm. Deconvolution of the CD spectrum of apoE$_{129-169}$ in the presence of heparin yields 49% α-helical and <1% β-strand. Thus, heparin induces a greater than 2-fold increase in α-helical content in apoE$_{129-169}$.

Figure 5B shows the CD spectra of $apoE_{211-243}$. In contrast to the effects of heparin on the CD spectrum of $apoE_{129-169}$ in which two pronounced minima are induced at 208 and 222 nm indicative of high helical content, the addition of heparin to $apoE_{211-243}$ induces an increase in the negative ellipticity between 190-240 mm with a broad band shape and minimum at ~220 nm i.e., $[\theta]_{220}$ = -4200 and $[\theta]_{220}$ = -12,000 deg.cm^2/dmol in the absence and presence of heparin, respectively. Deconvolution of the spectra yields 9% α-helical and 66% β-strand in the

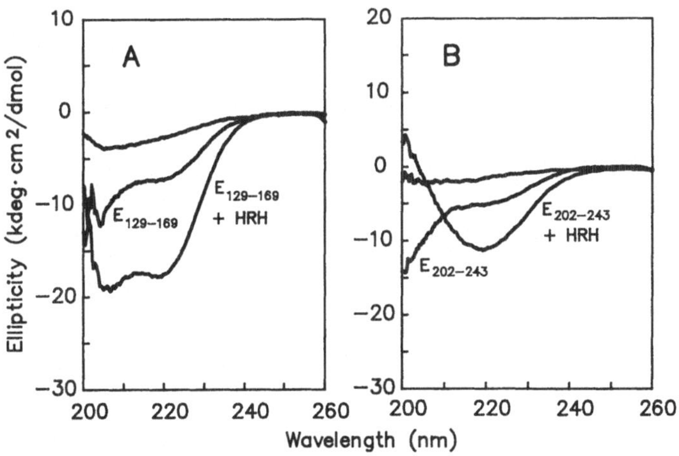

Fig. 5. The effect of heparin on the CD properties of $apoE_{129-169}$ (Panel A) and $apoE_{211-243}$ (Panel B). From top to bottom, the curves represent: (Panel A) heparin in standard buffer (10 mM Hepes, 0.15 M NaCl, pH 7.4), $apoE_{129-169}$, and heparin plus $apoE_{129-169}$; (Panel B) heparin in standard buffer, $apoE_{211-243}$, and heparin plus $apoE_{211-243}$. [peptide] = 100 µg/ml; [heparin] = 2 x 10^{-5} M.

presence of heparin compared to 11% α-helical and 16% β-strand in the absence of heparin. Thus, heparin induces a greater than 4-fold increase in β-strand structure.

In summary, only limited information is available on the effect of heparin on apoE structure. In this study, we have analyzed the conformations that the heparin-binding domains E-I and E-II might assume when associated with GAG. Both predictive methods (23,25) and solution studies with synthetic peptides indicate that heparin induces amphipathic α-helix in E-I and amphipathic β-strand in E-II.

ACKNOWLEDGMENTS

We wish to thank Ms. Susan Treadway for the preparation of this manuscript, Mr. Steve Biedenbach for the figures and James T. Sparrow and Doris A. Sparrow for the synthetic peptides.

REFERENCES

1. M. S. Brown and J. L. Goldstein, A receptor-mediated pathway for cholesterol homeostasis, Science 232:34 (1986).
2. J. D. Morrisett, R. L. Jackson and A. M. Gotto, Jr., Lipid-protein interactions in the plasma lipoproteins, Biochim. Biophys. Acta 472:93 (1977).
3. R. L. Jackson, J. D. Morrisett and A. M. Gotto, Jr., Lipoprotein structure and metabolism, Physiol. Rev. 56:259 (1976).
4. R. W. Mahley and T. L. Innerarity, Lipoprotein receptors and cholesterol homeostasis, Biochim. Biophys. Acta 737:197 (1983).
5. H. F. Hoff, C. L. Heideman, R. L. Jackson, R. J. Bayardo, H.-S. Kim and A. M. Gotto, Jr., Localization patterns of plasma apolipoproteins in human atherosclerotic lesions, Circ. Res. 37:72 (1975).
6. H. F. Hoff, W. A. Bradley, C. L. Heideman, J. W. Gaubatz, M. D. Karagas and A. M. Gotto, Jr., Characterization of low density lipoprotein-like particles in the human aorta from grossly normal and atherosclerotic regions, Biochim. Biophys. Acta 573:361 (1979).
7. B. Radhakrishnamurthy, S. R. Srinivasan, P. Vijayagopal, E. R. Dalferes, Jr. and G. S. Berenson, 1982, Mesenchymal injury and proteoglycans of arterial wall in atherosclerosis, in: "Glycosaminoglycans and Proteoglycans in Physiological and Pathological Processes of Body Systems," Karger, Basel.
8. G. Camejo, The interaction of lipids and lipoproteins with the extracellular matrix of arterial tissue. Its possible role in atherogenesis. Adv. Lipid Res. 19:1 (1982).
9. P.-H. Iverius, The interaction between plasma lipoproteins and connective tissue glycosaminoglycans. J. Biol. Chem. 247:2607 (1972).
10. M. Burnstein and H. R. Scholnick, Lipoprotein-polyanion-metal interactions, Adv. Lipid Res. 11:67 (1973).
11. A. D. Cardin, K. R. Witt and R. L. Jackson, Visualization of heparin-binding proteins by ligand blotting with ^{125}I-heparin, Anal. Biochem. 137:368 (1984).
12. W. Hollander, Unified concept on the role of acid mucopolysaccharides and connective tissue proteins in the accumulation of lipids, lipoproteins and calcium on the atherosclerotic plaque, Exp. Mol. Pathol. 25:106 (1976).
13. N. Hirose, D. T. Blankenship, M. A. Krivanek, R. L. Jackson and A. D. Cardin, Isolation and characterization of four heparin-binding cyanogen bromide peptides of human plasma apolipoprotein B, Biochemistry 26:5505 (1987).
14. A. D. Cardin, N. Hirose, D. T. Blankenship, R. L. Jackson, J. A. K. Harmony, D. A. Sparrow and J. T. Sparrow, Binding of a high reactive heparin to human apolipoprotein E: Identification of two heparin-binding domains, Biochem. Biophys. Res. Commun. 134:783 (1986).
15. A. D. Cardin, K. R. Witt, J. Chao, H. S. Margolius, V. H. Donaldson and R. L. Jackson, Degradation of apolipoprotein B-100 of human plasma low density lipoproteins by tissue and plasma kallikreins, J. Biol. Chem. 259:8522 (1984).
16. S. H. Gianturco, A. M. Gotto, Jr., S.-L. C. Hwang, J. B. Karlin, A. H. Y. Lin, S. C. Prasad and W. A. Bradley, Apolipoprotein E mediates uptake of S_f 100-400 hypertriglyceridemic very low density lipoproteins by the low density lipoprotein receptor pathway in normal human fibroblasts, J. Biol. Chem. 258:4526 (1983).

17. G. A. Martin, S. J. Busch, G. D. Meredith, A. D. Cardin, D. T. Blankenship, S. J. T. Mao, A. E. Rechtin, C. W. Woods, M. M. Racke, M. P. Schafer, M. C. Fitzgerald, D. M. Burke, M. A. Flanagan and R. L. Jackson, Isolation and cDNA sequence of human post-heparin plasma hepatic triglyceride lipase, J. Biol. Chem. (in press).
18. W. H. Burgess, T. Mehlman, D. R. Marshak, B. A. Fraser and T. Maciag, Structural evidence that endothelial cell growth factor α and acidic fibroblast growth factor, Proc. Natl. Acad. Sci. USA 83:7216 (1986).
19. J. Sommer, S. M. Gloor, G. F. Rovelli, J. Hofsteenge, H. Nick, R. Meier and D. Monard, cDNA sequence coding for a rat heparin-binding glia-derived nexin and its homology to members of the serpin superfamily: protease nexin I, Biochemistry 26:6407 (1987).
20. S. C. Bock, R. L. Wion, G. A. Vehar and R. M. Lawn, Cloning and expression of the cDNA for human antithrombin III, Nucleic Acids Res. 10:8113 (1982).
21. P. Berman, P. Gray, E. Chen, K. Keyser, D. Ehrlich, H. Karten, M. LaCorbiere, F. Esch and D. Schubert, Sequence analysis, cellular localization, and expression of a neuroretina adhesion and cell survival molecule, Cell (in press).
22. K. H. Weisgraber, S. C. Rall, Jr., R. W. Mahley, R. W. Milne, Y. L. Marcel and J. T. Sparrow, Human apolipoprotein E: Determination of the heparin binding sites of apolipoprotein E3, J. Biol. Chem. 261:2068 (1986).
23. D. Eisenberg, R. M. Weiss and T. C. Terwilliger, The hydrophobic moment detects periodicity in protein hydrophobicity, Proc. Natl. Acad. Sci. USA 81:140 (1984).
24. D. Osterman, R. Mora, F. J. Kezdy, E. T. Kaiser and S. C. Meredith, A synthetic amphiphilic β-strand tridecapeptide: a model for apolipoprotein B, J. Am. Chem. Soc. 106:6845 (1984).
25. M. Schiffer and A. B. Edmunson, Use of helical wheels to represent the structures of proteins and to identify segments with helical potential, Biophys. J. 7:121 (1967).
26. E. T. Kaiser and F. J. Kezdy, Amphiphilic secondary structure: Design of peptide hormones, Science 223:249 (1984).
27. J. B. Matthew and F. R. N. Gurd, Stabilization and destabilization of protein structure by charge interactions, Methods Enzymol. 130:437 (1986).

CHARACTERIZATION AND METABOLISM OF GLYCATED HIGH DENSITY LIPOPROTEINS IN DIABETIC PATIENTS

Carlos Calvo, Bang-Yao Luo, Florence Puygranier,
Gabriel Ponsin, and François Berthezène

Laboratoire de Métabolisme des Lipides, Service
d'Endocrinologie et des Maladies de la Nutrition, INSERM U.197
Hôpital de l'Antiquaille, 69321 Lyon Cédex 05, France

INTRODUCTION

One of the characteristics of diabetes consists in the non enzymatic glycation of plasma and cellular proteins, resulting from hyperglycaemia [1]. This process occurs in vivo by direct chemical reaction of glucose with the epsilon amino-group of lysine residues. The resulting labile intermediate Schiff base subsequently undergoes a slow Amadori rearrangement to form the more stable ketoamine and hemiketal adducts [2]. Several studies have clearly indicated that, in diabetes, low density lipoproteins (LDL) are glycated [3-5]. This non enzymatic glycation has functional consequences including : defective ability of LDL to be recognized by LDL receptor and alteration of LDL catabolism [6-8].

Although the occurence of glycated HDL in patients with type 1 diabetes has been reported [4,9], very little is known concerning the factors which could affect the HDL glycation process. Studies using HDL glycated in vitro have demonstrated that the modified lipoprotein is catabolized more rapidly than control HDL in guinea pigs [10], but the effects of glycosylation on HDL metabolism have not been studied in diabetic humans. The present study was planned to evaluate the characteristics of the glycation process of HDL apoproteins and the time of disappearance of glycated HDL after normalization of glycemia in diabetic patients.

MATERIALS AND METHODS

Subjects. The study on the glycation process was performed on patients with either type 1 or type 2 diabetes. The type 1 group (24 to 70 years old) consisted in 5 males and 5 females while the type 2 group (36 to 74 years old) included 14 males and 11 females. 14 healthy normoglycaemic subjects were used as controls. The study on the disappearance of glycated HDL has been performed in 9 patients (3 males, 6 females) with type 2 diabetes.

Lipoprotein isolation. Venous blood samples were collected in EDTA to a final concentration of 1 mg/ml after overnight fasting. Plasma were immediately obtained after centrifugation at 4°C. HDL were separated by sequential ultracentrifugation of plasma in the density interval of 1.063-1.21 g/ml adjusted by the addition of solid KBr. Isolated HDL fractions were extensively dialyzed against 0.01 mol/l Tris-HCl, pH 7.4, 0.15 mol/l NaCl, 0.01 mmol/l sodium azide, 1 mmol/l EDTA and then against 0.01 mol/l phosphate buffered saline (PBS), pH 7.4, containing 1 mmol/l EDTA.

Preparation of HDL apoproteins. HDL isolated from either diabetic or normal subjects were delipidated at 4°C in a mixture of diethylether : ethanol (3:1, V/V). The resulting apoproteins were solubilized in 0.2 mol/l Tris-HCl, pH 8.0, containing 5.4 mol/l urea and separated by chromatography on a column of Sephadex G-150 superfine (100 x 2.5 cm). The column was eluted with a 0.1 mol/l Tris-HCl buffer, pH 8.6, containing 5.4 mol/l urea and 0.01 % EDTA.

Preparation of phospholipid/Apo A-I complexes. POPC (1-palmitoyl-2-oleyl-sn-glycero-3-phosphocholine) and DMPC (1,2-dimyristoyl-sn-glycero-3-phosphocholine) were obtained from Avanti Polar-lipids, Birmingham, Alabama and from Calbiochem, La Jolla, California, respectively. Highly purified human apo A-I was a gift of Dr Henry J. Pownall, Baylor College of Medicine, Houston, Texas. POPC (or DMPC) and apo A-I (100:1, molar ratio) were mixed in the presence of sodium cholate at room temperature. The detergent was then separated from the lipid-protein complex by dialysis [11].

In vitro glycation of apoprotein A-I. The ability of apo A-I to undergo a non enzymatic glycation in vitro was studied as follows : apo A-I (1 mg/ml), either alone or in complex with POPC or DMPC was incubated in the presence of glucose at various concentrations for up to 7 days at 37°C, in PBS, PH 7.4, containing 50 mU/ml penicillin. At the end of the incubation periods, the samples were extensively dialyzed against PBS and used for glycation measurements.

Measurement of the non-enzymatic glycation of HDL apoproteins. The non-enzymatic glycation was measured after reduction in the presence of NaB^3H_4. Apoprotein (1 mg/ml) either alone, or complexed with POPC, or as a component of native HDL, was incubated in PBS containing 12 mmol/l NaB^3H_4 for 1 h on ice and 3 h at room temperature. Unbound label was removed by chromatography on a column of Sephadex G-25 followed by extensive dialysis against PBS. Duplicate aliquots of the sample were then counted for ^3H radioactivity.

The validity of this methodology was controlled in a series of preliminary experiments. Several samples of HDL from diabetics or of POPC/apo A-I complex previously incubated with glucose were treated as follows : aliquots of each sample were incubated with NaB^3H_4 for direct glycation measurement, while other aliquots were delipidated before glycation assay. No difference was observed between the results obtained with these two procedures. Moreover, when the samples were delipidated after the incubation with NaB^3H_4 less than 5 % of the radioactivity were recovered in the lipid phase. It was concluded that the presence of lipid did not cause any trouble in the glycation assay. Therefore, the measurements of HDL glycation were routinely achieved without prior delipidation of the samples. The validity of the NaB^3H_4 method was further studied in two other independent experiments. Firstly, samples of glycated apo A-I were submitted to affinity chromatography on a column of amino-phenyl boronic acid which separated the glycated protein from the bulk of apo A-I. Results indicated that only the fraction retained by the column was able to significantly incorporate radioactivity after incubation with NaB^3H_4. Secondly, the results obtained with the NaB^3H_4 procedure were compared to those obtained with the Thiobarbituric acid (TBA) method [12]. Both procedures gave comparable results.

RESULTS

- We compared the degree of glycation of HDL isolated from either type 1 or type 2 diabetic patients with that of control subjects.(Fig. 1).

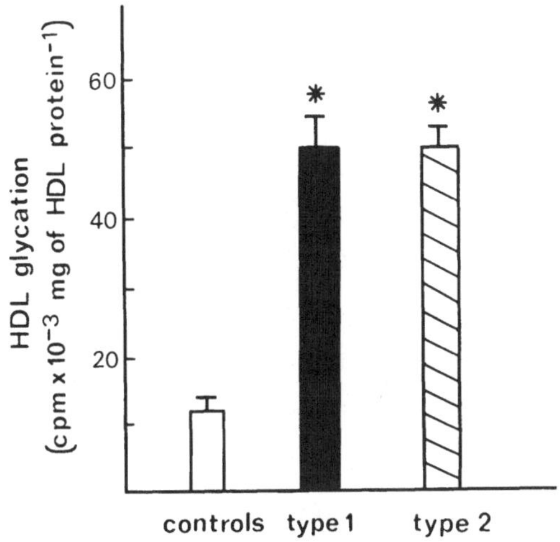

Fig. 1. Each bar represents the mean ± SEM. *$p < 0.001$ vs controls.[22]

- A significant positive correlation was found between the level of HDL glycation and blood glucose concentration of the 4th day before collection of blood sample. (Fig. 2).

- Samples of HDL from diabetic patients were delipidated and pooled. The apoproteins were separated and their glycation levels measured. Although all apoproteins were glycated, glycated apo A-I accounted for more than 80 % of the total HDL glycation.

- Highly purified apo A-I was incubated at 37°C in the presence of glucose at various concentrations. Apo A-I was glycated and this glycation was dependent upon both incubation time and glucose concentration.(Fig. 3).

- We studied the effects of phospholipid on the glycation of apo A-I in vitro using POPC/apo A-I complex. The glycation of apo A-I was approximately doubled in the presence of lipid. The results were comparable when DMPC was substituted for POPC.(Fig. 4).

- We studied the evolution of lipid composition of HDL in type 2 diabetic patients following intravenous infusion of insulin. Mean glycemia was 17 mmol/l before treatment ; it was almost normalized in all patients within 6 hours of insulin infusion and was maintained in the range of 5.5 to

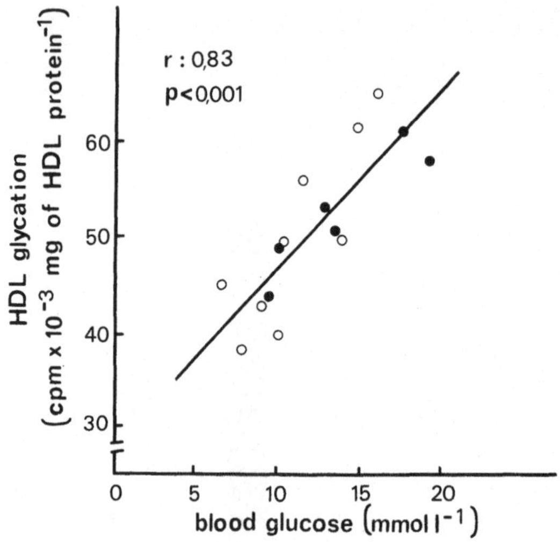

Fig. 2. (○) Type 1 ; (●) Type 2 diabetic patients.[22]

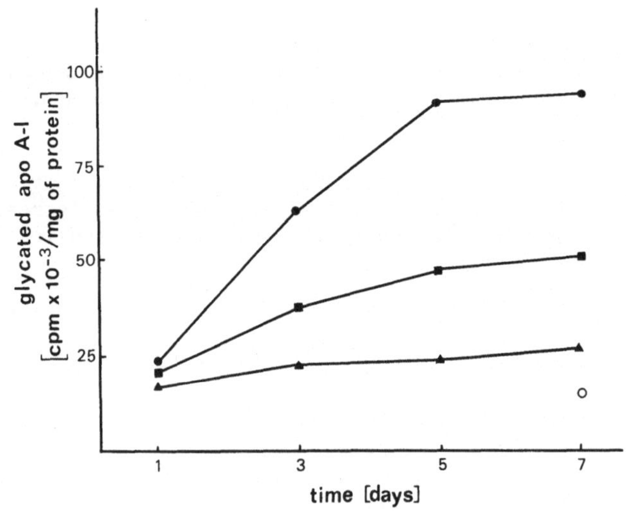

Fig. 3. Glucose (●) 80 ; (■) 20 ; (▲) 5 ; (○) 0 mmol/1 in PBS. Each value represents the mean of duplicate determinations.[22]

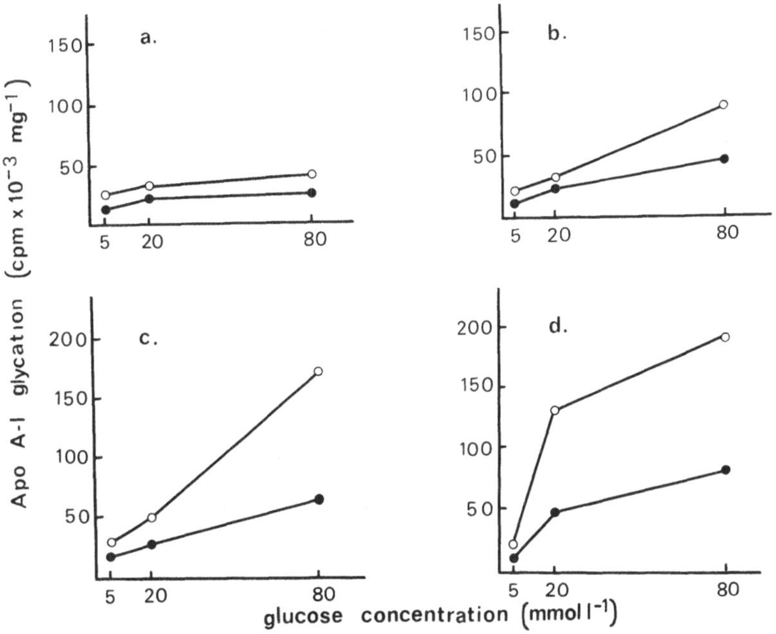

Fig. 4. Apo A-I alone (●) and apo A-I/POPC complex (○) were incubated for 1 (a), 3 (b), 5 (c), 7 (d) days. Each value represents the mean of duplicate determinations.[22]

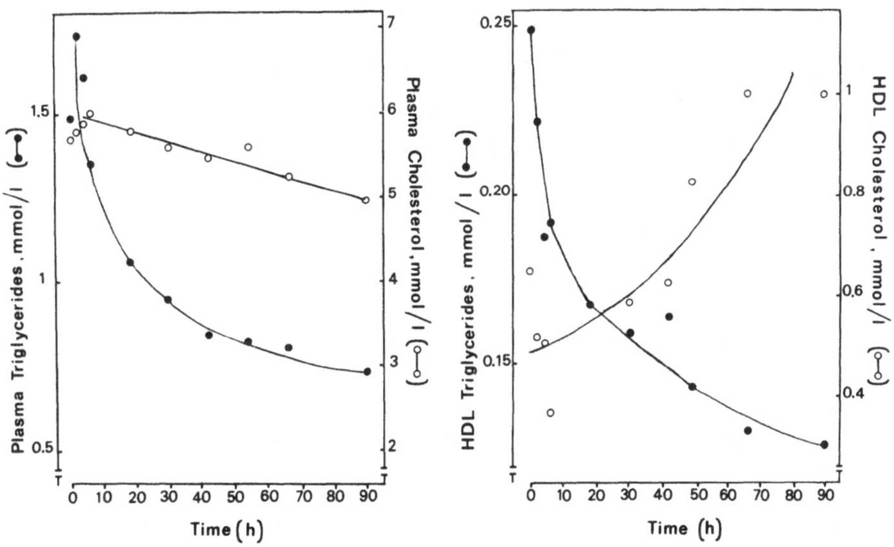

Fig. 5. Evolution of plasma and HDL lipids after normalization of glycaemia in type 2 diabetics infused with insulin.

6.6 mmol/l during 5 days. We observed a sharp decrease in plasma triglycerides and total cholesterol. HDL triglycerides fell significantly with a concomitant increase in HDL cholesterol. (Fig. 5).

Apo B levels did not change (1.06±0.15 vs 0.94±0.16 mg/ml, mean ± SE). Serum apo A-I concentration slightly decreased during the first eighteen hours (1.26±0.06 vs 1.09±0.11 mg/ml, $p < 0.02$) and then remained constant (1.04±0.11 mg/ml after 4 days). Apo A-II did not change.

- The disappearance curve of glycated HDL had two phases with different slopes. When the results are expressed as glycated HDL/apo A-I ratio, one curve can be drawn from which we can calculate an half-life of glycated HDL of 78 hours.

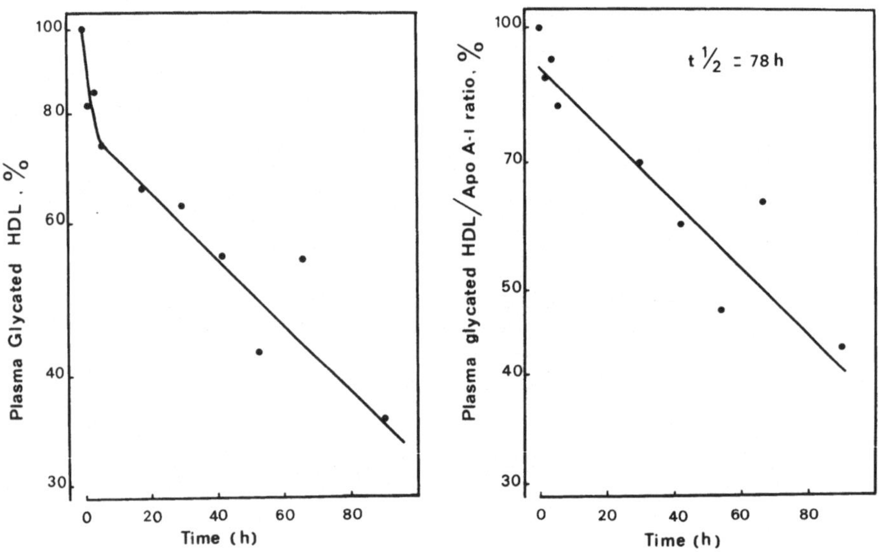

Fig. 6. The results are expressed as percent of the value obtained when euglycemia was reached.

DISCUSSION

The present report shows that HDL are somewhat glycated in normal subjects and that this glycation is about 4 fold higher in Type 1 and Type 2 diabetic patients. In both groups, the degree of HDL glycation was correlated with the plasma glucose concentration. These findings support and extend those of Curtiss and Witztum who found a similar increase in HDL glycation in patients with type 1 diabetes [4,9]. Apo A-I was the most glycated of HDL apoproteins. This preferential glycation could be expected, since apo A-I contains 21 lysine residues (8.6 %) per molecule. In vitro studies showed that apo A-I could be glycated after incubation in the presence of glucose. The glycation of the protein was largely facilitated when it was incubated as a component of phospholipid/apo A-I complex. This finding is of interest in view of the experimental evidences that, in HDL, the lipid-protein interaction is governed by the amphipathic helical theory [11,13-14]. The helical structure permits the non polar face to penetrate the lipid matrix and the polar face to remain in contact with the aqueous phase of plasma. In this structure, lysine residues may appear at the polar-non polar interface [15]. This particular position may improve the accessibility of certain lysine amino-groups to glucose. Therefore, in vivo, it would be tenable that the apoprotein glycation might be not only a function of glucose concentration but also a function of the lipid composition of HDL.

The changes observed in HDL composition during insulin infusion in diabetic patients were expected. An inverse relationship between HDL cholesterol and HDL triglyceride levels and between total triglycerides and HDL cholesterol has been found in non insulin dependent diabetics [16]. Our results show that hypertriglyceridemia and abnormalities in lipid composition of HDL disappear very quickly after normalization of glycemia. An unexplained finding was the slight but significant decrease in apo A-I concentration during the first eighteen hours after euglycemia was reached. It has been reported that, during insulin infusion in diabetic patients, HDL cholesterol particularly the HDL2 subfraction and apo A-I either did not change [17] or increased [18-19]. In all these data, lipids and apoproteins' assays have been performed several days or weeks after normalization of hyperglycemia. In these cases, insulin infusion gave changes in lipid production and lipoprotein and hepatic lipase activities which presumably did not occur in few hours in our study.

Glycated HDL disappeared from the plasma with an apparent half-life of 78 hours. This is slightly shorter than that of apo A-I measured in controls with radioiodinated apoprotein [20]. We are not aware of data in apo A-I metabolism in diabetic patients ; in rabbits with alloxan induced insulin deficiency, normal apo A-I disappears more slowly from the plasma than in control animals [21]. If the finding of a lower turnover of normal HDL during hyperglycemia can be extrapolated to human, it is of interest to find in diabetic patients an apparent half-life of glycated apo A-I slightly lower than that of apo A-I. We can hypothetize that non enzymatic glycation of apo A-I accelerates its catabolism, perhaps because of an increase in dissociation of the protein from the lipid moiety.

ACKNOWLEDGMENTS

We thank Dr Laurence Perrot for performing the apoprotein radioimmunoassays.

REFERENCES

1. L. Kennedy and J.W. Baynes, Non enzymatic glycosylation and the chronic complications of diabetes : an overview, Diabetologia 26:93 (1984).
2. G.E. Means and M.K. Chang, Non-enzymatic glycosylation of proteins. Structure and function changes, Diabetes 31 (suppl.3):1 (1982).
3. E. Schleicher, T. Deufel and O.H. Wieland, Non enzymatic glucosylation of human serum lipoproteins : elevated -lysine glycosylated low density lipoprotein in diabetic patients, FEBS Lett. 129:1 (1981).
4. L.K. Curtiss and J.L. Witztum, A novel method for generating region-specific monoclonal antibodies to modified proteins. Applications to the identification of human glucosylated low density lipoproteins, J. Clin. Invest., 72:1427 (1983).
5. T.J. Lyons, J.W. Baynes, J.S. Patrick, J.A. Colwell, and M.F. Lopes-Virella, Glycosylation of low density lipoprotein in patients with Type 1 (insulin-dependent) diabetes : correlations with other

parameters of glycaemic control, Diabetologia 29:685 (1986).
6. J.L. Witztum, E.M. Mahoney, M.I. Branks, M. Fisher, R. Elam, and D. Steinberg, Non enzymatic glucosylation of low-density lipoprotein alters its biological activity, Diabetes 31:283 (1982).
7. B. Gonen, J. Baenziger, G. Schonfeld, D. Jacobson, and P. Farrar, Non enzymatic glycosylation of low density lipoproteins in vitro: effects on cell-interactive properties, Diabetes 30:875 (1981).
8. U.P. Steinbrecher, and J.L. Witztum, Glucosylation of low-density lipoproteins to an extent comparable to that seen in diabetes slows their catabolism, Diabetes 33:130 (1984).
9. L.K. Curtiss, and I.L. Witztum, Plasma apolipoprotein A-I, A-II, B, C and E are glucosylated in hyperglycemic diabetic subjects, Diabetes 34:452 (1985).
10. J.L. Witztum, M. Fisher, T. Pietro, U.P. Steinbrecher,and R.L. Elam, Non enzymatic glucosylation of high-density lipoprotein accelerates its catabolism in guinea pigs, Diabetes 31:1029 (1982).
11. G. Ponsin, K. Strong, A.M. Gotto, J.T. Sparrow, and H.J. Pownall, In vitro binding of synthetic acylated lipid-associating peptides to high-density lipoproteins : effect of hydrophobicity, Biochemistry 23:5337 (1984).
12. T.D. Mehl, S.E. Wenzell, B. Russel, D. Gardner, and I.J. Merimee, Comparison of two indices of glycemic control in diabetic subjects : glycosylated serum protein and hemoglobin, Diabetes Care 6:34 (1983).
13. J.P. Segrest, R.L. Jackson, J.D. Morrisett, and A.M. Gotto, A molecular theory of lipid-protein interactions in the plasma lipoproteins, FEBS Lett. 38:247 (1974).
14. G. Ponsin, J.T. Sparrow, A.M. Gotto, and H.J. Pownall, In vivo interaction of synthetic acylated apopeptides with high density lipoproteins in rat, J. Clin. Invest. 77:559 (1986).
15. P. Kanellis, A.Y. Romans, B.J. Johnson, H. Kercret, R. Chiovetti Jr., T.M. Allen, and J.P. Segrest, Studies of synthetic peptide analogs of the amphipathic helix, Effect of charged amino-acid residue topography on lipid affinity, J. Biol. Chem. 255:11464 (1980).
16. R.C. Biesbroeck, J.J. Albers, P.W. Wahl, C.R. Weinberg, M.L. Bassett, and E.L. Bierman, Abnormal composition of high density lipoproteins in non-insulin-dependent diabetics, Diabetes 31:126 (1982).
17. F.V. Vlachokosta, A.C. Asmal, O.P. Ganda, and T.T. Aoki, The effect of strict control with the artificial β-cell on plasma lipid levels in insulin-dependent diabetes, Diabetes Care 6:351 (1983).
18. E. Helve, High density lipoprotein subfractions during continuous insulin infusion therapy, Atherosclerosis 64:173 (1987).
19. B.V. Howard, Lipoprotein metabolism in diabetes mellitus, J. Lipid Res. 28:613 (1987).
20. E.J. Schaefer, L.A. Zech, L.L. Jenkins, T.J. Bronzert, A.A. Rubalcaba, F.T. Lindgren, R.L. Aamodt, and H.B. Brewer, Jr., Human apolipoprotein A-I and A-II metabolism, J. Lipid Res. 23:850 (1982).
21. A. Golay, L. Zech, M.Z. Shi, C.Y. Jeng, Y.-A. M. Chiou, G.M. Reaven, and Y.-D. I. Chen, Role of insulin in regulation of high density lipoprotein metabolism, J. Lipid Res. 28:10 (1987).
22. C. Calvo, G. Ponsin, and F. Berthezène, Characterization of the non enzymatic glycation of high density lipoprotein in diabetic patients, Diabete Metab. (in press)

CONTROL OF SPONTANEOUS LIPID AND PROTEIN TRANSPORT

Henry J. Pownall

Baylor College of Medicine
The Methodist Hospital, M.S. A-601
6565 Fannin Street
Houston, Texas 77030, U.S.A.

INTRODUCTION

Lipids in plasma are solubilized by a specialized group of molecules known as the apolipoproteins. In addition to carrying plasma lipids, these proteins have other regulatory activities related to lipid metabolism. Apolipoproteins A-I and C-I are activators of plasma lecithin:cholesterol acyltransferase, apolipoprotein C-II stimulates lipoprotein activity, and apolipoproteins B and E contain ligands for the receptor-mediated endocytosis of lipoproteins. Specific transfer proteins can also move lipid molecules among lipoprotein classes. In addition, apoproteins and sparingly soluble lipids transfer among cell and lipoprotein surfaces by a mechanism that involves rate-limiting desorption from the surface of the lipoprotein followed by diffusion-controlled association with acceptor cells or lipoproteins. Following association with cells, lipids may be internalized by a flip-flop mechanism. Thus, plasma lipids and proteins can undergo a large number of physical transport processes that are independent of receptors and enzyme activities. Below I will summarize the current understanding of how these spontaneous transfer processes are regulated by specific functional groups in lipids and apolipoproteins.

SPONTANEOUS TRANSFER OF FLUORESCENT LIPID ANALOGS

Early studies demonstrated that pyrene-labeled analogs of lipids transfer between lipid surfaces (Pownall et al., 1983). These studies showed that each methylene unit in a series of fatty acids, methyl esters of fatty acids, long-chained alcohols or alkanes contributed approximately 750 cal/mole to the free energy of activation for spontaneous transfer. This value is close to that reported for the free energy of alkane transfer between water and hydrocarbons so it was concluded that the activated state for spontaneous lipid transfer was similar or identical to that of the monomeric lipid in water. Moreover, these data showed that the major determinant of lipid transfer was its hydrophobicity as determined by its alkyl chain length. These conclusions were extended by Massey et al. (1982a, 1982b, 1984), who showed that the spontaneous transfer of phospholipids among reassembled high density lipoproteins was a predictable function of the hydrophobicity of the lipid but relatively insensitive to the structure of the polar headgroup. Homan and Pownall (1988) studied the same group of phospholipids in single bilayer vesicles of phosphatidylcholine and observed a similar effect of acyl chain and headgroup structure on spontaneous transfer rates except that the transfer was slower by a factor of about 5. More importantly, they observed a second, slower component that was identified as lipid translocation from the inner leaflet of the vesicle to the outer one. This process, which is commonly called flip-flop, was quantified and it was shown that the acyl chain length has little or no effect on the rates and free energies of activation. In contrast, the structure of the polar headgroup was a major determinant with the flip-flop rate decreasing with the size and increased charge on the headgroup. For phosphatidylcholine, phosphatidic acid, phosphatidyl glycerol, and phosphatidyl ethanolamine, the rates are extremely slow; the respective halftimes for flip-flop were 347 h, 35 h, 69 h, and 10 h.

SPONTANEOUS TRANSFER OF NATURALLY OCCURRING LIPIDS

Massey & Pownall (1988) have measured the rates and activation energies of transfer of a series of fatty acids from single bilayer vesicles of 1-palmitoyl-2-oleoyl phosphatidylcholine (POPC) to albumin. Only one component was observed and its magnitude was a predictable function of the acyl chain length and the number of double bonds in the transferring species. The rate constant (sec^{-1}) for spontaneous fatty acid transfer is given by

$$\ln k_i = 23.5 - 1.31n + 1.86m \qquad 1$$

where n is the number of carbons in the acyl chain and m is the number of double bonds. This equation tells us that, as the number of double bonds is increased or the chain length shortened, the transfer rate increases. The respective halftimes for the transfer of palmitic and oleic acids between single bilayer vesicles of POPC are 41 ms and 140 ms at 30°.

The effect of chain length and structure on the rates of spontaneous transfer have not been extended to a series of naturally occurring PCs. We have measured the transfer of a series of PCs from reassembled high density lipoproteins to human low density lipoproteins. On the basis of the observed dependence of the transfer rate on the number of double bonds and the total number of carbons in both acyl chains, we derived the equations

$$\log \tau_{1/2}\ (\text{min}) = 0.234n - 0.189m - 5.76 \qquad 2a$$
$$\ln k_i^{2}(\text{min}^{-1}) = 12.7 - 0.53n + 0.44m \qquad 2b$$

where n and m have the same meaning as in Equation 1. These equations predict the rates with chain lengths up to 18 carbons and with up to 3 double bonds, e.g., for POPC $\tau_{1/2}$ = 115 min (observed) and 110 min (calculated).

SPONTANEOUS TRANSFER OF APOLIPOPROTEINS AND MODEL PEPTIDES

The C apolipoproteins transfer among VLDL, HDL, and the chylomicrons at a rate that is known to be fast compared to techniques used to isolate lipoproteins. During alimentary lipemia, some of the C proteins are transferred from HDL or VLDL to the newly secreted chylomicrons. Transfer of the apo C-II component of the C proteins is important to stimulation of lipolysis by lipoprotein lipase so that this process represents an important step in triglyceride metabolism. Conventional separation methods are too slow to quantify C protein transfer between lipoproteins and spectroscopic detection of protein transfer is not possible because of competing contributions by other protein components. As an alternative, we have measured the transfer of apo C-I, apo C-II, and apo C-III from single bilayer vesicles of POPC to vesicles that contain a fluorescence quencher. The rate of disappearance of fluorescence is a quantitative measure of the rate of apo C protein transfer. The rates of transfer calculated by extrapolation of Arrhenius plots to 37°C are much faster than those of phospholipid transfer and are composed of two components that differ by a factor of about 10. For apo C-I, apo C-II, and apo C-III, the respective halftimes for the fast (slow) components are 60 ms (540 ms), 74 ms (1,200 ms), and 71 ms (810 ms).

On the basis of these studies alone it was not possible to propose a reasonable model for the mechanism of apo C protein transfer. We therefore synthesized a series of model amphiphilic peptides that have the structure:

C_n-SerSerLeuLysGluTyrTrpSerSerLeuLysGluSerPheSer.

where C_n represents a saturated acyl chain of n carbons that is attached to the amino terminus. The peptide portion of this molecule is an amphiphilic helix. Previous studies have shown that affinity of the acyl lipid-associating peptide (C_n-LAP) for phospholipids increases as the acyl chain is increased and that the increased affinity correlates with an increase in its activation of plasma lecithin:cholesterol acyltransferase (Ponsin et al., 1984). Thus, the C_n-LAPs are a group of peptides in which there is a predictive relationship between structure and affinity for lipids. Therefore, if there is a correlation between hydrophobicity and the transfer mechanism, this should be apparent in the kinetics. As with the C-apoproteins, the kinetics of transfer of the C_n-LAPs were biexponential, with an equal fraction of the signal contributing to the fast and slow components. The rates of both components decreased as a function of n and the respective free energies of activation for transfer from single bilayer vesicles of POPC were linear with respect to n, suggesting that the peptide in the activated state is in an environment that is similar to that of water. We propose that the peptides form dimers on the lipid surface and that the initial fast component is due to the transfer of one component of the dimer out of the lipid surface and that the slow component is due to transfer of the remaining monomer in the lipid surface.

CONCLUSIONS

It is clear from the results given above that there is a pair of general rules that may be used to predict the spontaneous transport behavior of lipophiles:

1. The rate of transfer between lipid surfaces decreases as the hydrophobicity of the transferring species increases.

2. The rate of spontaneous lipophile translocation across model cell membranes decreases as the polarity and/or charge and size of the translocating species increases.

Table 1 provides some representative rates for the transfer of lipophiles between single bilayer vesicles of POPC or between reassembled high density lipoproteins. These numbers can be compared with those for other processes in the plasma compartment. There is little doubt that fatty acid and apo C protein transfer are much faster than the turnover of lipoproteins. This seems appropriate because, during lipolysis, it permits fatty acids to be rapidly transferred to albumin and the apo C-II component to be transferred to another triglyceride-rich lipoprotein where it activates lipoprotein lipase. The importance of free cholesterol transfer, although slow, is also clear, since there are no known plasma carrier proteins for free sterols and the transfer time is still faster than the turnover times of high and low density lipoproteins. Things are less clear with the phosphatidylcholines; although spontaneous transfer is faster than lipoprotein turnover, plasma contains phospholipid transfer proteins (Pownall et al., 1984; Via et al., 1985; Massey et al., 1985) that stimulate transfer presumably by function as carriers. In the case of the very slowly transferring lipids, the protein-mediated transfer is probably important. For the shorter-chained analogs, the protein-mediated process is slower than spontaneous transfer (Pownall et al., 1984). The cholesteryl esters and triglycerides are extremely hydrophobic and do not transfer between lipoproteins without a specific transfer protein (Morton and Zilversmit, 1983). Finally, we address the relative importance of lipid translocation across membranes. Cholesterol, diglycerides, triglycerides, and cholesteryl esters are nonpolar and, as a consequence, little energy is required to effect their translocation. In contrast, the translocation of phospholipids is slow; however, compared to the lifetime of a typical cell, these rates are fast. For example, the lifetime of an erythrocyte is on the order of months whereas the slowest rate of flip-flop, which was observed for phosphatidylcholine, was 14 days. Therefore, many spontaneous transfer processes provide an important background transport activity in which other activities such as lipoprotein endocytosis, lipolysis, and protein-mediated transfer or translocation must function.

Table 1. Representative Halftimes for Lipid Transport Processes

	DONOR	
Spontaneous Transfer of:	POPC SBV	R-HDL or HDL
POPC	8 h	1.5 h
oleic acid	90 ms	20 ms
cholesterol	15 min	2.9 min[1]
apo C proteins	~1 sec	200 ms
Other processes:		
HDL turnover	~5 days	
LDL turnover	~3 days	
erythrocyte lifetime	120 days	

[1]Lund-Katz et al. (1982).

REFERENCES

Homan, R. and Pownall, H.J., 1988, Transbilayer diffusion of phospholipids: Dependence on headgroup structure and acyl chain length, Biochim. Biophys. Acta, 938:155-166.

Lund-Katz, S., Hammerschlag, B., and Phillips, M.C., 1982, Kinetics and mechanism of free cholesterol exchange between human serum high- and low-density lipoproteins, Biochemistry, 21:2964-2969.

Massey, J.B., Gotto, A.M.,Jr., and Pownall, H.J., 1982a, Kinetics and mechanism of the spontaneous transfer of fluorescent phospholipids between apolipoprotein-phospholipid recombinants: Effect of the polar headgroup, J. Biol. Chem., 257:5444-5448.

Massey, J.B., Gotto, A.M.,Jr., and Pownall, H.J., 1982b, Kinetics and mechanism of the spontaneous transfer of fluorescent phosphatidylcholines between apolipoprotein-phospholipid recombinants, Biochemistry, 21:3630-3636

Massey, J.B., Hickson, D.L., She, H.S., Sparrow, J.T., Via, D.P., Gotto, A.M.,Jr., and Pownall, H.J., 1984, Measurement and prediction of the rates of spontaneous transfer of phospholipids between plasma lipoproteins, Biochim. Biophys. Acta, 794:274-280.

Massey, J.B., Hickson, D.L.M., Via, D.P., Gotto, A.M.,Jr., and Pownall, H.J., 1985, Fluorescence assay of the specificity of human plasma and bovine liver phospholipid transfer proteins. Biochim. Biophys. Acta, 835:124-131.

Massey, J.B. and Pownall, H.J., 1988, Kinetics of transfer of fatty acids and lysolecithin between phospholipid single bilayer vesicles and human serum albumin. Submitted to Biochim. Biophys. Acta.

Morton, R.E. and Zilversmit, D.B., Inter-relationship of lipids transferred by the lipid-transfer protein isolated from human lipoprotein-deficient plasma, J. Biol. Chem., 258(19):11751-11757, 1983.

Ponsin, G., Strong, K., Gotto, A.M.,Jr., Sparrow, J.T., and Pownall, H.J., 1984, In vitro binding of synthetic acylated lipid associating peptides to high density lipoproteins: Effect of hydrophobicity. Biochemistry, 23:5337-5342.

Pownall, H.J., Hickson, D.L., and Smith, L.C., 1983, Transport of biological lipophiles: Effect of lipophile structure, J. Amer. Chem. Soc. 105:2440-2445.

Pownall, H.J., Hickson, D.L.M., Gotto, A.M.,Jr., and Massey, J.B., 1984, In vitro transfer of phosphatidylcholines and their ether analogs by a human and rat plasma exchange factor, Biochem. Biophys. Res. Commun. 119:452-457.

Via, D.P., Massey, J.B., Vignale, S., Kundu, S.K., Marcus, D.M., Pownall, H.J., and Gotto, A.M.,Jr., 1985, Spontaneous and plasma factor-mediated transfer of pyrenyl cerebrosides between model and native lipoproteins, Biochim. Biophys. Acta, 837:27-34.

MODIFICATIONS IN THE CHEMICAL COMPOSITION AND THERMOMETRIC BEHAVIOR OF LDL AND HDL BY PROBUCOL IN TYPE IIa HYPERLIPOPROTEINEMIA

Christiane Dachet*, Claude Motta**, Danièle Neufcour* and Bernard Jacotot*

* INSERM U 32, Hôpital Henri-Mondor, 94010 Créteil Cedex France
** Laboratoire de Biochimie, Hôtel-Dieu 63000 Clermont-Ferrand, France

The role of low density (LDL) and high density (HDL) lipoproteins in cholesterol transport and the pathogenesis of coronary artery disease is widely known. Many studies have pointed out the importance of the physical chemical properties of these particles (1-2). In man, LDL from patients with familial hypercholesterolemia were larger and contained more cholesteryl esters and less triglycerides than the LDL from normal subjects (3, 4, 5). In animals, LDL from monkeys or swines fed an atherogenic diet were also large cholesteryl ester-enriched particles (6, 7). Probucol is a lipid-lowering drug used in the management of hypercholesterolemia. The drug is transported primarily in lipoproteins, mostly LDL (8). Its mode of action remains unclear. Probucol treatment has been reported to decrease LDL and HDL cholesterol (9, 10). The drug is effective in a high proportion of patients, including subjects with the heterozygous and homozygous forms of familial hypercholesterolemia (11, 12) and in the receptor-deficient WHHL-rabbit (13). Kinetic studies of lipoprotein metabolism in animals (13) or in man (14, 15) suggest that probucol might enhance LDL catabolism. This effect on the lipoprotein metabolism could be explained by a modification in lipoprotein composition and structure. This study is designed to investigate the effect of probucol treatment on the chemical composition and fluidity of LDL and HDL.

Three groups of subjects were studied :

1) Control group : 14 healthy normolipidemic subjects ranging in age from 20 to 48 years ; 2) HC group : 15 type IIa hypercholesterolemic patients aged from 17 to 55 years. None of these patients had been treated by lipid-lowering drugs in the last six weeks. Ten had associated xanthomas ; 3) HC-Prob group : 16 type IIa hypercholesterolemic patients ranging in age from 20 to 50 years, treated by probucol 5 mg twice daily, for at least six months. Eleven had associated xanthomas. Clinical data of the groups studied are shown in Table 1. Total cholesterol and LDL cholesterol levels were higher in the HC group than in the control group. There was no difference in the HDL cholesterol and plasma triglyceride levels. In the HC-Prob group, plasma cholesterol and LDL cholesterol levels were lower than in the HC group. Plasma triglycerides were slightly increased, while the HDL cholesterol was slightly decreased. Lipoproteins, LDL and HDL, were isolated by sequential ultracentrifugation from individual fasting plasma for physical and chemical analyses.

Table 1. Clinical data of the groups studied

Groups	Cholesterol mg/dl	Triglycerides mg/dl	HDL cholesterol mg/dl	LDL cholesterol mg/dl
Control (n = 14)	194 ± 36	70 ± 21	48 ± 11	131 ± 32
HC (n = 15)	363 ± 115	76 ± 28	53 ± 18	291 ± 116
HC-Prob (n = 16)	268 ± 72	119 ± 61	36 ± 10	203 ± 63

Lipoprotein fluidity was assessed by fluorescence polarisation of a fluorescent probe : the diphenilhexatriene, DPH. The DPH steady state fluorescence anisotropy (r) was determined from the intensities of emission polarized parallel ($I_{/\!/}$) and perpendicular ($I_{\perp}$) to the polarized excitation at 24° C by use of the standard formula : $r = \frac{I_{/\!/} - I_{\perp}}{I_{/\!/} + 2I_{\perp}}$

A structure order parameter for the lipid phase was calculated from the fluorescence anisotropy values.

Analysis of the lipid composition of LDL from various groups (Table 2), shows that LDL from the HC group appeared to contain significantly less triglycerides (TG) and slightly more cholesteryl esters than LDL from the control group. Consequently, the cholesteryl ester to triglyceride (CE/TG) ratio was significantly higher in the LDL from the HC group than in LDL from the control group. LDL from the HC-Prob group contained significantly more triglycerides (TG) than LDL from the HC group, and slightly less cholesteryl esters. The CE/TG ratio was decreased in LDL from the HC-Prob group. LDL from the HC-Prob group appeared to have the same lipid composition as LDL from the control group, except for the CE/TG ratio which was significantly higher than in the control or in the HC group.

There was no remarkable difference in the lipid composition of HDL from the control and hypercholesterolemic (HC) groups (Table 3) except for the CE/TC ratio which was lower in HDL from the HC group. HDL from the HC-Prob group contained significantly more triglycerides than HDL from the HC group. Therefore the CE/TG ratio was significantly lower in HDL from the HC-Prob group than in HDL from the HC group and even slightly lower than that of the control group.

Mean values of steady-state DPH fluorescence anisotropies at 24° C were significantly higher in LDL from the HC group than in LDL from the control group (Table 4), whereas the DPH fluorescence anisotropy in LDL from the HC-Prob group was significantly lower than in LDL from the HC group, and the same as the mean values obtained in LDL from the control group. The same results were observed concerning the calculated structure order parameter, but in this case the differences were not significant.

Table 2. Lipid composition of LDL from different groups (mg/100 mg of protein)

	Control (n = 14)	HC (n = 15)	HC-Prob (n = 16)
Total cholesterol (TC)	135 ± 25	153 ± 32	143 ± 22
Cholesteryl esters (CE)	96 ± 15	111 ± 23	107 ± 18
Triglycerides (TG)	23 ± 9	17 ± 4[b]	29 ± 9[ad]
Phospholipids (PL)	91 ± 9	97 ± 16	88 ± 11
CE/TC	0.734 ± 0.014	0.726 ± 0.014	0.749 ± 0.016[ad]
CE/TG	4.44 ± 0.99	6.57 ± 1.44[c]	4.03 ± 1.26[d]

a = $p < 0.05$ compared to the control group
b = $p < 0.01$ " " " " "
c = $p < 0.001$ " " " " "
d = $p < 0.001$ compared to the HC group.

Table 3. Lipid composition of HDL from different groups (mg/100 ml of protein)

	Control (n = 14)	HC (n = 15)	HC-Prob (n = 16)
Total cholesterol (TC)	29 ± 15	30 ± 10	26 ± 5
Cholesteryl esters (CE)	21 ± 2	25 ± 8	22 ± 5
Triglycerides (TG)	5.6 ± 2.8	4.7 ± 1.8	7.1 ± 2.9[b]
Phospholipids (PL)	48 ± 21	45 ± 17	40 ± 9
CE/TC	0.828 ± 0.032	0.808 ± 0.017[a]	0.821 ± 0.043
CE/TG	4.49 ± 1.33	5.41 ± 1.53	3.26 ± 1.25[ac]

a = $p < 0.05$ compared to the control group
b = $p < 0.01$ compared to the HC group
c = $p < 0.001$ " " "

Table 4. Mean values of DPH anisotropy and structure order parameter at 24° C in LDL from different groups.

	Control (n = 14)	HC (n = 15)	HC-Prob (n = 16)
Fluorescence anisotropy (r)	0.259 ± 0.010	0.272 ± 0.015[a]	0.261 ± 0.009[b]
Structure order parameter (S)	0.825 ± 0.024	0.849 ± 0.035	0.829 ± 0.023

a = $p < 0.05$ compared to the control group
b = $p < 0.05$ compared to the HC group

Table 5. Fluorescence anisotropy and structure order parameter of DPH in steady state at 24° C in HDL from different groups.

	Control (n = 14)	HC (n = 15)	HC-Prob (n = 16)
Fluorescence anisotropy (r)	0.229 ± 0.006	0.244 ± 0.014[a]	0.226 ± 0.009[b]
Structure order parameter (S)	0.752 ± 0.016	0.786 ± 0.033[a]	0.748 ± 0.022[b]

a = $p < 0.01$ compared to the control group
b = $p < 0.001$ compared to the HC group

The results observed in HDL particles were essentially the same than those observed in LDL (Table 5). The mean value of DPH anisotropies in HDL from the HC group was significantly higher than that from control group, while the mean value of DPH anisotropies in HDL from the HC-Prob group was lower than in HDL from the HC group. The mean values of anisotropies in HDL from the HC-Prob and control groups were identical. The same results were observed concerning the calculated structure order parameter. The differences were significant.

In conclusion, if we compare the lipoproteins from hypercholesterolemic patients to the particles from normolipidemic subjects, hypercholesterolemia LDL are found to be cholesterol-enriched and triglyceride-poor particles, in which the CE/TG ratio is significantly higher. Such large triglyceride-poor and cholesteryl ester enriched particles were observed in patients with familial hypercholesterolemia (4, 5). In addition, in animals, there is a strong correlation between LDL molecular weight and the development of coronary atherosclerosis (16). The atherogenic nature of the large LDL may be related to the chemical and physical properties of these particles. Large LDL may be more atherogenic because of the increased number of cholesteryl ester molecules they can deliver to cells (17). No difference was observed in the HDL chemical composition between particles from the HC and control groups. However, both HDL and LDL from the HC group were less fluid and more structured particles than lipoproteins from the control group.

LDL from the HC group treated by probucol contained more triglycerides and less cholesteryl esters than LDL from untreated hypercholesterolemic groups. Their CE/TG ratio was significantly lower. The chemical composition of LDL from the Probucol group was the same as that of the control group. HDL from hypercholesterolemic patients treated by probucol contained more triglycerides than HDL from the untreated hypercholesterolemic and control groups. Lipoproteins, LDL and HDL, from hypercholesterolemic patients treated by probucol were found to be more fluid and less structured particles. The presence of Probucol in lipoproteins induces changes in the chemical composition and structure of the particles.

The modifications in the lipoprotein chemical composition observed in hypercholesterolemic patients treated by Probucol, mainly the increase in LDL and HDL triglyceride content, may be due to an increase in plasma lipid transfer activities. Most of the HDL triglycerides come from triglyceride-rich particles through an exchange mechanism involving cholesteryl esters -- triglycerides (18). In addition, the activity of CETP provides a mechanism that remodels LDL and HDL into smaller particles (19). In plasma, CETP activity results in an increase in LDL and HDL triglycerides at the expense of core cholesteryl esters. The very hydrophobic character of probucol can induce modifications of the surface apoproteins, which can increase the binding of CETP to either donor or acceptor lipoproteins and therefore increase the lipid transfer activities (19). These structural modifications could also explain some earlier results : 1) Probucol is a highly effective inhibitor of endothelial cell-mediated oxidation of LDL (20). 2) Probucol enhances the metabolism of LDL by mechanisms independent of the native or modified LDL receptors (13). Similarly we propose that, in HDL, Probucol enhances their catabolism and consequently increases their turnover.

REFERENCES

1. A.R. Tall, D.M. Small, D. Atkinson, and J.L. Rudel, Studies on the structure of low density lipoproteins isolated from Macaca fascicularis fed an atherogenic diet, J. Clin. Invest. 62:1354 (1978).
2. H.J. Pownall, R.L. Jackson, R.I. Roth, A.M. Gotto, J.R. Patsch, and F.A. Kummerow, Influence of an atherogenic diet on the structure of swine low density lipoproteins, J. Lipid Res. 21:1108 (1980).
3. W. Patsch, R. Ostlund, I. Kuisk, R. Levy, and G. Schonfeld, Characterization of lipoprotein in a kindred with familial hypercholesterolemia, J. Lipid. Res. 23:1196 (1982).
4. A.V. Jardhav, and G.R. Thompson, Reversible abnormalities of low denstiy lipoprotein composition in familial hypercholesterolemia, Europ. J. Clin. Invest. 9:63 (1979).
5. B. Teng, G.R. Thompson, A.D. Snidermann, T.M. Forte, R.M. Krauss, and P.O. Kwiterovich, Composition and distribution of low density lipoprotein fractions in hyper-alpha-lipoproteinemia, normolipidemia and familial hypercholesterolemia, Proc. Ntl. Acad. Sci. USA, 80:6662 (1988).
6. L.L. Rudel, M.G. Bond, and B.C. Bullock, LDL heterogneity and atherosclerosis in non human primate, Ann. N.Y. Acad. Sci. 454:248 (1985).

7. A.R. Tall, D. Atkinson, D.M. Small, and R.W. Mahley, Characterisation of the lipoproteins of atherosclerotic swine, J. Biol. Chem. 252:7288 (1977).
8. C. Dachet, B. Jacotot, and J.C. Buxtorf, The hypolipidemic action of Probucol. Drug transport and lipoprotein composition in type IIa hyperlipoproteinemia, Atherosclerosis 58:261 (1985).
9. J. Lelorier, S. Dubreuil-Quidoz, S. Lussier-Cacan, Y.S. Huang and J. Davignon, Diet and Probucol in lowering cholesterol concentrations, Arch. Intern. Med. 137:1429 (1977).
10. M.J. Mellies, P.S. Gartside, L. Glatfelter, F. Vink, G. Guy, G. Schonfeld, and G.T. Glueck, Effect of probucol on plasma cholesterol, high and low density lipoprotein cholesterol and apoproteins A-I and A-II in adults with primary familial hypercholesterolemia, Metabolism 29:956 (1980).
11. S.G. Baker, B.I. Joffe, D. Mendelsohn, and H.C. Seftel, Treatment of homozygous familial hypercholesterolemia with probucol, S. Afr. Med. J. 62:7 (1982).
12. A. Yamamoto, Y. Matsuzawa, B. Kishino, R. Hayashi, K. Hirobe, and T. Kikkawa, Effects of Probucol on homozygous cases of familial hypercholesterolemia, Atherosclerosis 48:157 (1983).
13. M. Naruszewicz, T.E. Carew, R.C. Pittman, J.L. Witztum, and D. Steinberg, A novel mechanism by which Probucol lowers low density lipoprotein levels demonstrated in the LDL receptor deficient rabbit, J. Lipid Res. 25:1206 (1984).
14. P.J. Nestel, and T. Billington, Effects of probucol on low density lipoprotein removal and high density lipoprotein synthesis, Atherosclerosis 38:203 (1981).
15. Y.A. Kesaniemi, and S.M. Grundy, Influence of probucol on cholesterol and lipoprotein metabolism in man, J. Lipid Res. 25:780 (1984).
16. L.L. Rudel, and L.L. Pitts, Male female variability in the dietary cholesterol induced hyperlipoproteinemia of Cynomolgus monkeys (Macaca fascicularis), J. Lipid Res. 19:992 (1978).
17. R.W. St. Clair, P. Greenspan and M. Leight, Enhanced cholesterol delivery to cells in culture by low density lipoproteins from hypercholesterolemic monkeys. Correlation of cellular cholesterol accumulation with low density lipoprotein molecular weight, Arteriosclerosis 3:77 (1983).
18. S. Eisenberg, High density lipoprotein metabolism, J. Lipid Res. 25:1017 (1984).
19. J. Ihm, D.M Quinn, S.J. Busch, B. Chataing, and S.A.K. Harmony, Kinetics of plasma protein-catalyzed exchange of phosphatidylcholine and cholesteryl ester between plasma lipoproteins, J. Lipid Res. 23:1328 (1982).
20. S. Parthasarathy, S.G. Young, J.L. Witztum, R.C. Pittman, and D. Steinberg, Probucol inhibits oxidative modification of low density lipoprotein, J. Clin. Invest. 77:641 (1985).

SERUM AMYLOID A (SAA) - THE PRECURSOR OF PROTEIN AA IN SECONDARY AMYLOIDOSIS

G. Husby, A. Husebekk, B. Skogen, K. Sletten,
G. Marhaug, J. Magnus and V. Syversen

University Hospital of Tromsø, 9012 Tromsø
Norway and University of Oslo 3, Norway

In 1968, Pras and co-workers (1) published the water extraction technique which has become the standard method for purification of amyloid fibrils. Subsequent gel filtration of dissociated fibrils obtained from different clinical and experimental types of amyloidosis has in many instances revealed an elution pattern consisting of two main protein peaks (2). The second of these peaks represents a protein with molecular weight ranging from 4.2 to 31 kD in different amyloid preparations (2). It is the recognition and characterization of this low molecular weight amyloid protein that has enabled a chemical and immunologic classfication of amyloid fibrils (2,3). It has been shown that different, apparently non-related proteins can constitute the fibrils subunit is different cases of amyloidosis. Moreover, the nature of the protein is in most cases related to specific clinical types of amyloidosis (2,3). Certain serum proteins are precursors for the different fibril proteins in the various systemic forms of amyloidosis (2,3). Two important fibril proteins related to systemic amyloidosis are the amyloid L (AL) and amyloid A (AA) proteins. Protein AL, which consists of homogenous (monoclonal) immunoglobulin kappa or lambda light chains or fragments thereof (2,3) are seen in primary (idiopathic) and myeloma-associated amyloidosis (AL amyloidosis). Protein AA is associated with secondary (reactive) amyloidosis, the recessively inherited familial Mediterranean fever (FMF), and spontaneous and experimental amyloidosis in animals (2,3). These forms of amyloidosis can therefore, also be called AA amyloidosis. This presentation will concentrate on protein AA and its serum precursor SAA.

AA AND SAA PROTEINS RELATED TO REACTIVE AMYLOIDOSIS

The unique protein AA was first described by Benditt and co-workers in 1971 (4). The precursor for tissue AA is serum amyloid A (SAA) as evidenced by recent experiments performed in our laboratory (5). We introduced human HDL-SAA complexes from acute phase sera to mice during the induction of amyloidosis with endotoxin. Control animals received endotoxin only. Human AA detected with specific antisera (not cross-reacting with mouse AA) using immunodiffusion, immunoblot and radio-immunoassay techniques was present in amyloid fibrils isolated from the animals who has received human SAA, while only mouse AA was found in the fibrils obtained from the control animals. The mouse thus converted human SAA to AA and incorporated it in its own amyloid fibrils, and the precursor-product relationship between SAA and AA was thereby established. We have reproduced these experiments using different, human SAA-rich HDL preparations to ensure the significance of our findings (6,7) which have been questioned by others (8). However, very recent experiments by Tape and co-workers (9) confirm our observations on the precursor-product relationship between SAA and tissue AA.

STRUCTURE OF SAA AND AA

Human SAA consists of 104 amino acid residues corresponding to a molecular weight of about 11.5 kD and is larger, but otherwise essentially identical to protein AA in its primary structure (10,11,12). AA is most probably formed by proteolytic cleavage of SAA (Table I) and consists of a large N-terminal fragment of the latter protein (2,13). AA proteins of varying size have been found among different amyloid preparations as well as in amyloid extracted from a single organ, with molecular weights reportedly varying from 4.5 to 9.2 kD (2,13), corresponding to 45-83 amino acid residues. The sequence of SAA beyond position 45 is thus not absolutely necessary for fibril formation. However, most human AA proteins consist of the 76 N-terminal amino acid residues of SAA thus lacking a 28 amino acid C-terminal fragment which has been split off from the latter molecule (13).

It has been suggested that the size of the AA molecule may influence the tissue distribution of AA amyloidosis (14). However, the opposite could also be the case, namely that differential enzymatic handling of SAA at the local site may produce differently sized AA molecules. It has thus very recently been suggested (9) that the enzymatic cleavage of SAA takes place after the fibrils have been formed. In early stages, therefore, the fibrils may consist of intact SAA, either alone or together with AA, compatible with the observations in amyloid isolated from the duck (15), horse, cow and some human tissues (se ref. 16). The extreme hydrophobicity of the sequence corresponding to the 11 first N-terminal amino acids is the most striking feature of SAA and AA (Table I) (11,12,16,17,18). The sequence between residues 33-51 is highly conserved in the evolution pointing to an important biological function (10,11,12,16,20,21). Otherwise, the functional significance of SAA is largely unknown (13).

Computer analyses of SAA and wet gel x-ray studies of AA (17,18) confirm the hydrophobicity at the N-terminal residues 1-11. These residues form an α-helix and are probably involved in lipid binding, thus supporting previous suggestions (12) that this part of the molecule may be involved in the complexing of SAA with HDL and in the formation of AA fibrils. The authors (17) suggest that "the extremely hydrophobic N-terminal peptide and the hydrophobic sides of the secondary structural units exposed following cleavage of SAA, combine to produce a hydrophobic core in each AA fibril". Residues 50-74 also appear to form an α-helix possibly involved in lipid binding (12).

Apo SAA AND "Apo AA"

It is now established that the bulk of SAA is complexed to HDL in serum (22,23,24). Indeed, our studies have shown that SAA can constitute up to 50% of the total apolipoproteins of HDL in human acute phase sera (22). We have measured SAA concentrations of 1,280 mg/1 in such sera, while our mean concentration among normal controls is 1 mg/1 (25). SAA is a sensitive indicator of inflammation in diseases known to stimulate the acute phase response (25,26,27). For example, we have shown that measurements of SAA is of significance for the early detection of infectious complications in cystic fibriosis (28), opportunistic infections in AIDS (29), and has prognostic value with regard to determining the amount of tissue necrosis in myocardial infarction (30,31).

TABLE I HUMAN (H) AND MINK (M) SAA AND AA

					5					10					15					20
H SAA (11)	Arg	Ser	Phe	Phe	Ser	Phe	Leu	Gly	Glu	Ala	Phe	Asp	Gly	Ala	Arg	Asp	Met	Trp	Arg	Ala
H AA (10)																				
M SAA (20)		PCA	Trp				Phe				Ile/Val	Gln			Trp			Tyr		
M AA (19)		PCA	Trp				Phe				Val	Gln			Trp			Tyr		
					25					30					35					40
H SAA	Tyr	Ser	Asn	Met	Arg	Glu	Ala	Asn	Tyr	Ile	Gly	Ser	Asp	Lys	Tyr	The	His	Ala	Arg	Gly
H AA																				
M SAA										Lys	Asn									
M AA										Lys	Asn									
					45					50					55					60
H SAA	Asn	Tyr	Asp	Ala	Ala	Lys	Arg	Gly	Pro	Gly	Gly	Ala/Val	Trp	Ala	Ala	Glu	Ala/Val	Ile/Leu	Ser	Asn
H AA												Val*					Ala*	Ile		
M SAA				Gln								Ala				Lys	Val	Ile		Asp
M AA				Gln								Ala				Lys	Val	Ile		Asp
					65					70					75					80
H SAA	Ala	Arg	Glu	Asn	Ile	Glu	Arg	Phe	Phe	Gly	His	Gly	Ala	Glu	Asn	Ser	Leu	Ala	Asp	Glu
H AA																	-COOH			
M SAA					Gln			Ile/Val	Thr	Asp	Leu	Ile/Phe	Lys	Gly	Asp	Ser	Lys			
M AA					Glx-COOH															
					85					90					95					100
H SAA	Ala	Ala	Asn	Glu	Trp	Gly	Arg	Ser	Gly	Lys	Asp	Pro	Asn	His	Phe	Arg	Pro	Ala	Gly	Leu
M SAA																		Pro		
				104																
H SAA	Pro	Glu	Lys	Tyr-COOH																
M SAA		Asp		-COOH																

* Ala has been found in position 52 and Val in position 57 in another human AA (42).

By immunoblotting experiments using monospecific antisera to human SAA/AA we have detected small amounts of protein AA among the HDL apolipoproteins in acute phase sera containing large amounts of SAA (5,6,7). This means that SAA has been converted to AA when present on HDL in the circulation. It also shows that the 28 C-terminal amino acids of SAA which are not present in AA, are not essential for the binding to HDL although it may influence the affinity between SAA and HDL. This finding is in concert with recent computer analyses showing that the strongest lipid binding property resides among the 11 first N-terminal residues of SAA (12,17). Whether the existence of already formed AA molecules in the circulation is of importance for the formation of amyloid, is not clear at the present. However, we also studied the saturation kinetics between HDL and SAA/AA by adding different amounts of these proteins to purified, normal HDL in vitro followed by ultracentrifugation (7). The results showed clearly that SAA had a higher affinity for HDL than AA. The latter protein could therefore be more readily accessible for fibril formation. The lower affinity of AA for HDL may indicate that the C-terminus of SAA participates in the interaction with HDL, although residues 75-104 in SAA, as mentioned, contain no lipid-binding sequence (11,12,17). Another e planation may be that AA is formed by subspecies of SAA with low affinity for HDL, (see later). It could therefore be that apo SAA and "apo AA" molecules that have the lower affinity for HDL are those that form amyloid, and that "apo AA", although present in small amounts in the circulation, is of significance for amyloidosis.

Using the human-mouse model for AA amyloidosis we also showed that HDL-SAA complexes formed in vitro by adding purified SAA to normal HDL as well as similarly formed HDL-AA complexes, when given intravenously to the mice during amyloidogenesis, could be traced as human AA constituents of the amyloid fibrils formed in the mice (7,32). This showed that also circulating AA complexed to HDL can be amyloid precursor.

DISPLACEMENT OF SAA FROM HDL-SAA COMPLEXES BY Apo AI AND Apo AII

Others have shown that when SAA increases in serum during the acute phase in non-human primates, it displaces apo AI and apo AII or apo C protein from the HDL particle (33), and that human SAA displaces apo AI and apo AII from such complexes when added to normal HDL in vitro (34). We wanted to see if the opposite process, namely the displacement of SAA from acute phase HDL by apo AI or apo AII, can also take place (7). The results indeed showed that apo AI, but to an even greater extent apo AII effectively removed SAA from acute phase HDL when added in excess amounts in vitro (7,32). High concentrations of these apoproteins in vivo may thus displace SAA and AA from HDL, after which they can be incorporated in amyloid fibrils. It is interesting that tissues like the liver and intestines that actually produce apo AI and apo AII, are also predeliction sites for AA type amyloidosis.

"AMYLOID-PRONE" SAA MOLECULES: COMPARATIVE STUDIES IN MAN AND VARIOUS ANIMAL SPECIES

SAA is heterogenous, and amino acid sequence studies have revealed more than one amino acid at several positions of the protein (11,12,13, 20,21,35). The availability of complementary DNA proves for murine and human SAA has enabled more detailed studies of the induction, regulation and structure of SAA protein products (36,37,38,39), and these have confirmed and even extended the structural studies of the SAA protein products. Three or more genes for both murine and human SAA exist (36,37). In 1977 our amino acid sequence studies of murine SAA revealed two residues, namely valine and isoleucine at position 7, possibly representing two isotypes of this protein (35). This heterogeneity had not been observed in murine tissue AA (40) thus indicating that certain SAA molecules are genetically determined to be amyloidogenic. Hoffman et al. (40) confirmed that only one out of two isotypes of SAA, namely SAA2 forms AA and thereby amyloid, and the same group further showed that SAA2 is selectively removed from the circulation during amyloidogenesis (37). We explored this observation further (32) and could show that mouse SAA2 indeed had lower affinity for HDL than SAA1 in vitro. Furthermore, the expression of the "amyloid-prone" SAA2 gene has been shown to be defective in SJL mice that are resistant to amyloid induction (37). Structural diversity in SAA genes may therefore play a role in determining susceptibility to amyloidosis (36,37,38,39).

We have examined this hypothesis by studying the structure of SAA and AA in other species (Table I). Indeed, studies in the mink revealed a striking parallel to the situation in the mouse, with both valine and isoleucine at position 10, indicating two SAA isotypes, while only valine could be seen at this position in AA (19,20). Our preliminary studies in the horse indicate a similar selection (21).

WHAT IS THE SITUATION IN MAN?

In contract to mouse and mink, there is no obvious "amyloid-prone" sequences among the first 14 N-terminal amino acids of human SAA (10,11, 12). However, a heterogeneity at position 58 (leucine and isoleucine) in SAA does not occur in any human AA protein so far sequenced (Table I). But residue 58 of SAA clearly belongs to a part of the molecule which is not necessarily critical for fibril formation. On the other hand, it has been shown that the inherited, type AA amyloid associated with familial Mediterranean fever (FMF) which is made up by an AA protein consisting of 76 amino acids has threonine at residue 69 (42), whereas all AA proteins related to secondary amyloidosis has phenylalanine at this position. The latter is also the case with SAA from two single human individuals (11) as well as a pool of human sera (12). This substitution in FMF amyloid involving all three nucleotides, as demonstrated by the structure of a corresponding human gene (39) may be important for this inherited type of amyloidosis. Of interest in this regard is also that the amino acid sequence deduced from the structure of a human SAA gene recently determined (38) differed from any human AA proteins so far studied. This could be an example of a "non-amyloidotic" SAA gene.

SOME CURRENT HYPOTHESES REGARDING THE FORMATION OF AA AMYLOID

In all forms of systemic amyloidosis the precursor for the fibril protein is a serum protein, which in AA amyloidosis is SAA (5). Amyloid formation may be caused by excess amounts of such a protein, as the result of increased production and/or decreased clearence. However, only a minor proportion of patients with an underlying disease like rheumatoid arthritis develop amyloidosis in spite of chronically raised SAA. Additional factors are therefore needed for AA amyloidosis to occur, in other words the etiology of this disease is multifactorial. Genetically determined "amyloid-prone" SAA molecules is one such factor, together with diverse (37) or altered (43) expression of the SAA genes. The kinetics between apo SAA, "apo AA", apo AI and apo AII on HDL may facilitate amyloid deposition at certain sites (7). Two other factors, namely amyloid enhancing factor (AEF) and glycosaminoglycans (GAGs) have received much attention in recent years. AEF is capable of shortening the preamyloid phase, that is the lag period from start of induction to demonstration of amyloid fibrils in the tissue, in experimental amyloidosis in the mouse and hamster (44,45). AEF is produced by reticuloendothelial cells in the spleen and liver in close proximity to early amyloid deposition, and is also co-isolated with amyloid fibrils of different chemical composition (45). The precise chemical nature of AEF has not been established, but it has recently been suggested to consist of both carbohydrate and protein (44). GAGs, probably heparin and heparan sulfate, are also invariably present in amyloids of different types, and occurs in the tissues in close temporal relationship to amyloid deposition (44). The large negative charge of GAGs may indicate an involvement in protein folding and thereby in their incorporation in the fibrils (44). Our very recent experiments (J. Magnus et al., manuscript in preparation) confirm that GAGs are specifically associated with amyloid, in other words GAGs may be the "common denominator" in amyloids having different protein composition. The suggestion that AEF and GAGs are closely related is intriguing; perhaps AEF is a complex of GAGs and the actual amyloid protein making up the fibrils.

Another mechanism for amyloid formation may be defective enzymatic break-down of the precursor protein. Protein AA appears to be an intermediate fragment in the catabolism of SAA (46). Insufficient further break-down and removal of protein AA may therefore be important for amyloid formation. In support of this is the experimental evidence of defective reticuloendothelial (i.e. Kuppfer cell) function in pre-amyloidotic mice (47).

The ongoing reserach on these mechanisms for amyloidosis may not only shed new light on the formation of AA amyloid, but also on the pathogenesis of the various underlying diseases for secondary amyloidosis.

REFERENCES

1. M. Pras, M. Schubert, D. Zucker-Franklin, A. Rimon and E.C. Franklin, The characterization of soluble amyloid prepared in water. J. Clin. Invest. 47:923 (1968)
2. G. Husby and K. Sletten, Chemical and clinical classification of amyloid 1985. Scand. J. Immunol. 23:253 (1986)
3. G.G. Glenner, Amyloid deposits and amyloidosis. N. Engl. J. Med. 302:1283, 1333 (1980)
4. E.P. Benditt, M. Eriksen, M.A. Hermodsen and L.H. Ericsson, The major proteins of human and monkey amyloid substance: common properties including unusual N-terminal amino acid sequences. FEBS lett. 19:169 (1971)
5. A. Husebekk, B. Skogen, G. Husby and G. Marhaug, Transformation of amyloid precursor SAA to protein AA and incorporation in amyloid fibrils in vivo. Scand. J. Immunol. 21:283 (1985)
6. A. Husebekk, B. Skogen and G. Husby, Both SAA and AA are associated with HDL in human serum. Protides of the Biological Fluids (Ed. H. Peters), 34:367 (1986)
7. A. Husebekk, B. Skogen and G. Husby, Characterization of amyloid proteins AA and SAA as apolipoproteins of HDL. Displacement of SAA from the HDL-SAA by apo AI and apo AII. Scand. J. Immunol. 25:375 (1987).
8. M.L. Baltz, I.F. Rowe, D. Caspi, W.G. Turnell and M.B. Pepys, Is the serum amyloid A protein in acute phase plasma high-density lipoprotein the precursor of AA fibrils? Clin Exp. Immunol. 66:701 (1986)
9. C. Tape, R. Tan, M. Nesheim and R. Kisilevski, Apo SAA and AA: Precursor and product revisited. In Proceedings of the Vth International Symposium on Amyloidosis, Hakone, Japan (1987)
10. K. Sletten and G. Husby, The complete amino acid sequence of non-immunoglobulin amyloid fibril protein AS in rheumatoid arthritis. Eur. J. Biochem. 41:117 (1974)
11. K. Sletten, G. Marhaug and G. Husby, The covalent structure of amyloid related serum protein SAA from two patients with in flammatory disease. Hoppe-Zeyler's Z. Physiol. Chem. 364:1039 (1983)
12. D.C. Parmelee, K. Titani, L.H. Ericsson, N. Eriksen, E.P. Benditt and K.A. Walsh, Amino acid sequence of amyloid-related apoprotein (apo SAA_1) from human HDL. Biochemistry 21:3298 (1982)
13. G. Husby and K. Sletten, Amyloid proteins. Pp. 23-24 in J. Marrink and M.H. van Rijswijk (Eds). Amyloidosis. Martinus Nijhoff Publishers, Dordrecht (1986)
14. G.T. Westermark, P. Westermark, K. Sletten, Amyloid fibril protein AA. Characterization of uncommon subspecies from a patient with rheumatoid arthritis. Lab. Invest. 57:57 (1987)
15. L.H. Ericsson, N. Eriksen, K.A. Walsh and E.P. Benditt, Primary structure of duck amyloid protein A. FEBS LETT. 218:11 (1987)
16. A. Husebekk, G. Husby, K. Sletten, G. Marhaug and K. Nordstoga, Characterization of amyloid protein AA and its serum precursor SAA in the horse. Scand. J. Immunol. 23:703 (1986)

17. W. Turnell, R. Sarra, I.D. Glover, J.O. Baum, D. Caspi, M.L. Baltz and M.D. Pepys, Secondary structure prediction of human SAA1. Presumtive identification of calcium and lipid binding sites. Mol. Biol. Med. 3: 387 (1986)
18. W. Turnell, R. Sarra, J.O. Baum, D. Caspi, M.L. Baltz and M.B. Pepys, X-ray scattering and diffraction by wet gels of AA amyloid fibrils. Mol. Biol. Med. 3:409 (1986)
19. K. Waalen, K. Sletten, G. Husby and K. Nordstoga, The primary structure of amyloid fibril protein AA in endotoxin-induced amyloidosis of the mink. Eur. J. Biochem. 104:407 (1980)
20. V. Syversen, K. Sletten, G. Marhaug, G. Husby and B. Lium, The amino acid sequence of serum amyloid A (SAA) in mink. Scand. J. Immunol. 26:763 (1987)
21. K. Sletten, A. Husebekk and G. Husby, The aminoacid sequence of an amyloid fibril protein AA in horse. Scand. J. Immunol. 26:79 (1987)
22. G. Marhaug, K. Sletten and G. Husby,Characterization of amyloid related serum protein SAA complexed with serum lipoproteins (apo SAA). Clin. Exp. Immunol. 50:383 (1982)
23. E.P. Benditt and N. Eriksen, Amyloid protein SAA is associated with high density lipoprotein from human serum. Proc. Natl. Acad. Sci. USA 74:4025 (1977)
24. B. Skogen, A.L. Børresen, J.B. Natvig, K. Berg and T.E. Michaelsen, High-density lipoproteins as carrier for amyloid-related protein SAA in rabbit serum. Scand. J. Immunol. 10:39 (1979)
25. G. Marhaug, Three assays for the characterization and quantitation of human serum amyloid A. Scand. J. Immunol. 18:329 (1983)
26. K.P.W.J. McAdam, J. Li. J. Knowles, N.T. Foss, C.A. Dinarello, L.J. Rosenwasser, M.J. Selinger, M.M. Kaplan and R. Goodman, The biology of SAA: Identification of the inducer, in vitro synthesis, and heterogeneity demonstrated with monoclonal antibodies. Ann. NY Acad. Sci. 389:126 (1982)
27. M. Pepys, M.L. Baltz, Acute phase proteins with special reference to C-reactive protein and related proteins (pentaxins) and serum amyloid A protein. Adv. Immunol. 34:141 (1983)
28. G. Marhaug, H. Permin and G. Husby, Amyloid-related serum protein (SAA) as an indicator of lung infection in cystic fibrosis. Acta Pædiatr. Scand. 72:861 (1983)
29. A. Husebekk, H. Permin and G. Husby, Serum amyloid protein A (SAA) - an indicator of inflammation in AIDS and AIDS-related complex (ARC). Scand. J. Infect. Dis. 18:389 (1986)
30. G. Marhaug, L. Hårklau, B. Olsen, G. Husby, A. Husebekk and H. Wang, Serum amyloid A protein in acute myocardial infarction, Acta Med. Scand. 220:303 (1986)
31. G. Marhaug, M. Østensen, G. Husby S. Kolmannskog, T. Flægstad, T. Stokland and A. Husebekk, Clinical acute phase pattern of serum amyloid A. Protides of the Biological Fluids (Ed. H. Peters) 34:375 (1986)
32. A. Husebekk, B. Skogen and G. Husby, Replacement of SAA from the HDL-SAA complex by apo AI and apo AII. In: Proceedings of the Vth International Symposium on Amyloidosis, Hakone, Japan (1987)
33. J.S. Park and L.L. Rudel, Alteration of high density lipoprotein sub-fraction distribution with induction of serum amyloid A protein (SAA) in the non-human primate. J. Lip. Res. 26:82 (1985)
34. G.A. Coetzee, A.F. Strachan, D.R. van der Westhuyzen, H.C. Hoppe, M.S. Jeenan and F.C. deBeer, Serum amyloid A-containing human high density lipoproteins. J. Biol. Chem. 261:9644 (1986)
35. R.F. Anders, J.B. Natvig, K. Sletten, G. Husby and K. Nordstoga, Amyloid related serum protein SAA from three animal species: Comparison with human SAA. J. Immunol. 118:229 (1977)

36. J.D. Sipe, H.R. Colten, G. Goldberger, M.D. Edge, B.F. Tack, A.S. Cohen and A.S. Whitehead, Human serum amyloid A (SAA): Biosynthesis and postsynthetic processing of preSAA and structural variants defined by complementary DNA. Biochemistry, 24:2931 (1985)
37. K.-I. Yamamoto, M. Shiroo and S. Migita, Diverse gene expression for isotypes of murine serum amyloid A protein during acute phase reaction Science, 232:227 (1986)
38. B. Kluve-Beckerman, G.L. Long and M.D. Benson, DNA sequence evidence for polymorphic forms of human serum amyloid A (SAA) Biochem. Genet. 24:795 (1986)
39. P. Woo, J. Sipe, C.A. Dinarello and H.R. Colton, Structure of a human serum amyloid A gene and modulation of its expression in transfected L cells. J. Biol. Chem. 262:1570 (1987)
40. J.S. Hoffman, L.H. Ericsson, N. Eriksen, K.A. Walsh and E.P. Benditt, Murine tissue amyloid protein AA. NH2-terminal sequence identity with only one of two serum amyloid protein (ApoSAA) gene products. J. Exp. Med., 159:641 (1984)
41. R.L. Meek, J.S. Hoffmann and E.P. Benditt, Amyloidogenesis. One serum amyloid A isotype is selectively removed from the sirculation. J. Exp. Med. 163:499 (1986)
42. M. Levin, E.C. Franklin, B. Frangione and M. Pras, The amino acid sequence of a major non-immunoglobulin component of some amyloid fibrils. J. Clin. Invest. 51:2773 (1972)
43. H. Rokita, T. Shirahama, A.S. Cohen, R.L. Meek, E.P. Benditt and J. Sipe, Differential expression of the amyloid SAA3 gene liver and perinoneal macrophages of mice undergoing dissimilar inflammatory processes. J. Immunol. 139:3849 (1987)
44. A.D. Snow, J. Willmer and R. Kisilevsky, A close structural relationship between sulfated proteoglycans and AA amyloid fibrils. Lab. Invest. 57:687 (1987)
45. T.A. Niewold, P.R. Hol, A.C.J. van Andel, E.T.G. Lutz and E. Gruys, Enhancement of amyloid induction by amyloid fibril fragments in hamster, Lab. Invest. 56:544 (1987)
46. B. Skogen and J.B. Natvig, Degradation of amyloid proteins by different serine proteases. Scand. J. Immunol. 14:389 (1981)
47. A. Fuks and D. Zucker-Franklin, Impaired Kuppfer cell function precedes the development of secondary amyloidosis. J. Exp. Med. 161:1013 (1985)

MODULATION OF SERUM AMYLOID A GENE EXPRESSION BY CYOTKINES AND BACTERIAL CELL WALL COMPONENTS

Jean D. Sipe, Margaret A. Johns, Pietro Ghezzi* and Greta Knapschaefer

Boston University School of Medicine, Boston, MA 02118 and *Mario Negri Institute, Milan, Italy

INTRODUCTION

The serum amyloid A (SAA) proteins can be detected in the high density lipoprotein (HDL) fraction of plasma within a few hours after an organism has sustained injury (1-4). The amount and duration of SAA production depend upon the type of injury and its magnitude (2-6). ApoSAA proteins are cleared much more rapidly than other lipoproteins, with half lives of less than 2 hours (5). During homeostasis there is minimal, if any, SAA synthesis (2,4,7).

The SAA proteins were identified by their structural relationship to the amyloid A (AA) protein that is present as insoluble fibrils in the extracellular spaces of tissues when amyloidosis develops as a complication of chronic or recurrent acute inflammatory disorders (1,8). SAA concentrations are elevated over a 1000 fold range to concentrations approaching, for brief periods, 1 mg/ml in numerous conditions of injury, infection and malignancy. Amyloidosis is a relatively rare condition that occurs precipitously after a long course of inflammation, when SAA concentrations are significantly lower than during acute inflammatory episodes (9,10). The factors that cause the soluble apoSAA precursor to be converted to insoluble fibrils in specific anatomic sites are the subject of active investigation, as is the normal function SAA plays in restoring homeostasis.

THE ACUTE PHASE SAA RESPONSE

SAA synthesis is initiated as part of the acute phase response, the body's early reaction to injury and infection. It has been recognized for more than 40 years that the altered liver protein synthesis characteristic of the acute phase response is mediated by blood-borne leukocyte products resulting from tissue destruction (6). Investigations employing the acute phase SAA response as a model with which to dissect the cellular and biochemical events by which the reticuloendothelial system alters protein synthesis in liver have been fruitful. Rosenstreich and McAdam used the paired inbred C3H/HeJ and

C3H/HeN strains of mice to demonstrate that lymphoid cells are important to the acute phase SAA response (11). Unlike other strains, C3H/HeJ mice do not produce SAA after administration of lipopolysaccharide (LPS). C3H/HeJ mice have a defect in their ability to respond to LPS as the result of a mutation at the lps locus on chromosome 4 (12,13).

Endotoxin or lipopolysaccharide (LPS), the major component of the cell wall of gram-negative bacilli, is composed of lipid A attached through a basal core portion to a terminal polymeric oligosaccharide structure. Endotoxin preparations may contain varying amounts of protein, depending upon the method of isolation. The lipid A portion of LPS is thought to be primarily responsible for most of the biologic activities exhibited by endotoxin (14); however the protein associated with lipid A, the endotoxin associated protein (EAP), exhibits *in vitro* mitogenic and activating activities distinct from the lipid A moiety (15-20).

The lipid A portion of LPS initiates murine SAA synthesis in two steps, resulting in a brief period of high SAA concentration. The first event, which occurs in macrophages, is the production and release into the circulation of SAA inducer(s); the second, which occurs in liver, is the transcription of SAA 1, 2, and 3 mRNA followed by synthesis, secretion and uptake of SAA 1 and 2 by HDL (7,21-23) (Fig 1). The macrophages of C3H/HeJ mice do not produce SAA inducer(s) in response to LPS and thus LPS itself does not elicit an acute phase SAA response (Fig 2a) although the mice do respond normally to other stimulants such as casein (Fig 2b).

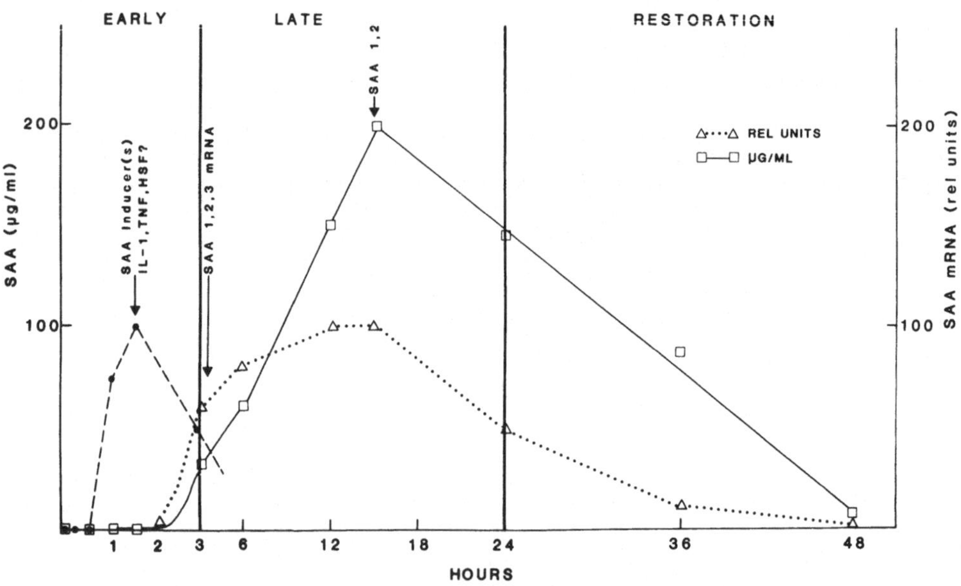

Figure 1. The humoral acute phase SAA response to endotoxin (LPS). (Composite of work from references 4, 7, 21,24). Reprinted with permission from "Immunophysiology: Role of cells and cytokines in immunity and inflammation", J.J. Oppenheim and E. Shevach, editors, Oxford Press, New York, (In press, 1988).

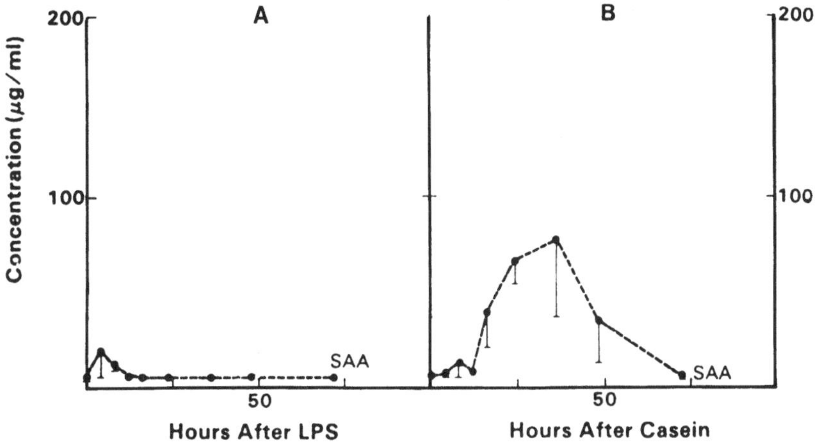

Figure 2. Response of C3H/HeJ mice to LPS and casein. Female C3H/HeJ mice were injected intraperitoneally, A, with 10 ug of K235 (ph) LPS, or B, with 50 mg of casein. (Reprinted with permission from Sipe, Vogel, Sztein, Skinner, Cohen, Ann. N.Y. Acad. Sci. 389:137, 1982).

SAA INDUCERS

Macrophages play a central role in mediating the physiologic effects of endotoxin (20). The SAA inducing activity present in serum 90 minutes after the administration of endotoxin (Fig 1) was originally attributed to IL-1 (21,24). However, subsequent studies implicated two additional monokines, tumor necrosis factor/cachectin (TNF) and hepatocyte stimulating factor/interleukin 6 (HSF) as modulators of hepatocyte protein synthesis (25-27). TNF as well as IL-1 has been shown to stimulate SAA synthesis both <u>in vivo</u> and <u>in vitro</u>, and the effect of IL-6 is currently being evaluated (28-30).

The time course for SAA production in LPS responder mice in response to IL-1 is different from that observed after LPS (Fig 3). In fact, whereas LPS and casein induce maximal SAA concentrations at 16-36 hours, the time of maximal SAA response to mouse IL-1 varies from 4 to 20 hours, according to the dose of IL-1 (25). Generally, the maximum <u>in vivo</u> SAA response to IL-1 is observed at 6 hours (Fig 3). Recently, we have observed that the time course of the SAA response to TNF is biphasic, with the maximal response at 6 hours, and a secondary response at 12 hours, presumably due to secondary IL-1 production, since TNF is a potent IL-1 inducer (31,32).

<u>Endotoxin associated protein</u>

Different profiles of LPS-initiated SAA synthesis in C3H/HeJ mice have been observed, which can be correlated with the protein content of the LPS preparation (Fig 4). Commercial LPS extracted by the Westphal method contains detectable protein, whereas LPS extracted by the hot phenol method does not (2,21,23). EAP induces a similar pattern of SAA synthesis in

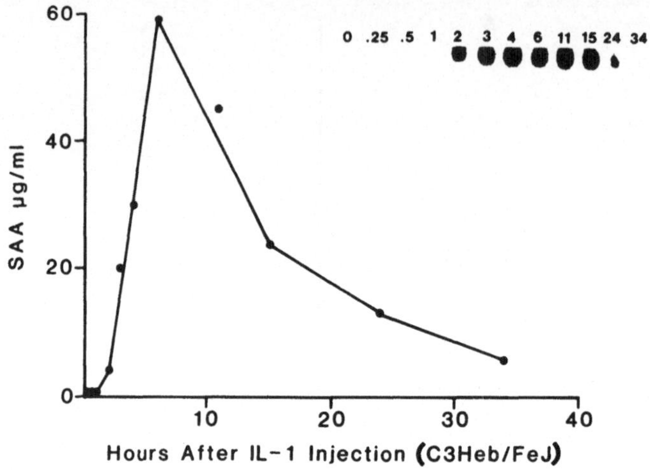

Figure 3. SAA gene expression in LPS responder C3Heb/FeJ mice after intraperitoneal injection of 2500 units of recombinant murine IL-1 as above. SAA was measured by radioimmunoassay and SAA mRNA (insert) was determined by Northern blot hybridization. (Reprinted with permission from Sipe and Ramadori, in Amyloidosis, J. Marrinck and M. Von Rijswijk, editors, Martinus Nijhoff, Utricht, 1986).

both LPS responder CF-1 and nonresponder C3H/HeJ mice and appears to act directly. EAP differs from LPS, in that incubation of EAP with macrophages fails to increase its SAA inducer activity (23). Furthermore, EAP exhibits IL-1 like activity in the thymocyte co-stimulator assay, and toward resting human T lymphocytes co-stimulated with Sepharose bound anti-CD3 monoclonal antibody 64.1. Unlike lipid A or LPS, the activity of EAP toward thymocytes was not blocked by polymyxin B (23). All of these observations strongly suggest that EAP may act as a direct SAA inducer.

REGULATION OF SAA GENE EXPRESSION

Glucocorticoids have long been known to protect against LPS-induced shock, and are potent inhibitors of IL-1, TNF, and IL-6 synthesis by macrophages (33-35). The synthetic antiinflammatory steroid dexamethasone (DEX) has been shown to inhibit LPS induced SAA production, but to enhance the response to IL-1 and TNF (31). DEX alone stimulates a low level of SAA production by as early as 4 hours reaching a maximum of 50 ug/ml at 6 hrs and with reduced, but detectable SAA concentration at 24 hours.

A comparison of the effect of DEX on SAA induction by cytokines and bacterial cell wall components is shown in Table I. CD-1 mice received an intraperitoneal injection of DEX 30 minutes prior to the ip injection of 50 ng of IL-1, 1 ug of TNF, 10 ug of protein-free LPS and 10 ug of EAP extracted from

TABLE I

SAA Induction in Mice Pretreated with Dexamethasone

SAA stimulus	SAA change relative to mice lacking DEX
LPS, 20 hrs	75% decrease
EAP, 20 hrs	33% decrease
IL-1, 6 hrs	67% increase
TNF, 6 hrs	50% increase

Male CD-1 mice, 6 weeks, were injected i.p. with 1 mg of dexamethasone phosphate 30 min prior to the i.p. injection of 10 ug of either LPS or EAP, or with 1 ug of human recombinant TNF or 10 ng of human recombinant IL-1 beta. SAA concentration was measured by radioimmunoassay in blood samples obtained at the indicated times (4).

Salmonella minnesota. The SAA concentration 20 hours after administration of LPS to mice pretreated with DEX was 75% lower than mice receiving saline pretreatment. When EAP was administered to DEX pretreated, the SAA concentration was 33 % lower than in mice receiving saline pretreatment. On the other

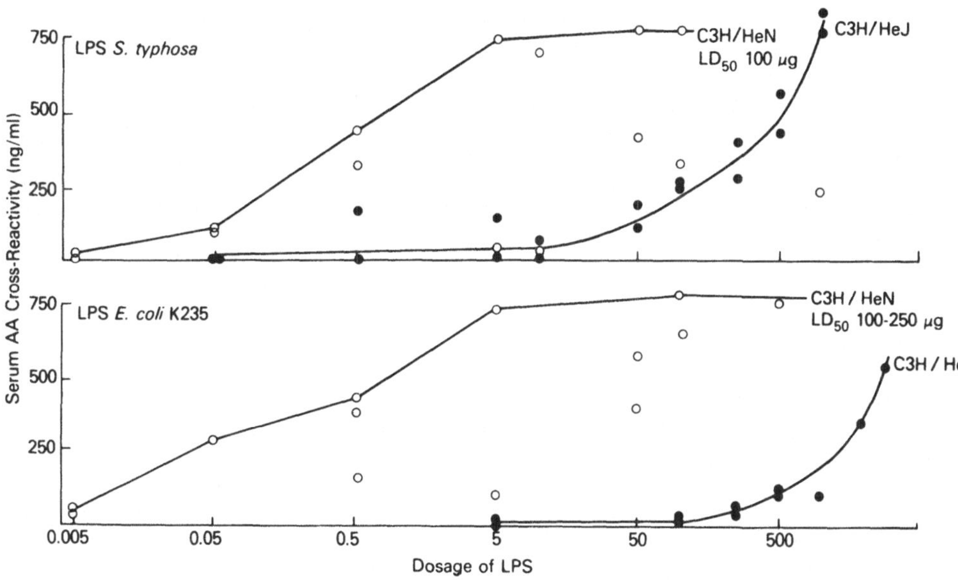

Figure 4. LPS dose-SAA response curves of C3H/HeN and C3H/HeJ animals after two types of LPS: Salmonella typhosa and Escherichia coli K235 (phenol extracted). Clinical manifestations of diarrhea and conjunctivitis were noted at dosages of LPS over 5 ug in C3H/HeN mice (LD50, 100-250 ug) but were not seen in C3H/HeJ mice, none of which died after LPS. (Reprinted with permission from McAdam and Sipe, J. Exp. Med. 139:1121, 1976).

hand, when either IL-1 or TNF was administered to dexamethasone treated mice, the SAA concentrations were 50 to 70% greater than in the saline pretreated control mice.

Thus EAP is both similar to and different from LPS and the cytokine SAA inducers. EAP differs from LPS in its ability to stimulate SAA synthesis in C3H/HeJ mice. EAP stimulated SAA synthesis is less sensitive to DEX inhibition than is LPS stimulated SAA production. SAA concentration following administration of EAP is elevated for as much as 48 hours longer than after administration of LPS. Future studies will be required to determine the clearance time of EAP. The EAP protein, similar to an outer membrane protein, may be of pathogenetic importance in chronic diseases where local events are important, such as arthritis and cardiovascular disease.

SUMMARY AND FUTURE DIRECTIONS

There is a marked difference in the time courses of SAA response to various inflammatory stimulants and to cytokines. The above studies have focused on in vivo production of the apoSAA1 and apoSAA2 isotypes by liver in response to bacterial cell wall components and the host derived mediators whose production they stimulate. A third gene, SAA 3, is expressed in liver coordinately with apoSAA 1 and apoSAA 2, but a corresponding protein has not been found in serum (7,28,36). Expression of this gene is markedly elevated in the macrophages of amyloidotic mice and may be of importance in amyloidosis (36).

Those factors that repress SAA gene expression remain to be determined, as do the biochemical events that occur between binding of the SAA inducer to the plasma membrane receptor and the transcription of the SAA genes. SAA mRNA has not been detected in liver until 2 hours after administration of IL-1 (Fig 3). The diverse physiological effects of IL-1 appear to be mediated by release of arachidonic acid from membrane phospholipids (37). It may be that the prolonged time course of SAA production in response to EAP relative to protein free LPS is due to the interaction of EAP with eukaryotic plasma membranes. It is not known how expression of the SAA genes is stopped, whether by an active process or by removal of the SAA inducers. Understanding of this issue is central to the pathogenesis of amyloidosis, where liver SAA gene expression is always reduced when amyloid A fibrils are deposited in peripheral organs.

ACKNOWLEDGMENTS

The authors thank Drs. Mark Weinstein, Alan Cohen, and Charles Dinarello for support.

REFERENCES

1. N. Eriksen and E. P. Benditt, Serum amyloid A (apoSAA) and lipoproteins, Methods Enzymol. 128:311 (1986).
2. K. P. W. J. McAdam and J. D. Sipe, Murine model for human secondary amyloidosis: Genetic variability of the acute phase serum protein SAA response to endotoxin and casein, J. Exp. Med. 144:1121 (1976).

3. M. D. Benson, M. A. Scheinberg, T. Shirahama, E. S. Cathcart, and M. Skinner, Kinetics of serum amyloid protein A in casein-induced amyloidosis, J. Clin. Invest. 59:412 (1977).
4. J. D. Sipe, Induction of the acute phase serum protein SAA requires both RNA and protein synthesis, Br. J. Exp. Path. 59: 305 (1978).
5. L. L. Bausserman, SAA kinetics in animals, "Amyloidosis", J. Marrinck and M.H. vanRijswijk, editors, Martinus Nijhoff, Dordrecht, (1986).
6. A. H. Gordon and A. Koj, The acute phase response to injury and infection. Research monographs in cell and tissue physiology, Vol 10, Elsevier, North Holland New York, (1985).
7. C. A. Lowell, D. A. Potter, R. S. Stearman, and J. S. Morrow, Structure of the serum amyloid A gene family, J. Biol.Chem. 261:8442 (1986).
8. M. Levin, E. C. Franklin, B. Frangione, and M. Pras. Immunologic studies of the major nonimmunoglobulin component of amyloid. I. Identification and partial characterization of a related serum component, J. Exp. Med. 138:373 (1973).
9. M. A. Gertz, M. Skinner, J. D. Sipe, A. S. Cohen and R. Kyle, Serum amyloid A and C-reactive protein in systemic amyloidosis. Clin. Exp. Rheum. 3:317 (1985).
10. M. Benson and A. S. Cohen, Serum amyloid A protein in amyloidosis, rheumatic and neoplastic diseases, Arthritis Rheum. 22:36, (1979).
11. D. L. Rosenstreich and K. P. W. J. McAdam, Lymphoid cells in endotoxin induced production of the amyloid-related serum protein SAA, Infect. Immun. 23:181 (1979).
12. B. J. Skidmore, D. C. Morrison, J. M. Chiller, and W. O. Weigle, Immunologic properties of bacterial lipopolysaccharide (LPS). II. The unresponsiveness of C3H/HeJ mouse spleen cells to LPS-induced mitogenesis is dependent on the method used to extract LPS. J. Exp. Med. 142:1488 (1975).
13. B. M. Sultzer and B. S. Nilsson, PPD tuberculin-a B-cell mitogen, Nature New Biol. 240:198 (1972).
14. D. Schlessinger, ed: American Society for Microbiology, Washington, D.C. Microbiology, pp3-167 (1980).
15. G. W. Goodman and B. M. Sultzer, Studies on the activation of lymphocytes by endotoxin protein, J. Immunol. 122:1329 (1979).
16. B. M. Sultzer and G. W. Goodman, Endotoxin protein: A B-cell mitogen and polyclonal activator of C3H/HeJ lymphocytes, J. Exp. Med. 144:821 (1976).
17. D. C. Morrison, S. J. Betz, and D. M. Jacobs, Isolation of a lipid A bound polypeptide responsible for "LPS-initiated" mitogenesis of C3H/HeJ spleen cells, J. Exp. Med. 144:840, (1977).
18. D. C. Morrison and S. J. Betz, Chemical and biological properties of a protein-rich fraction of bacterial lipopolysaccharides. II. The in vitro peritoneal mast cell response, J. Immunol. 119:1790 (1977).
19. W. F. Doe, S. T. Yang, D. C. Morrison, S. J. Betz, and P. M. Hensen, Macrophage stimulation by bacterial lipopolysaccharides. II. Evidence for differentiation signals delivered by lipid A and by a protein rich fraction of lipopolysaccharides, J. Exp. Med. 148:557 (1978).

20. S. N. Vogel, A. C. Weinblatt, and D. L. Rosenstreich, Inherent macrophage defects in mice, "Immunol. Defects in Laboratory Animals", M. E. Gershwin and B. Merchant, ed., Plenum Press, New York, pp. 327-357 (1981).
21. J. D. Sipe, S. N. Vogel, J. L. Ryan, K. P. W. J. McAdam, and D.L. Rosenstreich, Detection of a mediator derived from endotoxin-stimulated macrophages that induces the acute phase SAA response in mice. J. Exp. Med. 150:597 (1979).
22. E. P. Benditt, J. S. Hoffman, N. Eriksen, D. C. Parmelee, and K. A. Walsh, SAA, an apoprotein of HDL: Its structure and function, Ann. N.Y. Acad. Sci. 389:183 (1982).
23. M. A. Johns, J. D. Sipe, D. Melton, T. L. Strom, and W. R. McCabe, Endotoxin associated protein: Interleukin 1 like activity on serum amyloid A synthesis and lymphocyte activation. Infect. Immun. In press (1988).
24. G. Ramadori, J.D. Sipe, C.A. Dinarello, S.B. Mizel, and H. R. Colten, Pretranslation modulation of acute phase hepatic protein synthesis by recombinant generated mouse Interleukin 1 (IL-1) and purified human IL-1, J. Exp. Med. 162:930 (1985).
25. D. Perlmutter and H. R. Colten, Molecular biology of the complement proteins, Ann. Rev. Immunol. 4:231 (1986).
26. B. Woloski and G. Fuller, Identification and partial characterization of hepatocyte-stimulating factor from leukemia cell lines: comparison with interleukin 1, Proc. Natl. Acad. Sci. 82:1443 (1985).
27. H. Baumann, R. E. Hill, D. N. Sauder, G. P. Jahreis, Regulation of major acute phase plasma proteins by hepatocyte-stimulating factors of human squamous carcinoma cells, J. Cell. Biol. 102:370 (1986).
28. J. D. Sipe and G. Ramadori, Sites of SAA/AA synthesis, In Amyloidosis, J. Marrinck and M.H. VanRiswijck editors, Martinus Nijhoff, Dordrecht (1986).
29. J. D. Sipe, S. N. Vogel, S. Douches and R. Neta. Tumor necrosis factor/cachectin is a less potent inducer of serum amyloid A synthesis than interleukin 1, Lymphokine Res. 6:93, (1987).
30. P. Woo, J.D. Sipe, C.A. Dinarello, and H.R. Colten, Structure of a human serum amyloid A gene and modulation of its expression in transfected L cells, J. Biol. Chem. 262:15790 (1987).
31. P. Ghezzi and J.D. Sipe, Dexamethasone modulation of LPS, IL-1 and TNF stimulated serum amyloid A synthesis in mice. Manuscript submitted.
32. C. A. Dinarello, J. G. Cannon, S. M. Wolff, H. A. Bernheim, B. Beutler, A. Cerami, I.S. Figari, M. A. Palladino, J. V. O´Connor, Tumor necrosis factor (cachectin) is an endogenous pyrogen and induces production of interleukin-1, J. Exp. Med. 163:1433 (1986).
33. B. Beutler and A. Cerami, Cachectin: More than a tumor necrosis factor, New Engl. Jour. Med. 316:379 (1987).
34. D. S. Snyder and E. R. Unanue, Corticosteroids inhibit murine macrophage Ia expression and interleukin 1 production, J. Immunol. 129:1803 (1987).
35. B. M. R. N. J. Woloski, E. M. Smith, W. J. Meyer, G. M. Fuller, and J. E. Blalock, Corticotropin-releasing activity of monokines, Science 230:1035 (1985).

36. H. Rokita, T. Shirahama, A.S. Cohen, R.L. Meek, E.P. Benditt, and J.D. Sipe, Differential expression of the amyloid SAA 3 gene in liver and peritoneal macrophages of mice undergoing dissimilar inflammatory episodes, J. Immunol. 139:3849 (1987).
37. C. A. Dinarello, Interleukin 1, Rev. Infect. Dis. 6:51 (1984).

PROTEIN S AND SAA : GENETICS, STRUCTURE AND METABOLISM

ARE THEY APOLIPOPROTEINS AND IDENTICAL?

C.L. Malmendier and J-F. Lontie

Fondation de Recherche sur l'Athérosclérose
Brussels and Research Unit on Atherosclerosis
Université Libre de Bruxelles, Brussels, Belgium

The present paper intends first to welcome into the club of apolipoproteins serum amyloid A protein (SAA) and S, and second to make the first step for proposing a new nomenclature for a family of proteins integrated in the acute phase reactants and mostly associated to HDL in plasma.

GENETICS OF APOLIPOPROTEIN SAA

The murine SAA gene family is made up of three genes SAA1, SAA2 and SAA3 plus a pseudogene (1). The increase observed after endotoxin administration is specific, since the levels of the mRNAs encoding albumin and apolipoprotein A-I in liver decrease 2-fold by 24 h. This correlates with a 2-fold decrease of the serum concentrations of these two proteins as well as their in vitro protein synthesis in primary hepatocytes (1). SAA1+2 mRNAs maintain their maximum levels until 36 h after endotoxin administration.

The human SAA gene is localized on the short arm of chromosome 11 in the p11-pter region (2). It should be noted that this region of chromosome 11 also contains the gene sequences of other apoproteins, namely apo A-I, C-III and A-IV (3). However in humans there must be at least 3 genes as suggested by the existence of two major and four minor isotypes in serum. There are two allelic forms α and β with double substitution of valine for alanine at residues 52 and 57. Incomplete sequence of SAA2 shows that it lacks the N-terminal arginine. At least two AA amyloid deposits (chronic inflammatory arthritis and familial mediterranean fever) occur, corresponding to SAA1α and SAA1β with additional amino acid substitution (4). Woo stated that the organization of the SAA gene is similar to that of other apolipoprotein genes.

The genes of apo S have not yet been determined.

STRUCTURE AND PHYSICOCHEMICAL PROPERTIES OF APO SAA AND S

Molecular weight

The molecular weight of human apo SAA (5,6) was found similar to that of apo S (7) and threonine-poor apolipoproteins (8) (Table 1).

Isoelectric point

The pI of the major proteins was determined by isoelectric focusing (Table 1). In human subjects, pI's of SAA1 and SAA2 (5) were identical to those of apo S4 and apo S5 (7) but slightly lower than those reported by Shore et al. (8).

Amino acid composition

The different proteins have in common a similar amino acid composition (9). They are all relatively threonine- and valine-poor and arginine-rich.

Sequence

The complete amino acid sequence of apo SAA1 was realized by Parmelee et al.(10) and by Sletten et al.(11). It shows a protein of 104 residues with a tyrosine C-terminal and an arginine N-terminal. SAA1 differs from SAA2 only by the N-terminal arginine residue missing in the second one (5). Between the sequences proposed by the different authors, there are relatively important differences between Parmelee and Sletten (12) on one hand and Sipe (13) on the other hand (sequence deduced from cDNA) and major differences for amyloid AA between Levin (14) and Woo (4) and the other authors (Table 2).

Generally all amino acid substitutions comme from a single base change in the codon except for the position 69 when the

Table 1. Physico-Chemical Properties

		Apparent M. W. (Daltons)	pI
Apo SAA*	1	10,000-12,000	6.1
	2		5.6
Apo S**	1	9,000-11,000	8.00
	2		6.85
	3		6.55
	4		6.10
	5		5.7
	6		5.2
	7		5.0
Threonine-poor apolipoproteins***			
	1	10,000	6.5
	2		6.0

* ERIKSEN and BENDITT, 1980; MARHAUG et al.,1982
** MALMENDIER et al., 1979
*** SHORE et al., 1978

Table 2. Comparison of sequences of various human amyloid proteins.

	50		ALA					VAL	
(10)	GLY -	GLY -	VAL -	TRP -	ALA -	ALA -	GLU -	ALA -	ILE -
(13)									
(11)									LEU
(14)				ARG					
(4)			ALA					VAL	
		60							
(10)	SER -	ASP -	ALA -	ARG -	GLU -	ASN -	ILE -	LYS -	ARG -
(13)					LEU			GLN	
(11)		ASN						GLN	
(14)									
(4)		ASN							
			70						76
(10)	PHE -	PHE -	GLY -	HIS -	GLY -	ALA -	GLU -	ASP -	SER
(13)									
(11)								ASN	
(14)	LEU	THR	ARG						
(4)	LEU	THR							

References in parentheses.

substitution of threonine for phenylalanine requires the replacement of all bases of the codon. The major change was observed in patients with familial mediterranean fever.

The amino acid sequence of apo S4 and S5 was partially determined by Tartar et al. (15) and shows a similarity for the first 58 and 55 residues respectively with SAA1β : the only difference is that apo S5 lacks the N-terminal arginine.

Secondary and quaternary structures

The presence of an amphipatic helix of 26 residues responsible for the binding to lipids was shown in amyloid A protein (16). The same amphipathic helix was also demonstrated in apo S4 and S5 and suggested the likelihood of an association with lipids. Residues 1-24 and 50-74 would readily form amphipathic alpha helices of about 7 turns each. This may underlie the association of apo SAA with the HDL particle.

The existence of a stable association to HDL was reinforced by the observation that 125I radiolabeled apo S was recovered almost completely with HDL either in in vitro incubation or in vivo kinetic experiments (17). This in vivo association was maintained during the 15 days of the experiments (17).

Immunochemical properties

The determination of plasma SAA but also apo S was made using ELISA (18) or radioimmunoassay techniques (19) with antisera against SAA. Ouchterlony immunodiffusion showed that all other apoproteins (A, B, C, D..) did not cross-react with anti apo SAA except apo S (20). Total HDL and HDL-Fraction V of patients showed a cross-reaction.

METABOLISM OF APO SAA AND APO S

Benditt et al. in 1979 (21) have measured in an experiment of 60 min the plasma clearance rates of SAA-rich plasma and 125I SAA-rich HDL in 6-9 mice in which SAA level was increased after surgical removal of small intestine. In both cases, the plasma half-life approximates 38 min. In 1983, using similar material, Hoffmann and Benditt (22) found a plasma half-life of 75-80 min, as compared with a value of 11 h for mouse apo A-I, using only the fastest exponential calculated for the first 3 hours (Fig. 1). After injection of reconstituted 125I-apo SAA-HDL, the clearance kinetics paralleled that of similarly prepared 125I-apo A-I-HDL. They ignored these results, pleading a modification of SAA behavior. In 48 h experiments, Bausserman et al. (23) injected in monkeys 125I-apo SAA reassociated to HDL and concluded that apo SAA is cleared much more rapidly than apo A-I and apo C-III2.

A more careful study was performed in vervet monkey by Parks and Rudel (24)(Figure 1). Their experiments lasting for 6 days used 131I chylomicrons and 125I-HDL of chair-restrained animals. Their decay curve was biphasic with t1/2 of ∿0.45 days for the fast exponential and 2.5 to 4.3 days for the slowest component. The FCR differed for HDL (1.02 pools/day) and for chylomicrons (0.74 pools/day)(Table 3). They concluded that HDL-apo SAA is catabolized more rapidly than HDL-apo A-I and apo A-II. Recently we injected 125I-apo S and 13I-apo A-I into healthy volunteers with very low plasma concentrations of apo S (0.59 $\pm$ 0.12 mg/dl)(Table 3). The plasma decay curve, followed for 15 days, was 3-exponential for apo S and 2-exponential for apo A-I. However, the two slowest exponentials were common and a multicompartmental model was built taking into account the common subsystem.

Three conclusions derived from the comparison of these various kinetics:

1/ the clearance rates are not directly comparable between mice, monkeys and humans because of differences in the basal metabolic rates and in the lipoprotein compositions.

2/ the duration of experiments is often not adapted to the biological half-life of the molecule i.e. the apoprotein may be associated to different lipoprotein particles catabolized on different time scales.

3/ the apoprotein associates to other apoproteins on the same particle, implying that the slowest decay in the mammalian species is not artifactual but reflects the metabolism of common particles containing apo S, or apo SAA and apo A-I (17). In mice, half-life estimates from the slope of the terminal components of apo SAA (6-48 h) were not significantly different for 125I-apoSAA-rich-HDL and 125I-control HDL, although the authors estimated that this was due to denaturing conditions. Denaturation may more likeky lead to rapid plasma clearance than the reverse.

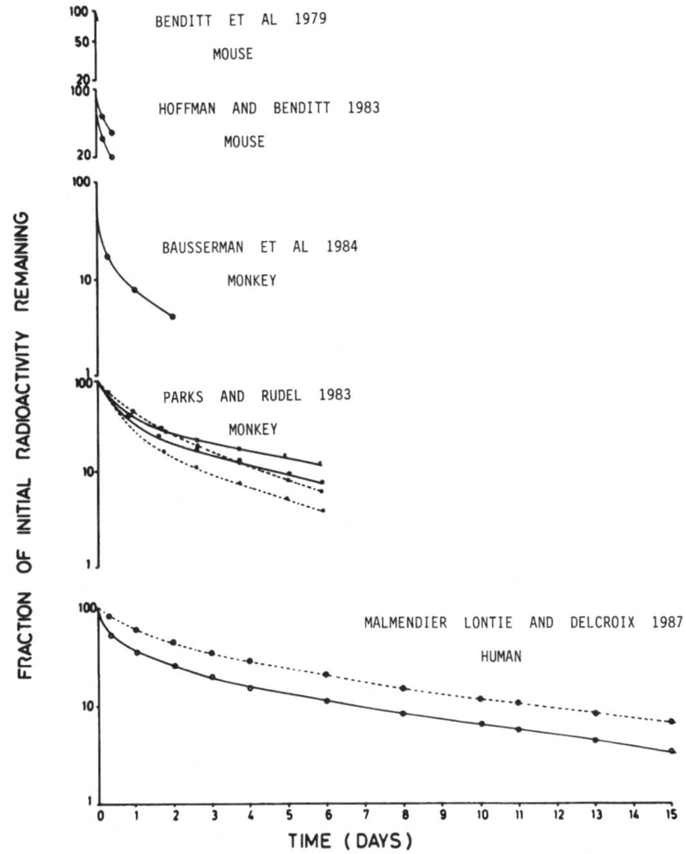

Fig. 1. Apo SAA or apo S radioactivity plasma decay curves, expressed as fraction of initial radioactivity remaining, on the same time scale (days)

Table 3. Protocol and kinetic parameters of experiments involving apo SAA or apo S

ANIMAL	INJECTED MATERIAL	DURATION OF EXPERIMENT	T ½ days	FCR pools/days	SYNTHETIC RATE mg/kg day
Mouse	SAA-rich plasma	60 min	38 min		
	^{125}I SAA-rich HDL		(0.026)		
Mouse	SAA-rich plasma	3 h	75 - 80 min (0.055)		
	^{125}I-SAA-rich HDL	48 h	0.58		
Cynomolgus monkey	^{125}I-apo SAA-HDL	48 h	∼ 1.0		
Vervet	^{131}I chylomicrons	6 days	0.45 2.5	0.74	
monkey	^{125}I-HDL		4.3	1.02	
Man	^{125}I-apo S	15 days	0.15 3.25	0.40	0.1

IDENTIFICATION AS AN APOLIPOPROTEIN

The criteria for a protein to be an apolipoprotein are: "An apolipoprotein is a homogenous protein composed of a single polypeptide chain which associates with or forms an integral part of a lipoprotein and/or lipoprotein particle in plasma" (25).

The fact that apo S or SAA concentration in plasma is very small in normal subjects casted some doubts on its existence as an apolipoprotein. However, in addition to the fact that it is always present, its value may increase up to 300 fold in pathological conditions (20,26), becoming a major apoprotein of HDL. There are several arguments in favor of the apolipoprotein nature of apo SAA and apo S:

a) the presence of amphipathic helices and the binding to lipids

Residues 1-24 and 50-74 would readily form amphipathic alpha helices of about 7 turns each. This may underlie the association of apo SAA with the HDL particle (10). Segrest et al. (16) have already noted this feature in the first segment of amyloid AA. A third region (residues 48-51) may also be of importance for the interaction with HDL (27). In fact, except for the residues 59-74, the known part of the sequence of apo S shows the same structure.

Association kinetics was shown between apo S and DMPC and a stable complex DMPC-apo S was isolated on Sepharose 6 B column (28). Binding to phospholipids was thus confirmed.

The binding to lipids is an essential condition to be an apolipoprotein (29): "An apolipoprotein is defined as a lipid-binding protein capable of forming a soluble, polydisperse lipoprotein family".

b) the constant association with HDL

The first paper mentioning the association of amyloid protein SAA to HDL was that of Benditt and Eriksen in 1977(30).

In 1978 Shore and Shore (8) isolated two threonine-poor proteins which they reported for the first time as apolipoproteins. Subsequently related proteins (apo S) were described and called apoproteins in subjects fed with glucose infusions (7) almost simultaneously with the demonstration that SAA protein was an HDL-apoprotein (5).

It is noteworthy that apo S was also found in HDL density range (7,17). Moreover, after its injection in vivo, 125I-apo S was repeatedly recovered with HDL even after repeated ultracentrifugations of plasma and after gel filtration of plasma or lipoproteins on FPLC chromatography. A maximum of 2% of radioactivity was found in the non-lipoprotein fraction as already observed for other apoproteins (A-I, A-II, C-II, C-III). Affinity chromatography on columns of anti-apo A-I linked to Sepharose 4B showed that 42.5% of apo S radioactivity was retained with lipoprotein particles containing apo A-I.

Furthermore, lipid composition of apo S-rich HDL of neurological and postsurgery patients showed a same percentage of protein (∿50%) and phospholipid (∿28%) as normal HDL but a lower cholesterol (∿16%) and a higher triglyceride (∿6%) content (20).

In mice, HDL from endotoxin-treated animals contained, on a percent weight basis, 10% more protein and 12% less phospholipid. No change in the relative content of other lipids were noted and no differences in the relative proportions of the various phospholipids were found (31).

c) the displacement by apo A-I and the reverse

Apo SAA may replace almost completely apo A-I when it is incubated with HDL (32) but no evidence of displacement of apo A-II or C was found. Concomittantly with the induction of apo SAA and apo S (20,33) biosynthesis, there is an apparent decrease in the percentage composition of the major apoproteins of HDL (A-I and A-II).

When apo S or SAA concentrations are increased more than 3-fold up to 100-fold, apo A-I, apo A-II and even apo C are decreased in large proportions (20,33)

Husebekk et al. (1987) illustrated that the reverse process could take place i.e. that both apo A-I and apo A-II may displace SAA from acute phase HDL when added in vitro to HDL-SAA complexes (34).

d) the proximity of the coding genes with other apolipoproteins

Another argument in favor of the apolipoprotein nature of apo SAA is the close proximity of the coding genes on chromosome 11 with apoproteins A-I, A-IV and C-III.

e) the common metabolic pathway

Finally, the demonstration of a common metabolic pathway of some lipoprotein particles containing both apo A-I and apo S emphasizes the apoprotein nature of apo S.

ARGUMENTS FOR IDENTITY OF APO SAA AND APO S

From the point of view of the structure and physicochemical properties, it appears that at least the major apo SAA (SAA1 and SAA2) and the major apo S (S4 and S5) are similar if not identical as suggested by identical isoelectric point, similar amino acid composition, identical sequence of the first 58 residues, isomorphism and a common immunoreactivity.

More, the plasma concentrations of both apoproteins increase in various clinical conditions and both proteins behave as acute phase reactants (chronic inflammatory diseases, bacterial infections, trauma,...).

Other similar proteins have been described. The exact nature of apo X mentioned by Douste-Blazy et al. (35), remains to be elucidated. These proteins increase in rabbits after irradiation or cyclophosphamide administration and are located mostly in VLDL. Their amino acid composition and sequence are unknown.

In contrast, the designation as apo T (36) of abnormal apoproteins in HDL of patients after surgery or trauma is not at all justified, as these T proteins are either SAA or S.

CONCLUSIONS

This paper represents the first attempt

1) to emphasize the similarities of apo SAA and apo S which represent variants of the same protein
2) to definitely introduce these proteins in the apolipoprotein family
3) to propose a common nomenclature in order to facilitate comprehensive approach in this field
4) to stress the link between the different forms and the clinical pathological manifestations.

It is likely that apparently minor structural changes may render some forms amyloidogenic and others not. Therefore to keep the term SAA protein may appear unappropriate. The same is true for apo S. The relation (timing, dependency) between these proteins and other acute phase proteins would require more careful clinical studies. In the future, the relationship between isoforms (S1 to S7, SAA1, SAA2,...) or the various forms of modified sequences on one hand and, on the other hand the specificity of pathological conditions leading (or not) to amyloidosis or atherogenesis, must be precised.

More metabolic studies are needed for this purpose in order to understand the function of these proteins.

REFERENCES

1 C.A. Lowell, R.S. Stearman, and J.F. Morrow, Transcriptional regulation of serum amyloid A gene expression, J.Biol.Chem. 261:8453 (1986).

2 B. Kluve-Beckerman, S.L. Naylor, A. Marshall, J.C. Gardner, T.B. Shows, and M.D. Benson, Localization of human SAA gene(s) to chromosome 11 and detection of DNA polymorphisms, Biochim.Biophys.Res.Comm. 137:1196 (1986).

3 S.W. Law, G. Gray, H.B. Brewer Jr., A.Y. Sakaguchi, and S.L. Naylor, Human apolipoprotein A-I and C-III genes reside in the p11--q13 region of chromosome 11, Biochim.Biophys. Res.Comm. 118:934 (1984).

4 P. Woo, Gene structure of a human serum amyloid A protein and comparison with amyloid A, In: Amyloidosis, J. Marrink & M.H. Van Rijswijk, eds., Martinus Nijhoff Publ., Dordrecht, 135 (1986).

5 N. Eriksen, and E.P. Benditt, Isolation and characterization of the amyloid-related apoprotein (SAA) from human high density lipoprotein, Proc.Natl.Acad.Sci.USA, 77:6860 (1980).

6 G. Marhaug, K. Sletten, and G. Husby, Characterization of amyloid related protein SAA complexed with serum lipoproteins (apoSAA), Clin.Exp.Immunol., 50:382 (1982).

7 C.L. Malmendier, J. Christophe, and J.P. Ameryckx, Separation and partial characterization of new apoproteins from human plasma high density lipoproteins, Clin.Chim.Acta 99:167 (1979).

8 V.G. Shore, B. Shore, and S.B. Lewis, Isolation and characterization of two threonine-poor apolipoproteins of human plasma high density lipoproteins, Biochemistry 17:2174 (1978).
9 C.L. Malmendier, and J.P. Ameryckx, L'apoprotéine S. Sa structure, son origine, sa fonction, son devenir, son induction, ses implications en pathologie, Exp.Ann.Bioch.Méd. 35:75 (1982).
10 D.C. Parmelee, K. Titani, L.H. Ericsson, N. Eriksen, E.P. Benditt, and K.A. Walsh, Amino acid sequence of amyloid-related apoprotein (apoSAA1) from human high-density lipoprotein, Biochemistry 21:3298 (1982).
11 K. Sletten, G. Marhaug, and G. Husby, The covalent structure of amyloid-related serum protein SAA from two patients with inflammatory disease, Hoppe-Seyler's Z.Physiol.Chem. 364:1039 (1983).
12 K. Sletten, and G. Husby, The complete amino-acid sequence of non-immunoglobulin amyloid fibril protein AS in rheumatoid arthritis, Eur.J.Biochem. 41:117 (1974).
13 J.D. Sipe, H.R. Colten, G. Goldberger, M.D. Edge, B.F. Tack, A.S. Cohen, and A.S. Whitehead, Human serum amyloid A (SAA): biosynthesis and postsynthetic processing of preSAA and structural variants defined by complementary DNA, Biochemistry 24:2931 (1985).
14 M. Levin, E.C. Franklin, B. Frangione, and M. Pras, The amino acid sequence of a major nonimmunoglobulin component of some amyloid fibrils, J.Clin.Invest. 51:2773 (1972).
15 A. Tartar, P. Maes, J-F. Lontie, and C.L. Malmendier, Microsequences of SA-4 and SA-5, Unpublished observations.
16 J.P. Segrest, H.J. Pownall, R.L. Jackson, G.G. Glenner, and P.S. Pollock, Amyloid A: amphipatic helixes and lipid binding, Biochemistry 15:3187 (1976).
17 C.L. Malmendier, J-F. Lontie, and C. Delcroix, In vivo metabolism of apolipoprotein S in humans. Comparison with apolipoprotein A-I metabolism, Clin.Chim.Acta 170:169 (1987).
18 D.Y. Dubois, and C.L. Malmendier, Noncompetitive enzyme linked immunosorbent assay for human apolipoprotein SAA or S, J.Immunol.Meth. In press, 1988.
19 N.L. Godenir, M.S. Jeenah, G.A. Coetzee, D.R. Van der Westhuyzen, A.F. Strachan, and F.C. De Beer, Standardisation of the quantitation of serum amyloid A protein (SAA) in human serum, J.Immunol.Meth. 83:217 (1985).
20 C.L. Malmendier, and J.P. Ameryckx, Apoprotein S versus SAA protein, In: Lipid Metabolism and Its Pathology, M.J. Halpern, ed., Plenum Press, 31 (1985).
21 E.P. Benditt, N. Eriksen, and J.S. Hoffman, Origin of protein AA, Symposium on amyloidosis, Portugal 23-28 september 1979.
22 J.S. Hoffmann, and E.P. Benditt, Plasma clearance kinetics of the amyloid-related high density lipoprotein apoprotein, serum amyloid protein (ApoSAA), in the mouse, J.Clin.Invest. 71:926 (1983).
23 L.L. Bausserman, P.N. Herbert, R. Rodger, and R.J. Nicolosi, Rapid clearance of serum amyloid A from high-density lipoproteins, Biochim.Biophys.Acta 792:186 (1984).
24 J.S. Parks, and L.L. Rudel, Metabolism of the serum amyloid A proteins (SAA) in high-density lipoproteins and chylomicrons of nonhuman primates (vervet monkey), Am.J.Pathol. 112:243 (1983).

25 P. Alaupovic, Structure and function of plasma lipoprotein with particular regard to hyperlipoproteinemias and atherosclerosis, Ann.Biol.Clin. 38:83 (1980).
26 C.J. Rosenthal, and E.C. Franklin, Variation with age and disease of an amyloid A protein-related serum component, J.Clin.Invest. 55:746 (1975).
27 W.G. Turnell, R. Sarra, J.O. Baum, and M.B. Pepys, Structural studies of human AA fibrils and of their precursor SAA, In: Protides of the Biological Fluids, H. Peeters, ed., vol. 34, Pergamon Press, 1986.
28 C.L. Malmendier, and M. Rosseneu, Unpublished results.
29 P. Alaupovic, The concepts, classification systems and nomenclatures of human plasma lipoproteins, In: Electrophoresis, L.A. Lewis, and J.J. Opplt, eds., Vol 1, Lipoproteins, CRC Press (1979).
30 E.P. Benditt, and N. Eriksen, Amyloid protein SAA is associated with high density lipoprotein from human serum, Proc. Natl.Acad.Sci.USA 74:4025 (1977).
31 J.S. Hoffman, and E.P. Benditt, Changes in high density lipoprotein content following endotoxin administration in the mouse, J.Biol.Chem. 257:10510 (1982).
32 G.A. Coetzee, A.F. Strachan, D.R. van der Westhuyzen, H.C. Hoppe, M.S. Jeenah, and F.C. de Beer, Serum amyloid A-containing human high density lipoprotein 3. Density, size, and apolipoprotein composition, J.Biol.Chem. 261:9644 (1986).
33 J.S. Parks, and L.L. Rudel, Alteration of high density lipoprotein subfraction distribution with induction of serum amyloid A protein (SAA) in the nonhuman primate, J.Lipid Res. 26:82 (1985).
34 A. Husebekk, B. Skogen, and G. Husby, Characterization of amyloid proteins AA and SAA as apolipoproteins of HDL. Displacement of SAA from the HDL-SAA complex by apo AI and apo AII, Scand.J.Immunol. In press.
35 R. Feliste, N. Dousset, M. Carton, and L. Douste-Blazy, Changes in plasma apolipoproteins following whole body irradiation in rabbit, Radiation Res. 87:602 (1981).
36 L.A. Carlson, L. Holmquist, and M. Lindholm, Abnormal apolipoproteins in high density serum lipoproteins in dyslipoproteinemia of severe trauma, Lancet 2:760 (1982).

PLASMA CHOLESTERYL ESTER AND PHOSPHOLIPID TRANSFER PROTEINS AND THEIR REGULATION

John J. Albers[a,b,c], John H. Tollefson[a,c], Russell A. Faust[b,c] and Toshio Nishide[a,c]

Departments of Medicine[a] and Pathology[b]
University of Washington School of Medicine and Northwest Lipid Research Center[c], Harborview Medical Center, Seattle, Washington, U.S.A

INTRODUCTION

Recently, there has been intense interest in the role of plasma lipid transfer proteins in lipoprotein metabolism. It has been postulated that a cholesteryl ester transfer protein (CETP) plays a key role in "reverse cholesterol transport" by modulating the transfer of CE from high density lipoprotein (HDL) to chylomicrons, very low density lipoprotein (VLDL), lipoprotein remnants and low density lipoprotein (LDL).[1] These lipoprotein acceptors transport the CE to the liver. On the other hand, these CE acceptor lipoproteins are considered atherogenic particles and thus this CE transfer process could conceivably enhance atherosclerosis.

ISOLATION AND CHARACTERIZATION OF THE PLASMA LIPID TRANSFER PROTEIN

We have isolated from human plasma and characterized two distinct lipid transfer proteins designated LTP-I and LTP-II.[2,3] LTP-I facilitates the exchange and net mass transfer of cholesteryl ester (CE), triglyceride (TG), and phospholipid (PL) among lipoproteins. Preliminary data also suggests that LTP-I promotes the transfer of CE from and to cells.[4,5] LTP-I is an acidic glycoprotein with an apparent molecular weight of 64,000 by SDS polyacrylamide gel electrophoresis and about 65,000 by molecular exclusion chromatography, has an isoelectric point of 5.0 by isoelectric focusing, binds poorly to heparin-sepharose, and is relatively resistant to elevated temperatures, 58 degrees for 1 hour.[2] The cholesteryl ester transfer protein (CETP), with a reported molecular weight of 74,000 daltons,[6,7] is identical to LTP-I since it has been shown to have sequence homology to LTP-I[8] and polyclonal antibody, against either whole LTP-I or a decapeptide of CETP, reveals reactive proteins in identical positions on SDS gel electrophoresis by the immunoblot technique.[9] Anti-LTP-I completely inhibits the plasma CE and TG transfer activity, but only about half of the plasma PL transfer activity.[3] This result suggests that there is only one plasma CE transfer protein, but that there is a second phospholipid transfer protein.

Two distinct lipid transfer proteins can be separated by heparin-sepharose.[2] All the CE and TG transfer activity and about half of the plasma PL transfer activity does not bind to a heparin sepharose column. However, about half of the PL transfer activity binds to the column and is eluted with 0.5M NaCl. LTP-II is purified from plasma by ultracentrifugation and a series of chromatography steps[3] patterned after the isolation procedure initially reported for LCAT,[10] and subsequently modified for the purification of LTP-I.[2] LTP-II has many properties similar to LTP-I. Although LTP-II has an apparent molecular weight and isoelectric point similar to LTP-I, unlike LTP-I, LTP-II also has a strong affinity for heparin-sepharose, is heat labile, does not bind anti-LTP-I, and does not transfer CE and TG, but does transfer PL (Table 1). Purified LTP-II promotes the net mass transfer of phospholipid from VLDL to HDL. LTP-II also has been shown to markedly enhance CE transfer mediated by LTP-I.[3] Addition of purified LTP-II to a mixture of lipid donor and acceptor lipoproteins and LTP-I results in a pronounced enhancement of CE transfer.

CELLULAR SECRETION AND REGULATION OF THE CHOLESTERYL ESTER TRANSFER PROTEIN LTP-I

LTP-I is synthesized and secreted by a variety of human cells, including monocyte derived macrophages[11] HepG2 and Hep3B hepatocarcinoma cells,[12] and CaCo-2 enterocyte epithelial cells.[13] Human macrophages in vitro synthesize and secrete LTP-I, and secretion is enhanced about 70% over control cultures by the stimulant phorbol myristate acetate (PMA).[11] Loading macrophages with cholesterol in either chemical or lipoprotein form results in a concentration dependent increase in secretion of CE transfer activity. The increase in regulation of LTP-I secretion is correlated with macrophage cholesteryl ester content.[14] Thus, LTP-I secretion may be a protective mechanism against excess lipid accumulation.

Table 1. Comparison of LTP-I and LTP-II

PROPERTY	LTP-I	LTP-II
Molecular Weight ($x10^{-3}$)		
Gel Filtration	65	70
SDS-Gel Electrophoresis	64	69
Isoelectric Point	5.0	5.1
Heat Stable	Yes	No
Binds Heparin	No	Yes
PPT by Anti-LTP-I	Yes	No
Transfers CE, TG	Yes	No
Transfers PL	Yes	Yes

When human CaCo-2 enterocytes are cultured on permeable membranes, they secrete a cholesteryl ester transfer protein with molecular identity to LTP-I.[13] Cholesteryl ester transfer activity secreted by CaCo-2 cells, as well as HepG2 cells, is completely inhibited by anti-LTP-I. LTP-I is detected in the basolateral, but not in the apical cell medium. Thus, CaCo-2 cells vectorially sort and secrete LTP-I from the basolateral cellular domain. Over a 24 hour period, CaCo-2 cells linearly secrete CE transfer activity, at approximately twice the rate of HepG2 cells. Furthermore, the CaCo-2 enterocyte, but not HepG2 hepatocyte, regulates LTP-I secretion in response to fatty acid concentration in the culture medium.[14] Thus, the intestine may be the principle regulated source of human plasma LTP-I.

ISOLATION AND CHARACTERIZATION OF A PLASMA LIPID TRANSFER INHIBITOR PROTEIN

Some animal species such as the pig, rat, and mouse have little CE transfer activity. However, chromatography of the lipoprotein free plasma fraction over phenyl-sepharose significantly enhances CE transfer activity. Because of these observations, we have hypothesized that plasma inhibitors are responsible for the masking of plasma lipid transfer activity in some species with low CE transfer activity.[15]

We have partially purified a protein component from pig plasma which completely inhibits cholesteryl ester transfer activity obtained from pig plasma.[16] We have also isolated from human plasma a unique HDL subclass which contains a lipid transfer inhibitor protein (LTIP).[17] LTIP has an apparent molecular weight of 29,000 daltons by SDS gel electrophoresis. Purified LTIP inhibits CE, TG, and PL transfer mediated by LTP-I and PL transfer mediated by LTP-II. The HDL subclass containing LTIP, isolated from human HDL by anti-LTIP affinity chromatography, inhibited CE transfer activity to a similar extent before and after delipidation. Some preparations of A-I also inhibited CE transfer but after chromatography on an anti-LTIP column all inhibitory properties were removed. As little as 1 μg of purified LTIP completely inhibited CE, TG, and PL transfer mediated by LTP-I, while 3 μg was required for complete inhibition of PC transfer mediated by LTP-II. Passage of either human, pig, rat, or mouse plasma over the anti-LTIP column significantly enhanced CE transfer activity in each of these species, supporting the hypothesis that LTIP may be a major determinant of the differences in plasma lipid transfer activity between and among animal species.

CONCLUSIONS

In conclusion, human plasma CE and TG transfer activity is proportional to the amount of LTP-I, and is influenced by the ratio and composition of the lipoprotein lipid donors and acceptors. Furthermore, the lipid transfer activity may be regulated by 1) lipid challenge, 2) a phospholipid transfer protein designated LTP-II, and 3) a unique HDL subclass containing a lipid transfer inhibitor protein. The differences in plasma lipid transfer activity between animal species is largely determined by a plasma lipid transfer inhibitor protein.

ACKNOWLEDGEMENTS

This work was supported by a grant from the National Institutes of Health, U.S.A. HL30086.

REFERENCES

1. Albers JJ: Role of HDL, LCAT, and lipid transfer protein in lipoprotein metabolism. J Jap Athero So 13:751 (1985)
2. Albers JJ, Tollefson JH, Chen C-H, and Steinmetz A: Isolation and characterization of human plasma lipid transfer proteins. Arteriosclerosis 4:49 (1984)
3. Tollefson JH, Ravnik S, and Albers JJ: Isolation and characterization of a phospholipid transfer protein (LTP-II) from human plasma. J Lipid Res (in press)
4. Stein O, Halperin G, and Stein Y: Cholesteryl ester efflux from extracellular and cellular elements of the arterial wall: Model systems in culture with cholesteryl linoleyl ether. Arteriosclerosis 6:70 (1986)
5. Granot E, Tabas I, and Tall AR: Human plasma cholesteryl ester transfer protein enhances the transfer of cholesteryl ester from high density lipoproteins into cultured HepG2 cells. J Biol Chem 262:3482 (1987)
6. Jarnagin AS, Kohr W, and Fielding C: Isolation and specificity of a Mr 74,000 cholesteryl ester transfer protein from human plasma. Proc Nat Acad Sci USA 84:1854 (1987)
7. Hesler CB, Swenson TL, and Tall AR: Purification and characterization of a human plasma cholesteryl ester transfer protein. J Biol Chem 262:2275 (1987)
8. Drayna D, Jarnagin AS, McLean J, Henzel W, Kohr W, Fielding C, and Lawn R: Cloning and sequencing of human cholesteryl ester transfer protein cDNA. Nature 327:632 (1987)
9. Faust RA and Albers JJ: Unpublished observations.
10. Albers JJ, Cabana VG, and Stahl YDB: Purification and characterization of human plasma lecithin-cholesterol acyltransferase. Biochemistry 15:1084 (1976)
11. Tollefson JH, Faust R, Albers JJ, and Chait A: Secretion of a lipid transfer protein by human monocyte-derived macrophages. J Biol Chem 260:5887 (1985)
12. Faust RA and Albers JJ: Synthesis and secretion of the plasma cholesteryl ester transfer protein by the human hepatocarcinoma cell line, HepG2. Arteriosclerosis 7:267 (1987)
13. Faust RA and Albers JJ: Regulated vectorial secretion of cholesteryl ester transfer protein (LTP-I) by the CaCo-2 model of human enterocyte epithelium. Submitted
14. Faust RA, Tollefson JH, Chait A and Albers JJ: Regulation of cholesteryl ester transfer protein secretion from human monocyte-derived macrophages by cell maturation and cell cholesteryl ester content. Submitted
15. Tollefson JH and Albers JJ: Isolation, characterization, and assay of plasma lipid transfer proteins. IN Methods in Enzymology (Plasma Lipoproteins), edited by Albers JJ, Segrest J. Academic Press Inc., Orlando (1986)

16. Tollefson JH, Liu A, and Albers JJ: Regulation of plasma lipid transfer by the plasma high density lipoproteins: A look at human and animal lipoprotein physiology. Submitted
17. Nishide T, Tollefson JH, and Albers JJ: Inhibition of lipid transfer by a specific high density lipoprotein subclass containing an inhibitor protein. Circulation Part II 76:IV-416 (1987)

THE HUMAN PLASMA CHOLESTERYL ESTER TRANSFER PROTEIN: STRUCTURE, FUNCTION AND PHYSIOLOGY

Christopher J. Fielding

Cardiovascular Research Institute
University of California Medical Center
San Francisco, California 94143

The first report of net transfer of preformed cholesteryl esters between plasma lipoprotein species was by Rehnborg and Nichols in 1964 (1). The identification of a protein factor catalysing the exchange of cholesteryl esters between plasma lipoproteins was first made by Zilversmit and colleagues in 1975 (2). A factor with similar properties was identified in human plasma by the same laboratory in 1978 (3). Since that time there has been a rapid increase both in interest and information concerning cholesteryl ester transfer protein (CETP). A summary of earlier attempts to isolate plasma cholesteryl ester transfer activity is contained in a recent review (4). The recent isolation of human plasma CETP, cloning of its cDNA, and the expression of the cloned gene in transfected cells (5,6) provides an appropriate opportunity to review the nature of this protein and its significance in plasma lipid metabolism.

Structure of human plasma CETP

An approximately 100,000- fold purification of CETP from normal human plasma carried out by this laboratory (Table 1, from ref 5) yielded a single protein species which comigrated with CETP activity (assayed as the transfer of ^{3}H-cholest- eryl ester radioactivity in HDL to unlabeled LDL) by both preparative gel electrophoresis and isoelectric focussing (5). The apparent molecular weight of this activity was 74 kDa, a value in close agreement with a value obtained by radiation inactivation analysis (7).

Partial aminoacid sequence was obtained from the purified protein. Oligonucleotide probes were synthesized corresponding to these sequences, and used to probe a human adult liver cDNA library (6). Hybridizing plaques were purified and cloned into M13 vectors for sequencing. As the sequence obtained did not extend in the 5' direction sufficiently to code for a complete signal prepeptide, this information was obtained by screening a human genomic library, using a partial cDNA clone. The complete sequence of human plasma CETP indicates that it consists of a 17 aa signal prepeptide, followed by a mature protein of 476 aa

TABLE 1

Purification of Cholesteryl Ester Transfer Protein

Step	Volume	Total Protein	Total Activity*	Specific Activity	Fold Purification	Recovery %
	ml	mg	μg CE/hr	μg CE/hr mg protein	relative to Plasma	
Plasma	900	47745	41445	0.868	1	100.
Middle Fraction**	335	439	19200	43.7	50	46.
Phenyl Sepharose	120	107	18386	172.	198	44.
DEAE-Sepharose	25	63	12676	201.	232	30.6
Hydroxylapatite	30	4.5	2229	495.	570	5.4
CM Cellulose	10	0.0076	715	94079.	108386	1.7

*Assayed as transfer of cholesteryl ester from ^{3}H-CE-HDL to unlabeled LDL, 2 hour incubation, 37°C, pH 7.4, as described under methods.
**Intermediate fraction of plasma centrifuged at density = 1.24 g/ml.

residues, and a predicted protein molecular weight of 53,108. Four N-linked glycosylation sites are evident from the sequence, suggesting that the difference between this value and the observed 74,000 total molecular weight of the protein may be due in large part to a large carbohydrate residue. However, a contribution from other post-translational modification, such as acylation, is not excluded.

Two pieces of evidence strongly support the concept that the sequence obtained is that of authentic human plasma CETP. Firstly, the aminoacid composition predicted from the full length cDNA obtained matches closely that of the purified native CETP purified by Jarnagin et al. (5) and independently by Hesler et al. (8)(Table 2). CETP shows a highly hydrophobic composition overall, with several extended sequences of nonpolar aminoacids (6). However, the sequence found shows no significant homology with other reported aminoacid sequences, suggesting that CETP is unrelated to the major plasma apolipoproteins, or to the lipase family of which plasma lecithin:cholesterol acyltransferase (LCAT) is a representative (9).

Secondly, when nonhuman cells in culture which expressed no endogeneous CETP activity were transfected with the full length cDNA of CETP, transfer activity detected in the cell culture medium had the properties of the native plasma protein (6).

These data, taken together, indicate clearly that human CETP is the 74 kDa protein purified along with cholesteryl ester transfer activity from normal human plasma (5).

A plasma inhibitor of CETP has been reported (10). The potential significance of this factor in native plasma was investigated by observing the increment of total CETP activity when CETP from transfected cells was added to plasma. There was no significant decrease in the activity of the exogeneous protein, suggesting that such an inhibitor does not reduce CETP activity in native human plasma, and

TABLE 2

COMPOSITION OF HUMAN PLASMA CHOLESTERYL ESTER TRANSFER PROTEIN (CETP)

	Jarnagin et al.[1]	Predicted from cDNA[2]	Hesler et al.[3]
Asx	40	40	53
Thr	29	31	19
Ser	37	46	18
Glx	67	51	60
Pro	18	16	21
Gly	33	27	33
Ala	29	23	33
Cys	5	7	2
Val	40	39	39
Met	10	12	7
Ile	25	36	29
Leu	56	51	56
Tyr	11	10	10
Phe	27	30	29
His	10	13	14
Lys	25	26	31
Arg	19	12	17
(Correlation	r = 0.93	r = 0.82)	
Total MW	74 kDa	-	74 kDa
Protein MW	-	53 kDa	-

[1] Jarnagin et al., Proc. Natl. Acad. Sci. USA 84:1854-1857, 1987.
[2] Drayna et al., Nature 327:632-634, 1987.
[3] Hesler et al., J. Biol. Chem. 262:2275-2282, 1987.

that the activity of CETP in plasma assayed with ^{3}H-labeled HDL cholesteryl ester reflects its concentration in plasma (Table 3). On these grounds, for a purification of approximately 100,000-fold from plasma (Table I) the predicted mass of CETP in human plasma should be about 0.5 -1.0 ug/ml. A comparable mass is predicted from the purfication described by Hesler et al. (8). These estimates will require confirmation from direct immunoassay.

TABLE 3

EFFECT ON PLASMA FACTORS/ INHIBITORS ON CETP ACTIVITY

Incubation	Activity[a]
CETP alone	10.4 ± 1.0
Plasma alone	4.0 ± 1.1
CETP + plasma	14.0 ± 2.1

[a]CETP was assayed as the transfer of (^{3}H-cholesteryl ester)-HDL to unlabeled LDL. Specific activity was 932 cpm/μg cholesteryl ester. Purified CETP was from transfected CHO cells. Plasma was from normolipemic fasting blood.

Functions of human plasma CETP

As pointed out by Morton and Zilversmit (11), CETP can catalyse the transfer of both cholesteryl esters and triglycerides, although isolated CETP shows a strong preference for cholesteryl ester transfer between the major lipoprotein classes (Table 4, from ref. 5).

Four potential functions of CETP can be postulated, based on our current knowledge of plasma cholesterol metabolism:

(a) The exchange of cholesteryl esters containing different fatty acid moieties between HDL, VLDL and LDL Present evidence, based largely on studies with rat liver (12) indicates that cholesteryl esters derived from hepatic acyl CoA:cholesterol acyltransferase are likely to have a different acyl spectrum than those generated in the plasma by LCAT activity.

(b) The exchange of triglycerides for cholesteryl esters This is usually considered the major function of CETP, based on studies of the incubation of lipoproteins in vitro, particularly mixtures of triglyceride- rich VLDL and cholesteryl ester- rich HDL. It should be borne in mind, however, that a fraction of VLDL in human plasma (VLDL without apo E, VLDL-E), which has the kinetic and biochemical properties of nascent or early VLDL (13,14) has the same cholesteryl ester/apo B mass ratio as product VLDL (VLDL containing apo E, VLDL+E). Comparison of the rates of mass net transport of cholesteryl ester from HDL (in exchange for triglyceride) and unidirectional isotopic transfer of labeled HDL cholesteryl esters suggests that in fasting whole plasma in vitro, a major part of the activity catalysed by CETP involves "nonproductive" exchange of cholesteryl ester for cholesteryl ester, rather than of cholesteryl ester for triglyceride.

(c) Transfer of cholesteryl esters newly generated by LCAT to HDL There is little evidence that LCAT activity in native plasma is associated with HDL in general. The major part of HDL consists of particles containing both apolipoprotein A-I (apo A-I) and apolipoprotein A-II, while LCAT was found by immunoaffinity chromatography to be

TABLE 4

Lipid Specificity of Cholesteryl Ester Transfer Protein

	HDL to LDL	LDL to HDL	HDL to VLDL	VLDL to HDL
Cholesteryl Ester	1.0	0.96 ± 0.19	0.73 ± 0.04	0.81 ± 0.24
Triglyceride	0.05 ± 0.06	0.10 ± 0.03	0.08 ± 0.05	0.11 ± 0.15

Values are nmoles lipid transferred/hr/ml of purified cholesteryl ester transfer protein relative to the transfer of cholesteryl ester from HDL to LDL assayed in parallel, which was 24 to 35 nmoles CE/hr/ml in different experiments. Values are the means ± standard deviations for three determinations. Details of the assay are in the text. Amounts of lipoproteins used per 0.5 ml incubation volume were: HDL, 50μg CE, LDL, 300 μg CE, and VLDL, 100 μg CE, which were saturating for transfer (3). The cholesteryl ester to triglyceride ratios for the lipoproteins were HDL, 7.7, LDL, 6.7, and 0.15 for VLDL.

largely associated with particles containing apo A-I and the minor apolipoprotein, apo D but no apo A-II (15). These data suggest that the major part of HDL cholesteryl ester in native plasma is likely to be derived by cholesteryl ester transfer from LCAT- containing particles, rather than by the direct activity of LCAT on HDL.

(d) Cholesteryl ester transfer between cells and plasma lipoproteins Studies by several laboratories over the last decade have demonstrated the transfer of intact cholesteryl esters between plasma lipoproteins and cells. The systems studied include that of chylomicron cholesteryl ester to endothelial cells (16) and liver (17); LDL cholesteryl ester to steroidogenic tissues such as the ovary (18); and HDL cholesteryl ester to liver (19) and a variety of cultured cells including adrenal, fibroblast and cultured hepatoma (Hep G2) cells (20-22). In the latter two cell types some stimulation by CETP was reported (22). However, the quantitative significance of the phenomenon, compared to the endocytosis of intact lipoproteins, particularly in vivo, remains to be established.

These considerations emphasize that while the chemical identity of CETP has been recently determined, its functions remain to be clarified and perhaps are as yet unidentified.

REFERENCES

1. Rehnborg, C.S. & Nichols, A.V. Biochim. Biophys. Acta 84:596-603, 1964.
2. Zilversmit, D.B., Hughes, L.B. & Balmer, J. Biochim. Biophys. Acta 409:393-398, 1975.
3. Pattnaik, N.M., Montes, A., Hughes, L.B. & Zilversmit, D.B. Biochim. Biophys. Acta 530:428-438, 1978.
4. Tall, A.R. J. Lipid Res. 27:361-367, 1986.
5. Jarnagin, A.S., Kohr, W. & Fielding, C.J. Proc. Natl. Acad. Sci. (USA) 84: 1854-1857, 1987.
6. Drayna, D., Jarnagin, A.S., McLean, J., Henzel, W., Kohr, W., Fielding, C. & Lawn, R. Nature 327:632-634.
7. Loudet, A.-M., Dousset, N., Potier, M., Manent, J., Carton, M. & Douste-Blazy, L. Med. Sci. Res. 15:251-252.
8. Hesler, C.B., Swenson, T.L. & Tall, A.R. J. Biol. Chem. 262:2275-2282, 1987.
9. McLean, J., Fielding, C., Drayna, D., Dieplinger, H., Baer, B., Kohr, W., Henzel, W. & Lawn, R. Proc. Natl. Acad. Sci. (USA) 83:2335-2339, 1986.
10. Morton, R.E. & Zilversmit, D.B. J. Biol. Chem. 256:11992-11995, 1981.
11. Morton, R.E. & Zilversmit, D.B. J. Biol. Chem. 258:11751-11757, 1982.
12. Goodman, D.S., Deykin, D. & Shiratori, T. J. Biol. Chem. 239:1335-1345, 1964.
13. Nestel, P., Billington, T., Nada, N., Nugent, P. & Fidge, N. Metab. Clin. Exp. 32:810-817, 1983.
14. Ishikawa, Y., Fielding, C.J. & Fielding, P.E. J. Biol. Chem. 263:2744-2749, 1988.
15. Fielding, P.E. & Fielding, C.J. Proc. Natl. Acad. Sci. (USA) 77:3327-3330, 1980.
16. Fielding, C.J., Vlodavsky, I., Fielding, P.E. & Gospodarowicz, D. J. Biol. Chem. 254:8861-8868, 1979.
17. Fielding, C.J. & Fielding, P.E. Biochim. Biophys. Acta 620:440-446, 1980.

18. Reaven, E., Chen, Y.-D.I., Spicher, M., Hwang, S.-F., Mondon, C.E. & Azhar, S. J. Clin. Invest. 77:1971-1984, 1986.
19. Stein, Y., Dabach, Y., Hollander, G., Halperin, G. & Stein, O. Biochim. Biophys. Acta 752:98-105, 1983..
20. Gwynne, J.T. & Hess, B. J. Biol. Chem. 255:10875-10883.
21. Pittman, R.C., Knecht, T.P., Rosenbaum, M.S. & Taylor, C.A. J. Biol. Chem. 262:2443-2450, 1987.
22. Granot, E., Tabas, I. & Tall, A.R. J. Biol. Chem. 262:3482-3487, 1987.

CHOLESTEROL ESTER TRANSFER PROTEIN. CHARACTERIZATION OF MONOCLONAL ANTIBODIES AGAINST THE HUMAN ANTIGEN

Y.L. Marcel, R.W. Milne, P.K. Weech, H. Czarnecka,
C.B. Hesler* and A.R. Tall*

Clinical Research Institute of Montreal, 110, Pine Ave West
Montreal, Quebec H2W 1R7, Canada
and *Department of Medicine, Columbia University College of
Physicians & Surgeons, New York, NY 10032, USA

INTRODUCTION

Early work by Rehnborg and Nichols (1) showed that upon incubation of normolipemic plasma, cholesteryl ester (CE) concentration increased in VLDL, LDL and HDL whereas triglycerides (TG) concentration decreased in VLDL and increased in LDL and HDL. These experiments (1,2) were the first to document CE transfer between lipoproteins but the concept was really developed with the demonstration by Zilversmit et al (3) that the d > 1.21 g/ml fraction contained a protein that catalysed CE transfer between lipoproteins, a protein which is now identified as cholesterol ester transfer protein (CETP) and which transfers CE and TG between lipoproteins.

Several groups have purified the lipid transfer proteins which have a specificity for CE, TG and phospholipids, and first estimated their molecular weights to be between 60 and 70,000 (4-7). In one of these reports evidence was presented for the existence of a separate phospholipid transfer protein (PTP) which does not mediate tranfer of CE or TG (7).

More recently two groups have purified to homogeneity the CETP which is characterized by a Mr on SDS gel electrophoresis in reducing or non-reducing conditions of 74,000 (8,9). One study reported the existence of CETP isoforms with pI between 5.3 and 5.6 (8), and the other gave 5.2 as the pI of the purified CETP (9). CETP is characterized by an unusually high content of non polar amino acids and its calculated hydrophobicity is greater than any other apoprotein (8). A puzzle in the characterization of CETP is the different apparent molecular weights reported by different groups. Albers et al (6) reported an Mr of 64,000 on SDS-gel electrophoresis under reducing conditions, in agreement with the Mr of 65,000 recently reported by Busch et al (10) under the same conditions. In contrast as cited above, Hesler et al (8) and Jarnagin et al (9) noted a Mr of 74,000 also by SDS gel electrophoresis. This difference in apparent molecular weight could perhaps be the result of proteolytic cleavage or could represent partially enriched material. Hesler et al (8) have reported CETP to be very sensitive to oxidative cleavage while Busch et al (10) proposed spontaneous decomposition. A strong argument (11) for the identity of the 74 kDa protein purified by

Hesler et al (8) with the cloned cDNA for CETP (11) is that the sequence derived from the cloned cDNA shows a high homology (92%) with the N-terminal sequence of 74 kDa protein.

The cDNA for CETP encodes a 476 aminoacid mature protein (predicted Mr 53,000) which has 4 potential asparagine-linked glycosylation sites and 6 cysteines. Removal of the N-linked sugard reduces the apparent Mr in SDS gels from 74 Kd to 60 Kd, showing that carbohydrate accounts for most of the difference in Mr between the cDNA-encoded peptide and the mature protein (12). However, the cDNA encoded peptide runs with Mr 53 Kd, indicating the presence of an additional post-translational modification such as D-linked sugard (unpublished).

Morton and Zilversmit have shown that the purified CETP mediates both net transfer and exchange of TG and CE; net transfer results from an heteroexchange of TG for CE, while simple exchange is a bidirectional transfer of homologous molecules (13). This mechanism of action explains the results obtained earlier by ourselves and others on the CETP mediated transfer between lipoproteins. Indeed only exchange of CE occurs between HDL and LDL (14,15) but net transfer of CE takes place from HDL to VLDL or chylomicrons (16,17) as well as from LDL to VLDL (18,19). The net transfer of CE to TG-rich lipoproteins is a function of their size which reflects their relatively higher TG content and their higher TG/CE ratio. We have shown that chylomicrons are better acceptors for net CE transfer than VLDL (16,20) and Eisenberg has demonstrated that HDL_2 in the presence of CETP could transfer CE more efficiently first to $VLDL_1$, second to $VLDL_2$ and last to $VLDL_3$ (21). While LCAT activity drives more CE transfer from HDL to VLDL (22), the initial rate of CE transfer is not influenced by LCAT inhibition (23). Consequently it would not appear that the LCAT reaction could directly stimulate CE transfer but that it favors net CE transfer by increasing the CE/TG ratio in HDL. In conclusion CETP activity may represent an essential and limiting factor in the reverse cholesterol transport to the liver.

Several authors have hypothesized that CETP participate in the centripetal transport of cholesterol from the peripheral (or extrahepatic) tissues to the liver (24-26). This concept stems from the logical integration of CETP activity with the hypothesis of an LCAT-mediated reverse cholesterol transport (27). In this pathway the cholesterol present in HDL, which originates from the nascent hepatic HDL, or which is transferred to HDL from various cells and cell membranes (27), or from chylomicron and VLDL surface constituents (28) is esterified by LCAT. In the absence of any acceptor lipoproteins, the accumulation of CE in HDL has been proposed to inhibit LCAT reaction (23) but the addition of CETP neutralizing antibodies to plasma has been found to have no effect on the LCAT reaction (F.T. Yen et al, unpublished results). Whereas equilibration of CE but no net transfer occurs between HDL and LDL (14,15), net mass transfer of CE take place from HDL to TG-rich lipoproteins (16,17) and the rate of transfer appears to increase with the size of acceptor lipoprotein and is fastest with chylomicrons (16,20) and large VLDL (21). In vivo this CE transfer to TG-rich lipoproteins coupled with lipolysis by lipoprotein lipase and transfer of surface constituents to the HDL fraction is presumably responsible for the formation of the smaller apo B-containing lipoproteins that are rich in CE and which constitute the IDL and LDL.

Some elements of this pathway seem to represent a futile cycle. For example, the unesterified cholesterol of TG-rich lipoproteins is derived from the liver for the primary purpose of transporting triglycerides, is subsequently transferred to HDL, esterified, and then transferred back to triglyceride-rich, apo B-containing lipoproteins. These become IDL that are taken up by hepatic apo E and apo B/E receptors (29). The cholesterol initially associated with the nascent HDL, secreted by the liver to supply the intravascular lipid transport system with most of its apolipoproteins, also appears to follow a similar futile pathway which returns its cholesterol to the liver. Therefore under normal conditions, the economy-minded organism keeps recycling the cholesterol needed for TG transport and for apolipoprotein secretion back to the liver where it controls hepatic cholesterol synthesis. The third component of the pathway which does not follow a futile cycle, and which is the reverse-cholesterol transport system, presumably should not amount to a very high net transport under normal steady state conditions where little excess cholesterol need to be returned from peripheral pools to the liver.

As many aspects of the physiological role of CETP and of its mode of action remain speculative or unknown, we have undertaken to prepare monoclonal antibodies (Mabs) specific for this protein which will provide important reagents to further study its function. We review here the production of 3 Mabs specific for CETP and their preliminary characterization and application to the immunoassay of the plasma antigen.

Production and characterization of Mabs against CETP

For the first series of Mabs, one Balb/c mouse was immunized by 3 subcutaneous injections of 10 μg of purified CETP emulsified in complete Freund's adjuvant. CETP used for immunization and screening was purified from plasma by centrifugation at 1.21 g/ml, by phenyl-Sepharose and CM-Sepharose chromatographies and by binding to a lipid emulsion and gel filtration (11). Four days before the fusion, the mouse was given an additional intravenous boost with 10 μg of CETP in saline. The spleen cells were fused with the murine myeloma cell line SP20 as previously described (30). After the fusion, the cells were distributed into 672 microculture wells. When sufficient cell growth had occured, aliquots (50 μl) of the supernatant were tested by solid phase radioimmunoassay for the presence of specific antibody.

In this initial screening, 22 of the supernatants contained antibody activity for the antigen preparation. The cells in all positive wells were recloned twice by limiting dilution. All hybridoma supernatants after recloning, were tested for their ability to immunoprecipitate CETP activity from a crude CETP preparation. The immunoprecipitation was carried out as described earlier (30) Briefly, Pansorbin was armed with rabbit antimouse IgG (0.25 ml rabbit antiserum/g Pansorbin) and 100 μl of the washed suspension was preincubated overnight at 4°C with 1 ml of culture supernatant from the appropriate Mab. After centrifugation, washing and resuspension, 200 μl of the complex, Pansorbin-antimouse IgG - Mab was added to 3 μl of the CETP preparation obtained after CM-cellulose chromatography, incubated for 4 hours and the suspension was centrifuged for 15 min at 2,000 x g. The supernatant which represents the immunodepleted preparation was assayed for remaining CETP activity as previously described (31) and the results are summarized in Table I. The antibodies present in the supernatants of 3 of the clones (2H4, 5C7 and 7E1) were able to remove 100% of the transfer activity. Immunoprecipitation with supernatants from the other clones gave results which were similar to those obtained with control irrelevant Mabs. In other experiments reported elsewhere (31), addition of increasing levels of the purified IgG resulted in the progressive inhibition of the lipid transfer activity. Antibodies 2H4, 5C7 and 7E1 could inhibit 100% of both triglyceride and cholesteryl ester transfer activity but only about 50% of the phospholipid transfer which corroborates earlier results indicating that triglycerides and cholesteryl ester are transferred by the same protein which exhibit also phospholipid transfer properties (7,10) but that a different phospholipid transfer factor also exists (7).

TABLE 1. Screening of Mabs against CETP by immunoprecipitation of the transfer activity.

	Incubation of *CE-HDL and LDL in the presence of						
	added CETP	no CETP	rabbit anti CETP	Mabs 2H4	Mabs 7E1	Mabs 5C7	Irrelevant Mab
Radioactivity remaining in HDL after incubation (cpm)	1270 ± 120	5180 ± 360	4520 ± 260	4895 ± 400	4865 ± 150	5010 ± 600	3150 ± 100

While the immunoprecipitation of the transfer activity does not rule out the possibility that the antibodies could bind to a protein associated with but different from the transfer protein, the immuno inhibition of the transfer reaction by the 3 Mabs does provide definitive evidence that they react with CETP. Additional evidence that each antibody does react with CETP is provided by immunoblots of partially purified CETP preparations analysed by SDS gel electrophoresis which show only one immunoreactive band with Mr of 74,000 (Figure 1). Likewise immunoblot of the same preparation analysed by 2D-gel electrophoresis indicated the presence of immunoreactive spots with the pI and Mr described for CETP (8,9). A further proof of the specificity of these antibodies is their reaction with the fusion protein obtained by expression of the cDNA coding for CETP (14) in a bacterial system (M.Brown, A. Tall et al). Supernatants of clones which did not immunoprecipitate CETP activity gave no reaction on Western blots of either the purified CETP preparations or the fusion proteins.

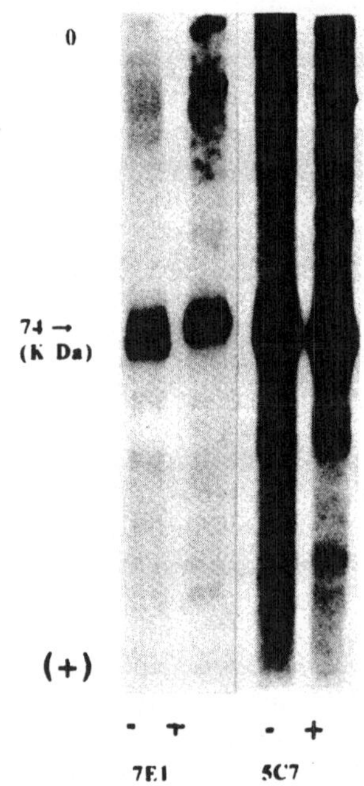

Figure 1: Immunoblots with Mabs 7E1 and 5C7 of a partially purified CETP preparation separated by SDS-gel electrophoresis.

The IgG fractions of the Mabs 2H4, 5C7 and 7E1 were isolated from the ascitic fluid of hybridoma-bearing mice by affinity chromatography on Protein A-Sepharose. The IgG was radioiodinated and the 3 Mabs were tested in cross-competition assays. In all combinations the Mabs showed reciprocal competition which would indicate that they recognize epitopes which are close together on CETP. Nevertheless, 2H4 can be distinguished from the other 2 Mabs by its higher affinity for rabbit CETP.

Development of a radioimmunoassay (RIA) for CETP

In initial experiments, each of the 3 Mabs was evaluated for the immunoassay of CETP and MAB 7E1 was chosen on the basis of its higher reactivity with the solid phase antigen which is the partially purified CETP obtained after CM-Sepharose chromatography (11). The antigen was coated on plastic wells at the concentration of 10 μg protein/ml. After saturation with albumin, the immobilized antigen competed with the soluble antigen for a limiting dilution of 7E1. Normal and complete displacement curves have been obtained under these conditions with normal serum or plasma as well as with plasma lipoprotein subfractions. ED50 were typically obtained with dilutions of serum between 1/30 and 1/60. In a small pilot experiment, we found that the same plasma or serum stored at 4°C with preservatives (antiproteases and antibacterial agents) varied between 3 and 4%. Serum stored for 3 weeks at -80°C was more immunoreactive than serum stored at 4°C.

Treatments of normolipemic sera with chaotropic agents, with detergents or with freezing and thawing failed to cause any significant change in CETP immunoreactivity. This indicates that the epitope for antibody 7E1 is not affected by the lipid environment or by the association of CETP with lipoproteins. In a preliminary series of determination of CETP levels in normolipemic subjects, there was no significant difference between males and females but there was a positive correlation between CETP and both HDL cholesterol and plasma apo AI. These initial results imply that CETP and apo AI-containing lipoproteins may be under the same control for synthesis, secretion and/or catabolism. Alternatively, the CETP could play a role in determining plasma HDL levels, for example, by promoting efflux of tissue cholesterol into HDL. These results are compatible with those of others who have shown that most of plasma CETP is found in HDL (26,31) and in association with apo AI (32).

ACKNOWLEDGEMENT

This work is supported by grants from the Quebec Heart Foundation and NIH #HL22682.

REFERENCES

1. Rehnborg C.S. and Nichols, A.V. (1964) Biochim. Biophys. Acta 84, 596-603.
2. Nichols, A.V. and Smith L. (1965) J. Lipid Res. 6, 206-210.
3. Zilversmit D.B., Hughes, L.B. and Balmer J. (1975) Biochim. Biophys. Acta 409, 393-398.
4. Morton R.E. and Zilversmit D.B. (1982) J. Lipid Res. 23, 1058-1067.
5. Ihm I, Ellsworth J.L., Chataing B. and Harmony J.A.K. (1982) J. Biol. Chem. 257, 4818-4827.
6. Albers J.J., Tollefson J.H., Chen C. and Steinmetz A. (1984) Arteriosclerosis 4, 9-58.
7. Tall A.R., Abreu E. and Shuman J. (1983) J. Biol. Chem. 258, 2174-2180.
8. Hesler C.B., Swenson T.L., and Tall A.R. (1987) J. Biol. Chem. 262, 2275-2282.

9. Jarnagin A.S., Kohr W. and Fielding C.J. (1987) Proc. Natl. Acad. Sci. USA 84, 1854-1857.
10. Busch S.J., Stuart, W.D., Hug B., Mao S.J.T. and Harmony J.A.K. (1987) J. BIol. Chem. 262, 17563-17571.
11. Drayna D., Jarnagin A.S., McLean J., Henzel W., Kohr W., Fielding C.J. and Lawn R. (1987) Nature 327, 632-634.
12. Swanson T.L., Simmons J.S., Hesler, C.B., Bisgaier, C. and A.R. Tall. (1987) J. Biol. Chem. 262, 16271-16274.
13. Morton R.E. and Zilversmit (1983) J. Biol. Chem. 258, 11751-11757.
14. Sniderman A., Teng B., Vezina C. and Marcel Y.L. (1978) Atherosclerosis 31, 327-333.
15. Barter P.J. and Jones M.E. (1979) Atherosclerosis 34, 67-74.
16. Marcel Y.L., Vezina C., Teng B. and Sniderman A. (1980) Atherosclerosis 35, 127-133.
17. Hopkins G.J. and Barter P.J. (1980) Metabolism 29, 546-550.
18. Barter P.J., Gorjatschko, L. and Calvert G.D. (1980) Biochim. Biophys. Acta 619, 436-439.
19. Deckelbaum R.J., Eisenberg S., Oschry Y., Butbul E., Sharon I. and Olivecrona T. (1982) J. Biol. Chem. 257, 6509-6517.
20. Noël S.P., Dupras R, Vezina C. and Marcel Y.L. (1984) Biochim. Biophys. Acta 796, 277-284.
21. Eisenberg S. (1985) J. Lipid Res. 26, 487-494.
22. Glomset J.A., Norum K.R. and King W. (1970) J. Clin. Invest. 49, 1827-1837.
23. Fielding C.J. and Fielding P.E. (1981) J. Biol. Chem. 256, 2101-2104.
24. Marcel Y.L. (1982) Adv. Lipid Res. 19, 85-135.
25. Tall A.R. (1986) J. Lipid Res. 27, 361-367.
26. Fielding C.J. (1987) Am. Heart J. 113, 532-537.
27. Glomset J.A. and Norum K.R. (1973) Adv. Lipid Res. 1, 1-65.
28. Tam S.P. and Breckenridge W.C. (1983) J. Lipid Res. 24, 1343-1357.
29. Mahley R.W., Hui D.Y., Innerarity T.L. and Weisgraber K.H. (1981) J. Clin. Invest. 68, 1197-1206.
30. Milne R.W., Blanchette L., Théolis R., Weech P.K., and Marcel Y.L. (1987) Molec. Immunol. 24, 435-447.
31. Hesler C.B., Tall A.R., Swenson T.L., Weech P.K., Marcel Y.L. and Milne R.W. (1988) J. Biol. Chem. 263, 5020-5023.
32. Tall A.R., Forester L.R., and Bongiovanni G.L. (1983) J. Lipid Res. 24, 277-289.
33. Cheung M.C., Wolf A.C., Lum K.D., Tollefson S.H., and Albers J.J. (1986) J. Lipid Res. 27, 1135-1144.

CHOLESTEROL ESTERIFICATION AND NET MASS TRANSFER OF CHOLESTERYLESTERS AND TRIGLYCERIDES IN PLASMA FROM HEALTHY SUBJECTS AND HYPERLIPIDEMIC CORONARY HEART DISEASE PATIENTS

A. Van Tol, L.M. Scheek and J.E.M. Groener

Dept. of Biochemistry I
Erasmus University Rotterdam
P.O. Box 1738, 3000 DR Rotterdam
The Netherlands

INTRODUCTION

Plasma from a variety of mammals, including man, contains lipid transfer proteins (LTP's) that promote the exchange/transfer of cholesterylesters (CE), triglycerides (TG) and phospholipids (PL) between the plasma lipoproteins VLDL, LDL and HDL. One plasma LTP has been purified to homogeneity[1,2]. The gene for this protein has been cloned and sequenced[3].

The cholesterylesters present in plasma are derived from two sources. A minor part is secreted into the blood from the intestine and the liver in nascent lipoproteins: chylomicrons and VLDL. The major part is synthesized in plasma by the enzyme lecithin:cholesterol acyltransferase (LCAT). Several reviews on the roles of LTP's and cholesterol esterification by LCAT in plasma lipoprotein metabolism have been published recently[4-7].

The present paper describes net mass transfer of CE from (VLDL+LDL) to HDL in plasma from normolipidemic healthy subjects. This pathway is defective in plasma from hyperlipidemic coronary heart disease (CHD) patients.

MATERIALS AND METHODS

Collection of Blood and Isolation of Plasma

Blood was drawn after a 12-14 hour overnight fasting period and collected in tubes containing EDTA (final concentration 1.5 mg/ml). The blood was cooled immediately to 0-4 °C and centrifuged at 3000 rpm for 15 min in a cooled tabletop centrifuge. The resulting plasma was kept on ice and the experiments were performed within 2 hours after blood sampling.

Assay of the Initial Rate of Cholesterol Esterification

The chemical composition of plasma lipoproteins changes substantially during prolonged incubation of plasma. These changes will affect the measured rates of cholesterol esterification and lipid transfer. Therefore we measured the initial rate of cholesterol esterification by LCAT by incubation of plasma at 37 °C for one hour. Plasma was used without dilution in order to work with physiological concentrations of endogenous lipoproteins, including endogenous activators and inhibitors. It was shown in separate experiments that iodo-acetate (1 mM) gives more than 95 % inhibition of cholesterol esterification. The reaction is linear for up to 2 hours of incubation.

Assay of the Initial Rate of Lipid Mass Transfer

After incubation of plasma for 1 hour, the apo B-containing lipoproteins were precipitated by Mg^{++}/phosphotungstate and the rate of lipid mass transfer between (VLDL+LDL) and HDL was determined assaying changes in TG and CE content of the supernatant. Incubation of plasma did not affect the efficiency of the precipitation procedure. This was checked by measuring apo B in the supernatants, as well as apo A-I in the precipitates. CE were determined by measuring the difference between total cholesterol and unesterified cholesterol. Lipid analyses were performed using enzymatic methods.

Assay of the Activity of CETP in Plasma

The amount of active LTP in plasma was determined exactly as described by Groener et al.[8], using an isotope assay detecting the exchange/-transfer of radioactive cholesterylesters from exogenous LDL to HDL (CETP activity). The measured activity is independent of endogenous lipoproteins.

RESULTS AND DISCUSSION

Plasma Lipids and Lipoproteins in the Studied Groups

Plasma from two groups of blooddonors were used. Firstly, a group of nine healthy laboratory workers (5 males, 4 females) with plasma lipid levels in the normal range. Plasma TG concentrations were 0.5-1.6 mM and total cholesterol ranged from 3.9-6.2 mM. Secondly, a group of fifteen hyperlipidemic males treated for coronary heart disease (CHD). Plasma samples were obtained more than three months after treatment. Most patients used calcium antagonists and/or beta-blockers. Various types of hyperlipidemia were present. Plasma TG levels ranged from 2-9 mM and plasma total cholesterol from 6-10 mM. As expected, the HDL-cholesterol level in plasma was significantly lower in CHD patients, compared to the normolipidemic subjects. Table 1 shows the mean values

Table 1. Plasma lipids in healthy subjects and in hyperlipidemic coronary heart disease (CHD) patients

	Healthy subjects	CHD patients
Plasma triglycerides (mM)	1.0 ± 0.3	3.5 ± 2.0
Plasma cholesterol (mM)	5.0 ± 0.8	7.6 ± 1.4
HDL-total cholesterol(mM)	1.3 ± 0.3	0.7 ± 0.2

Table 2. Initial rates of net mass transfer and synthesis of CE in plasma from healthy subjects and CHD patients*

	Increase in HDL-CE	Cholesterol esterification	CE transfer from HDL to (VLDL+LDL)
Healthy subjects	128 ± 52	30 ± 12	- 97 ± 57
CHD patients	76 ± 49	121 ± 26	45 ± 63

* Rates are given as nmoles/ml plasma/h

of plasma triglycerides, plasma cholesterol and HDL-cholesterol ± SD in both groups.

Cholesterol Esterification and Lipid Mass Transfer in Plasma

The initial rates of CE mass transfer and CE synthesis (means ± SD) are given in Table 2. All plasma samples from the healthy subjects showed a substantial increase in HDL-CE during incubation. In all samples tested, the rate of increase in HDL-CE was higher than the rate of cholesterol esterification (measured simultaneously in the same plasma samples). It can be concluded that substantial net mass transfer of CE occurs from (VLDL+LDL) to HDL. This is indicated in Table 2 as negative transfer from HDL to (VLDL+LDL). Incubation of the plasma samples from the hyperlipidemic CHD group also results in increases in HDL-CE, in 14 out of the 15 samples tested. The average rate of increase in HDL-CE is slower than in the normolipidemic group. The initial cholesterol esterification rate, however, is 4-fold higher in the CHD patients than in the normolipidemic group. As a result, the increase of CE in plasma due to esterification is higher than the observed increase in HDL-CE. If it is assumed that all CE formation in human plasma takes place on HDL, it can be concluded that a substantial part of the CE formed by LCAT action on HDL are transferred from HDL to (VLDL+LDL) in patient plasma. This was found to be the case in 12 out of the 15 hyperlipidemic plasma samples tested. 3 hyperlipidemic samples showed a "negative" transfer comparable with the healthy normolipidemic subjects.

The initial rates of TG transfer, measured by the increase in HDL-TG, was not significantly different between the healthy subjects and the CHD patients. In both groups TG were transferred from (VLDL+LDL) to HDL in all plasma samples (see Table 3 for means ± SD and ranges).

In addition to TG and CE, the surface lipids (unesterified cholesterol (UC) and PL) also accumulate in HDL during incubation of normal plasma. It is therefore conceivable that HDL acquires both core and surface lipids during incubation of plasma at 37 °C. Because the HDL protein component (apo A-I) remains constant, these data suggest an increase in size of one or more of the HDL subfractions. The changes of the different HDL lipids during incubation of normal plasma are given in Table 4 (mean values ± SD and ranges).

Table 3. Initial rates of net mass transfer of TG from (VLDL+LDL) to HDL in plasma from healthy subjects and CHD patients

	nmoles/ml plasma/h	
Healthy subjects	36 ± 14	(range 20-62)
CHD patients	39 ± 18	(range 14-74)

Table 4. Increase in HDL lipids on incubation of plasma from normolipidemic subjects

	nmoles/ml plasma/h	
HDL-CE	128 ± 52	(range 50-220)
HDL-TG	36 ± 14	(range 20-62)
HDL-UC	20 ± 15	(range 7-53)
HDL-PL	77 ± 49	(range -14-154)

CETP activity was measured in plasma samples from a group of healthy males and the hyperlipidemic CHD patients. Table 5 shows that this activity does not differ significantly in both groups, indicating that the differences observed in direction and activity of net mass transfer of CE (shown in Table 2) are not due to differences in the plasma levels of active LTP. This strongly suggests that the differences in CE transfer between the healthy subjects and the hyperlipidemic CHD patients are caused by differences in level and/or chemical composition of the acceptor and donor lipoproteins.

Table 5. CETP activity in plasma from healthy males and CHD patients measured by the exogenous substrate method

	Arbitrary units	
Healthy males	149 ± 33	(range 109-210)
CHD patients	151 ± 44	(range 78-242)

Conclusions and Summary of the In Vitro Incubation Experiments

It is concluded that in vitro incubation of plasma reveals a profound difference in net mass transfer of CE between plasma from normolipidemic healthy subjects and plasma from hyperlipidemic CHD patients. The differences are also evident in the initial rates of cholesterol esterification. The site of accumulation of CE during incubation of freshly isolated, undiluted normolipidemic control plasma is solely the HDL fraction. If plasma from hyperlipidemic CHD patients is used, the CE formed by LCAT action accumulate on (VLDL+LDL) as well as on HDL. Consequently, with the assumption that all CE formed in human plasma are synthesised on HDL, these observations predict that CE are transferred from HDL to (VLDL+LDL) in hyperlipidemic patient plasma. Also, even without this assumption, it is clear that substantial net mass transfer of CE occurs from (VLDL+LDL) to HDL in normolipidemic control plasma. Our observations on CE and TG transfer are summarized in Table 6.

Table 6. Changes in CE and TG in plasma lipoprotein classes during incubation of freshly isolated plasma

	Healthy subjects (normolipidemic)	**CHD patients (hyperlipidemic)**
HDL-TG	**Increase**	**Increase**
(VLDL+LDL)-TG	**Decrease**	**Decrease**
HDL-CE	**Increase**	**Increase**
(VLDL+LDL)-CE	**Decrease**	**Increase**

As shown in Table 3, HDL-TG increase during incubation of plasma in both groups at essentially the same rate. This indicates that the concentration of plasma TG does not solely determine the rate of plasma TG transfer. Net mass transfer of TG always proceeds in the direction of HDL. Since in this study there is no correlation between direction and/or rate of CE and TG transfer, our results do not support the theory that CE and TG transfer occurs reciprocally and in a equimolar way. A substantial part of CE transfer in plasma must be independent of TG transfer. However, our results cannot exclude a limited exchange of TG for CE on a 1:1 molar basis between different lipoprotein classes. In this study HDL were separated from (VLDL+LDL) by Mg^{++}/phosphotungstate precipitation of apo B-containing lipoproteins. This method works instantaneous and complete and is therefore very suitable for the assay of mass lipid transfer activities. A disadvantage of this method is that VLDL is not separated from LDL. Therefore it cannot be decided from the present experiments whether VLDL or LDL is the donor lipoprotein in the observed transfer of CE from (VLDL+LDL) to HDL in normal plasma. Experiments using methods suitable for the separation of VLDL and LDL (ultracentrifugation and gelfiltration) are in progress.

Possible Consequences for In Vivo Plasma Lipoprotein Metabolism

If the CE fluxes between (VLDL+LDL) and HDL measured during in vitro incubation of plasma are operating in vivo at the same rates, it is clear that the CE shifts may profoundly influence the distribution of plasma cholesterol between the different lipoprotein classes. In fact it can be calculated that the CE transfer observed in normal plasma only has to operate for a limited number of hours/day in order to have a profound influence on lipoprotein metabolism. Our results suggest a new pathway for plasma CE metabolism in healthy subjects. In this group with low plasma VLDL and LDL concentrations, a substantial part of the CE in (VLDL+LDL) may be metabolized by way of HDL. This could result in an elevated steady state level of HDL and a relatively low steady state concentration of LDL. It must be kept in mind that this situation may only occur during the fasted state, as all our observations were made in plasma samples lacking chylomicrons or chylomicron remnants. CE are not transferred from (VLDL+LDL) to HDL in plasma from hyperlipidemic CHD patients. This defect may contribute to the relatively high (VLDL+LDL) concentrations and relatively low HDL levels in patient plasma. As our hyperlipidemic patient group consisted of subjects with hypercholesterolemia, hypertriglyceridemia or both, further investigation is necessary in order to reveal the possible importance of the chemical composition of specific plasma lipoproteins in the defective mass transfer of CE in CHD. The observations described in this paper may have important consequences for basic metabolic pathways, like VLDL/IDL/LDL interconversion and reverse cholesterol transport, in health and disease.

ACKNOWLEDGEMENTS

The experiments were performed at the Dept. of Biochemistry I of the Erasmus University Rotterdam. Members of the laboratory staf of the Depts. of Biochemistry I and II (Chemical Endocrinology), as well as medical students of the Erasmus University volunteered as blooddonors. Dr. J.M. Hartog (Thoraxcenter) is thanked for providing the bloodsamples from the CHD patients. The expert technical assistance of E.M. Van Ramshorst and J.C. Van Geel is gratefully acknowledged. Financial support was obtained from the Netherlands Heart Foundation (Grant nr. 87.060).

REFERENCES

1. A. Stephens Jarnagin, W. Kohr and C. Fielding, Isolation and specificity of a M_r 74,000 cholesteryl ester transfer protein from human plasma, Proc. Natl. Acad. Sci. USA 84: 1854.
2. C.B. Hesler, T.L. Swenson and A.R. Tall, Purification and characterization of a human plasma cholesteryl ester transfer protein, J. Biol. Chem. 262: 2275 (1987).
3. D. Drayna, A. Stephens Jarnagin, J. McLean, W. Henzel, W. Kohr, C. Fielding and R. Lawn, Cloning and sequencing of human cholesteryl ester transfer protein cDNA, Nature 327: 632 (1987).
4. P.J. Barter, HDL metabolism in relation to plasma cholesteryl ester transport, in "Clinical and Metabolic Aspects of High Density Lipoproteins", N.E. Miller and G.J. Miller. eds., Elsevier, Amsterdam-New York-Oxford (1984) p. 167.

5. A.R. Tall, Metabolism of postprandial lipoproteins, in "Methods in Enzymology", Vol. 129, J.J. Albers and J.P. Segrest, eds., Acad. Press Inc., London (1986) p. 469.
6. C.J. Fielding, Mechanisms of action of lecithin-cholesterol acyltransferase, in "Methods in Enzymology", Vol. 129, J.J. Albers and J.P. Segrest, eds., Acad. Press Inc., London (1986) p. 783.
7. C.J. Fielding, Factors affecting the rate of catalyzed transfer of cholesteryl esters in plasma, Amer. Heart J. 113: 532 (1987).
8. J.E.M. Groener, R.W. Pelton and G.M. Kostner, Improved estimation of cholesteryl ester transfer/exchange activity in serum or plasma, Clin. Chem. 32: 283 (1986).

LECITHIN:CHOLESTEROL ACYLTRANSFERASE AND ITS ACTION ON DIFFERENT SUBSTRATES

Gabriele Knipping, Andrea Birchbauer,
Ernst Steyrer and Gerhard M. Kostner

Institute of Medical Biochemistry
University of Graz, A-8010 Graz, Austria

INTRODUCTION

Lecithin:cholesterol acyltransferase (LCAT) is synthesized by the liver and is responsible for the production of most of the cholesteryl esters (CE) transported in serum of man and of many animals (1). HDL, especially HDL_3 are assumed to represent the main substrate for LCAT, and apoA-I was found to be a necessary cofactor for the LCAT reaction (2). In recent publications we could demonstrate that the properties of LCAT purified from either pig or human plasma are similar. Using liposomes consisting of phosphatidylcholine (PC) and unesterified ^{3}H-cholesterol (FC) as substrate the enzymes of both species esterified FC at the same rate, if the molar PC/FC ratio was more than 3:1. The esterification rate was significantly reduced, if the molar ratio of PC/FC was below 3:1 (3). Since such differences in the molar PC/FC ratios of HDL- and LDL-fractions may play a role in the substrate properties of these lipoproteins, we further tested the action of LCAT from both species on isolated native HDL- or LDL-fractions. Here we could demonstrate that the esterification rate of FC in LDL was 30-40% that in HDL (4). Furthermore, these LDL-fractions became even better substrates, if the molar PC/FC ratio was raised to >3:1 by incubation with liposomes in the presence of cholesteryl ester transfer/exchange protein (CETP).

In man, the cholesteryl esters formed in HDL are believed to be transferred to VLDL or LDL by the action of CETP (5). Pig Plasma lacks or contains only small amounts of CETP activity. Since the fatty acid composition of LDL-CE is not significantly different from those of the HDL-CE in pig plasma (6,7), we reinvestigated the substrate properties of the individual lipoproteins in <u>native pig serum</u> to elucidate the role of LCAT on lipoproteins other than HDL. In the present report we studied the direct action of pig LCAT on the different lipoproteins by incubating freshly isolated pig serum in the presence or absence of active LCAT.

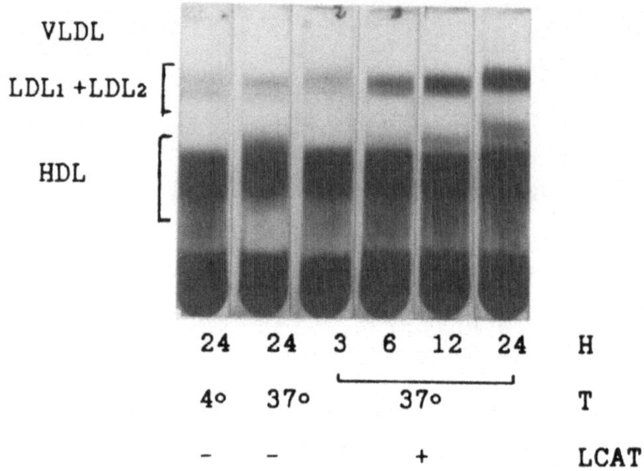

FIGURE 1: Flotation pattern of pig serum lipoproteins by density gradient ultracentrifugation. Pig serum was incubated in the presence and absence of the LCAT inhibitor sodium iodoacetate at 37°C at different time intervals. For control, one aliquot of the LCAT-inactive sample was stored at 4°. Samples were prestained with Coomassie-blue.

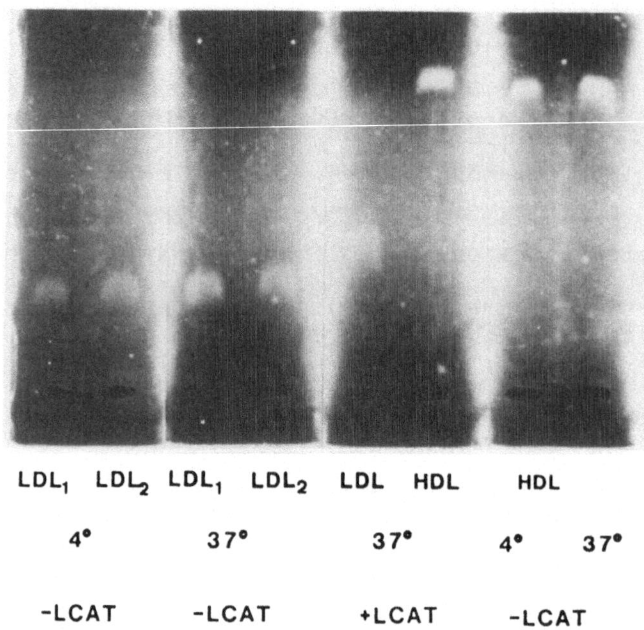

FIGURE 2: Agarose gel electrophoresis of LDL- and HDL-fractions after incubation in the presence and absence of active LCAT under conditions as in Fig. 1. Lipoproteins were isolated by a single density gradient ultracentrifugation run.

METHODS

Plasma was prepared from individual fasting pigs (sus domesticus) at 4°C. For inactivation of LCAT in some cases, 10 mM sodium iodoacetate was immediately added to whole blood. Fresh pig plasma was then divided into LCAT-active and LCAT-inactive samples. Aliquots of both active and inactive samples were incubated at 37°C; another aliquot of the LCAT-inactive sample was stored at 4°C as control. After incubation at various time intervals, LCAT was inhibited in the LCAT-active samples by the addition of 5 mM sodium iodoacetate, and the lipoproteins were isolated by density gradient ultracentrifugation (4,8), dialyzed against 0.15 M NaCl, containing 10 mM Tris-HCl (pH 7.4) and chemically analyzed for FC, CE and phospholipids (PL). Alternatively, the VLDL+LDL fraction was removed after incubation by precipitation with specific antibodies against pig apoB. After precipitation, the samples were centrifuged. The amount of free and esterified cholesterol was determined from total serum and the supernatant (="HDL"). The difference of these values was taken as "VLDL+LDL" free and esterified cholesterol.

RESULTS

Pig serum was incubated in the presence and absence of active LCAT for various periods of time and the individual lipoprotein classes were separated by density gradient ultracentrifugation. Fig. 1 shows the flotation pattern of the lipoproteins. The most striking result was the fusion of both pig LDL-subfractions, LDL_1 and LDL_2, upon prolonged incubation in the presence of active LCAT. This fusion process was time dependent, started after 3 h and was finished between 18 and 24 h of incubation. It was obvious that this fusion process was due to the action of LCAT, since no change of the LDL-flotation pattern occurred in the LCAT-inactive samples incubated for 24 h at 37°C.

Alterations in the HDL-flotation profile were also obtained. As seen in Fig. 1, in comparison to the LCAT-inactive sample, incubated at 4°C, the previously homogenous HDL broadened to some extent upon a 24 h-incubation at 37°C in the absence of active LCAT. In contrast, in the presence of active LCAT the previous more or less homogenous HDL splitted in a socalled HDL_2, HDL_3 and very high density lipoprotein fraction (VHDL). However, the mass of this HDL_2-fraction amounted maximally 10% of the total HDL-class after 24 h of incubation, whereas the mass of HDL_3 and that of VHDL represented 70-80% and 14-28%, respectively. This is in accordance to our previous finding (9), where we demonstrated that larger amounts of HDL_2-like particles only arise in the presence of active LCAT, CETP and triglyceride-rich particles.

In order to show that LCAT acts directly on LDL and not on α-lipoproteins or HDL-like particles, which might have floated in the LDL-density region after incubation with LCAT, we investigated the isolated LDL- and HDL-fractions by lipoprotein electrophoresis (Fig. 2). Whereas the mobility of total HDL was hardly altered regardless, whether LCAT was active or not, the mobility of LCAT-modified LDL was changed markedly. This was most probably due to the uptake of apoA-I, apoC-II and apoE upon incubation with LCAT. It is noteworthy to point out that these apoproteins resided on the modified LDL-particle as revealed by immunoelectrophoresis (data not shown).

In order to quantitate the absolute lipid changes in each lipoprotein class, the lipoproteins isolated by one single gradient ultracentrifugation run were chemically analyzed. Fig. 3 displays the absolute changes in the lipid composition after incubation of pig serum in the presence and absence of active LCAT. A 24 h-incubation in the absence of active LCAT resulted in a decrease of FC up to 25% in LDL, whereas the absolute amount of CE and PL increased between 4 and 8%. In HDL FC increased up to 42%, whereas CE and PL decreased between 4 and 9%. There was only a slight increase in FC and PL in VLDL. But due to the low content of VLDL in pig serum the values for VLDL must be viewed with caution.

Incubation for various periods of time in the presence of active LCAT resulted in a loss of FC in all lipoprotein classes. This decrease of FC started immediately and was most pronouned in LDL. In each fraction the loss of FC was equivalent to the newly formed CE. In contrast, the loss of PL in

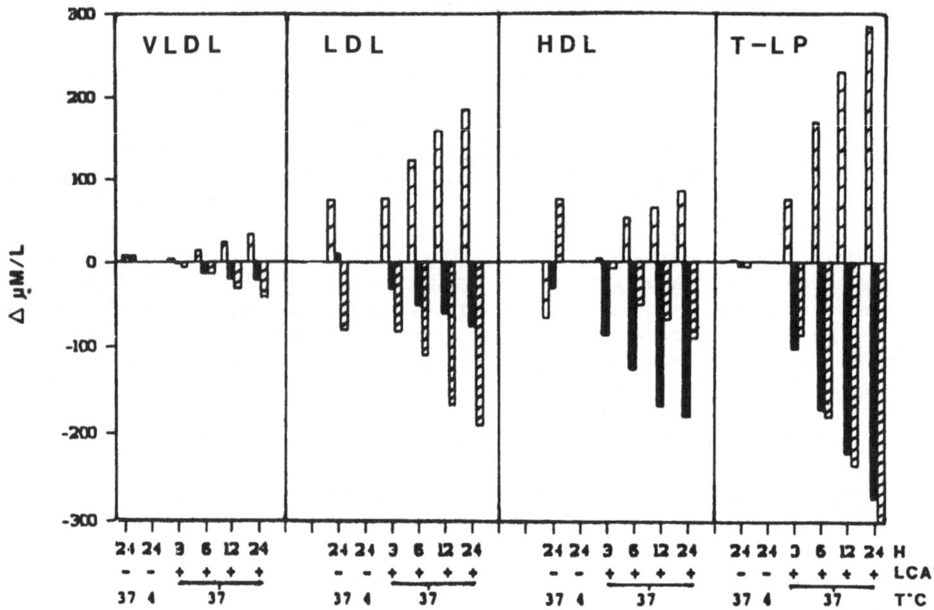

FIGURE 3: Absolute changes in the lipid composition after incubation of serum in the presence and absence of active LCAT. Conditions and isolation procedure as in Fig. 1. The values represent the absolute changes (Δ,uM/L) of FC [▨], CE [▨] and PL [■] obtained for VLDL, LDL, HDL and total lipoproteins (T-LP) in relation to the values obtained for the control sample stored at 4°C in the absence of LCAT (=base line). The original lipoprotein composition was as follows: VLDL, 36 uM/L FC, 15 µM/L CE, 32 µM/L PL; LDL, 279 µM/L FC, 823 µM/L CE, 292 µM/L PL; HDL, 146 µM/L FC, 734 µM/L CE, 606 µM/L PL.

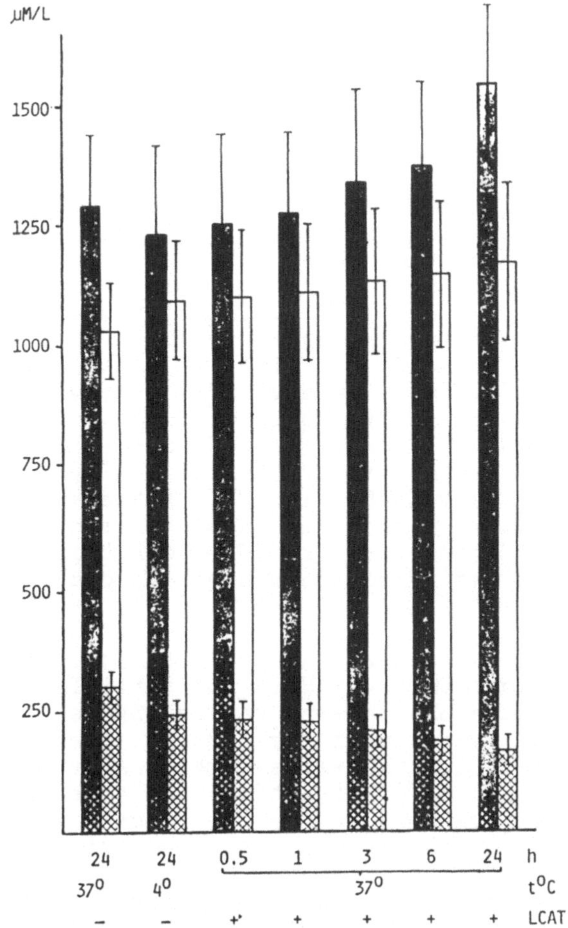

FIGURE 4: Absolute changes in lipid composition of lipoproteins after incubation in the presence or absence of active LCAT. Serum was incubated in the presence and absence of active LCAT at 37°C at different time intervals. As control one aliquot of the LCAT-inactive sample was stored at 4°C. FC and CE were estimated from total serum and from the supernatant obtained after precipitation of VLDL+LDL with anti-apoB. Grey bars, VLDL+LDL-CE, grey-hatched bars, VLDL+LDL-FC; white bars, HDL-CE, white-hatched bars, HDL-FC.

each lipoprotein fraction was not equivalent to the synthesis of CE. For the LCAT reaction it seemed that most of the PL originated from HDL, since this fraction lost more than twice as much PL as VLDL and LDL together. It is furthermore noteworthy that at any time interval studied, 60-70% of the newly synthesized CE were found in the LDL fraction.

Similar results were found with the second method used to quantify the absolute changes of FC and CE in HDL and apoB-containing lipoproteins. Here the apoB-containing lipoproteins were precipitated with anti-apoB after incubation of serum for various periods of time and the amount of unesterified and esterified cholesterol was determined from total serum and from the HDL-fraction. The results from these experiments are shown in Fig. 4. Even here most of the newly synthesized CE were incorporated in the VLDL+LDL fraction.

DISCUSSION

According to current concepts, LCAT acts preferentially on HDL particles that contain mainly apoA-I. With our experimental design we could demonstrate that this enzyme also acts directly on apoB-containing particles even in the presence of physiological amounts of HDL: both pig LDL-subclasses fused to one more or less homogenous fraction upon prolonged in vitro incubation in the presence of active LCAT, resulting in LDL-particles with molecular sizes in between those of LDL_1 and LDL_2. In addition, at any time interval tested, the enrichment with newly synthesized CE was most prominent in the LDL-fractions and less expressed in the HDL-fraction. We could exclude the possibility that CETP plays a key role in the enrichment of pig LDL with CE, since a 24 h-in vitro incubation of native pig plasma in the absence of active LCAT resulted in a CE-transfer from HDL to LDL in an amount of maximally 8% and was therefore minimal in comparison to human serum.

We could further ascertain that LCAT acts directly on LDL and not on α-lipoproteins or HDL-like particles, which might have floated in the LDL-density region, by testing the isolated LDL- and HDL-fractions by agarose gel electrophoresis. It is clearly visible that HDL-like particles or α-lipoproteins are not present in the LDL-fractions and thus are not responsible for the alterations in the LDL-fraction. These results are in accordance to our previous findings (4), where we showed that the amount of apoA-I on the LDL-particle itself is sufficient to activate LCAT and where we further demonstrated that even apoA-I-depleted LDL may serve as a substrate for LCAT to a certain extent. However, the slightly faster migration of fused LDL was due to the uptake of apoA-I, apoC-II and apoE as revealed by immunoelectrophoresis (data not shown). Such an uptake of apolipoproteins by LDL during the action of LCAT was already noticed by others (10).

We think that the methods used here (separation of lipoproteins by a single density gradient ultracentrifugation run or by precipitation with anti-apoB) allowed an absolute quantitation of the time dependent compositional changes in each lipoprotein class upon in vitro incubation in the presence and absence of active LCAT. We further think that these data are more reliable than those obtained by the use of isotopic labeling of cholesterol in whole serum, a method performed in previous studies (11). In preliminary experiments we found that ^{3}H-FC distributed unevenly among the different lipoproteins, e.g. at zero time and after 1 h of incubation the labeled FC was mostly found in the HDL-fraction. Thus newly synthesized ^{3}-CE must be preferentially found in the HDL-fraction under those conditions.

In conclusion, we have demonstrated that in pig, an animal, which lacks CETP, LCAT acts significantly - if not preferentially - on apoB-containing lipoproteins even in the presence of physiological concentrations of HDL. We suggest that such a mechanism may also act in human plasma. It could explain, why the LCAT activity of fisheye disease patients is nearly normal. Additionally, the amount of serum CE of such patients is 60% of normal, and these CE are predominantly found in LDL and not in HDL (12).

ACKNOWLEDGMENT

This work was supported by the Österreichischer Fonds zur Förderung der wissenschaftlichen Forschung Nr. P 6141B and the Österreichische Nationalbank.

REFERENCES

1) J.A. Glomset, An exercise in comparative biology, In: Progr. Biochem. Pharmacol. S. Eisenberg ed, Karger Basel (1979) 41 pp.
2) C.J. Fielding, V.G. Shore and P.E. Fielding, A protein cofactor of lecithin:cholesterol acyltransferase, Biochem. Biophys. Res. Comm. 46:1493 (1972).
3) G. Knipping, Isolation and properties of porcine lecithin: cholesterol acyltransferase, Eur. J. Biochem. 154:289 (1986).
4) G. Knipping, A. Birchbauer, E. Steyrer, J. Groener, R. Zechner and G.M. Kostner, Studies on the substrate specificity of human and pig lecithin:cholesterol acyltransferase: the role of low density lipoproteins, Biochemistry 25:5242 (1986).
5) P.E. Fielding and C.J. Fielding, A cholesteryl ester transfer complex in human plasma, Proc. Natl. Acad. Sci. U.S.A. 77:3327 (1980).
6) V. Nöthig-Laslo and G. Knipping, Surface structure of the two porcine low density lipoprotein subclasses. A spin labeling study, Int. J. Biol. Macromol. 6:255 (1984).
7) V. Nöthig-Laslo and G. Knipping, Comparative study of the lipid dynamics in the surface layer of porcine and human high density lipoprotein subclasses by spin labeling, Chem. Phys. Lipids 36:373 (1985).
8) G. Knipping, A. Birchbauer, E. Steyrer and G.M. Kostner, The action of lecithin:cholesterol acyltransferase on low density lipoproteins in native pig plasma, Biochemistry 26:7945 (1987).
9) G. Knipping, R. Zechner, G.M. Kostner and A. Holasek, Factors affecting the conversion of high density lipoproteins: experiments with pig and human serum, Biochim. Biophys. Acta 835:244 (1985).
10) R. Zechner, H. Dieplinger, A. Roscher and G.M. Kostner, The LDL pathway of native and chemically modified LDL isolated from plasma incubated in vitro, Biochem. J. 224:569 (1984).
11) O.V. Rajaram and P.J. Barter, Reactivity of human lipoproteins with purified lecithin:cholesterol acyltransferase during incubations in vitro, Biochim. Biophys. Acta 835:41 (1985).
12) L.A. Carlson and L. Holmquist, Paradoxical Esterification of plasma Cholesterol in fish eye disease, Acta med. Scand. 217:491 (1985).

IN VIVO EVIDENCE FOR CHOLESTEROL ESTER AND TRIGLYCERIDE EXCHANGE BETWEEN HIGH DENSITY LIPOPROTEIN AND INFUSED TRIGLYCERIDE RICH PARTICLES IN ABETALIPOPROTEINEMIA

D.W. Erkelens and R.P.F. Dullaart

Department of Internal Medicine
University Hospital
Utrecht, Netherlands

ABSTRACT

Abetalipoproteinemia is characterized by the absence of chylomicrons, very low density and low density lipoproteins from the plasma. To study neutral lipid exchange between lipoproteins in vivo a chylomicron model, Intralipid[R], was infused in an abetalipoproteinemic patient. During a three hour infusion of 250 mg/kg/hr after a priming dose of 100 mg/kg triglyceride, 8% of the triglyceride mass of Intralipid was replaced by cholesterolester, while 8% of the cholesterylester mass was replaced by triglyceride in high density lipoproteins. Thus, the exchange of cholesterylester and triglyceride between high density lipoprotein and triglyceride rich particles in vivo was directly demonstrated in the absence of apolipoprotein B.

INTRODUCTION

The plasma from patients with recessive abetalipoproteinemia contains only high density lipoprotein (HDL) with apolipoproteins A, E and C, but no apolipoprotein B carrying lipoproteins such as chylomicrons, very low density lipoproteins (VLDL) or low density lipoproteins (LDL) (1). This abnormality offers a unique model to study exchange of neutral lipids between lipoproteins(2) particularly since it has been shown in vitro that cholesterylester transfer from HDL occurs preferentially to triglyceride rich lipoproteins(3,4). In order to evaluate mass transfer quantitatively an artificial emulsion of triglyceride rich particles, Intralipid[R], was infused in an abetalipoproteinemic patient and triglyceride and cholesterylester were quantified in HDL and the re-isolated triglyceride rich particles at the end of the infusion.

PATIENT AND METHODS

The patient is a 30 year old female in whom the diagnosis of abetalipoproteinemia was made at the age of 25. She was the sole offspring from a consanguineous marriage and exhibited slight spinocerebellar ataxia, impaired vision from retinitis pigmentosa, and 60-70% acanthocytosis. Plasma total cholesterol was 1.3-1.6 mmol/l and triglyceride undetectable to 0.1 mmol/l. Apolipoprotein B levels were below the detection limit (1 mg/dl) of the kinetic immunoturbidimetry assay with polyclonal sheep anti-human LDL antiserum (Boehringer Mannheim, 726-494). Chylomicrons, VLDL and LDL

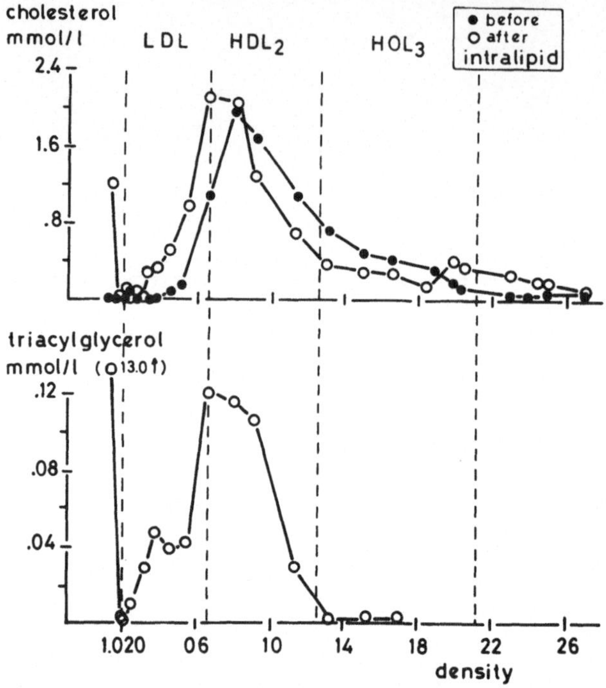

Fig. 1. Density gradient ultracentrifugation of plasma from an abetalipoproteinemic patient before (o) and after (o) infusion of artificial triglyceride particles, Intralipid[R]. Before infusion cholesterol is only present in the HDL density region while after infusion it is also present in the top fraction (upper pannel) . Triglyceride is not present before infusion but is distributed between the top fraction and the particles in the LDL-HDL density region (lower panel).

were absent from her plasma as shown by one step gradient ultracentrifugation (5) in Fig. 1. It has been described previously that monoclonal antibodies against apo B100 and apo B48 gave positive staining on immuno-histochemistry and immuno electron microscopy, proving local synthesis of apo B in intestinal mucosal and hepatic cells(6).

Infusion procedure

The patients regularly received infusion of an artificial triglyceride emulsion (Intralipid[R]) mixed with vitamin A (Vitralipid, KabiVitrum), and vitamin E (Ephynal, Hoffman-La Roche) for treatment of vitamin A, E and essential fatty acid deficiency. On the day of the test Intralipid[R] 10% only was infused for 3 hours at a speed of 250 mg triglyceride/kg/hr after a primary dose of 100 mg triglyceride/kg given within 2 minutes. Intralipid[R] contains artificial triglyceride particles comparable in size and lipid composition to chylomicrons(7) but does not contain (apo)proteins. Blood was sampled from the contralateral cubital vein at -1, 0, 1, 2, 3, 8, 24 and 25 hours, through an intravenous line, kept open with 0.9% saline drip. Except for a taste of glue shortly after the priming dose, the patient experienced no adverse effects.

Laboratory measurement

In each plasma sample (EDTA 1 mg/ml) total triglycerides and cholesterol were measured by enzymatic methods (Boehringer Mannheim kits 701912 and 237574 respectively). Chylomicrons were separated by slicing after ultra-

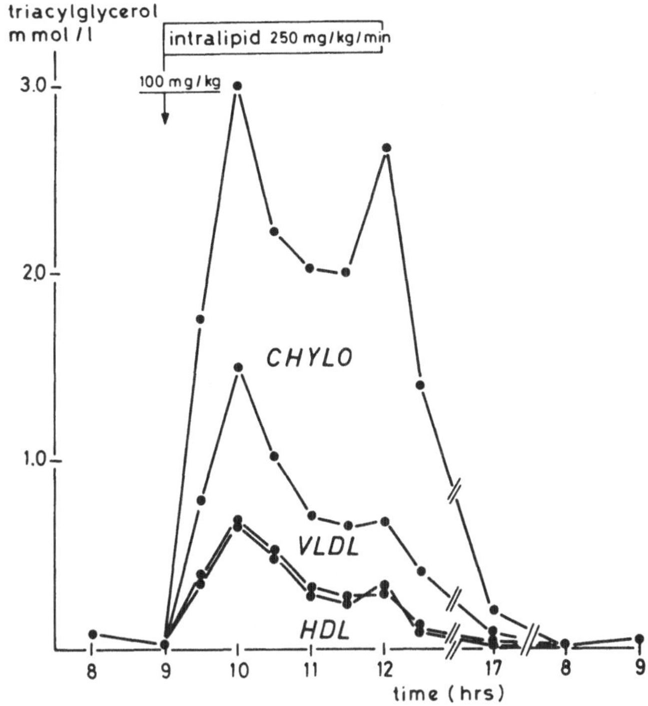

Fig. 2. Triglyceride distribution between chylomicron (chylo) very low density lipoprotein (VLDL) and high density lipoprotein (HDL) after infusion of Intralipid[R] in an abetalipoproteinemic patient.

centrifugation at 10^6 9.min in an SW 40.1 Beckman rotor. VLDL were separated after 20 hr d 1.006 ultracentrifugation at 40.000 rpm in a 40.3 fixed angle Beckman rotor. LDL or precipitable lipoprotein were separated from HDL or nonprecipitable lipoprotein by heparin manganese treatment of the infranatant.

Triglycerides, total cholesterol and free cholesterol were measured in each fraction as described above. Cholesterylester was calculated as the difference between total and free cholesterol.

Cholesterylester transfer activity of each sample was measured after delipidation in transfer assay between 14C cholesterylester containing HDL and cold LDL as described before(8). The 0, 1, 2 and 3 hour samples were subjected to a one step gradient ultracentrifugation(5) and cholesterol, free cholesterol and triglyceride were measured in each 0.5 ml fraction.

RESULTS

The triglyceride (TG) level increased from 0.0 mmol/l to 2.28 mmol/l during the 3 hrs infusion of triglyceride rich particles or TGRP (Fig. 2). Almost three quarters (74.7%) of total TG was located in the chylomicron + VLDL fraction which represents the infused particles and their early catabolic products, while the rest was located in the HDL fraction. The so called "LDL" fraction isolated by precipitation contained 0.01-0.05 mmol/l of TG.

Total cholesterol did increase slightly during infusion from 1.25 to 1.51 mmol/l. The amount of cholesterol in the VLDL + chylomicron fraction increased from 0.11 to 0.38 mmol/l (Fig. 3). In the d<1.019 fraction, or the VLDL infranatant the total cholesterol concentration remained unaltered, while the heparin manganese precipitable cholesterol increased from 0.68 to 0.96 mmol/l. This fraction is designated "LDL" since in normal patients it contains apo B carrying lipoproteins. In our patient, who has no circu-

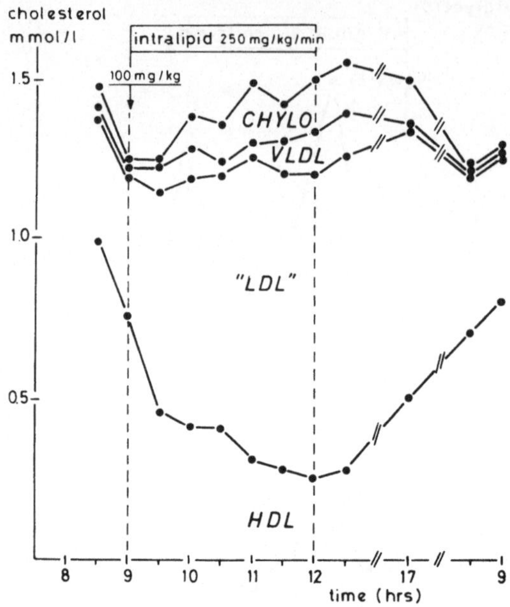

Fig. 3. Cholesterol distribution between lipoproteins during Intralipid[R] infusion. "LDL" designates the fraction precipitated by heparin-manganese. This probably represents apo E rich HDL since apo B containing LDL are absent from the plasma.

Table 1. Distribution of triglyceride and cholesterylester between "chylomicrons + VLDL" and HDL after 3 hrs infusion of triglyceride rich particles in an abetalipoproteinemic patient.

	Chylomicrons + VLDL		HDL + "LDL"	
Triglyceride mmol/l (%)	1.70	(74.7)	0.58	(25.3)
Cholesterylester mmol/l (%)	0.27	(25.4)	0.78	(74.6)

lating apo B, it probably represents apo E rich HDL precipitable by heparin manganese(1). The free cholesterol varied between 25 and 35% of total cholesterol in the different fractions, and cholesterol ester complementary between 65 and 75%. The distribution of triglyceride and cholesterylester among the lipoprotein fractions is given in Table I. When lipoproteins were separated by density gradient ultracentrifugation no, that is less than 0.02 mmol/l, TG was detected before infusion. At the end of the infusion the d 1.016 fraction contained 13.0 mmol/l. TG was not detectable from d 1.018-1.020 but increased to a maximum of 0.12 mmol/l in the LDL and HDL 2 density region (Fig. 1), with a peak at d 1.070 to 1.100.
Cholesterol was not detectable in the top fraction before infusion but was 1.2 mmol/l at the end of the infusion. The cholesterylester content of the top fraction was 0.69, 0.54 and 1.12 mmol/l or 23, 17 and 37% of total cholesterol ester present in the serum at 1, 2 and 3 hours of infusion respectively. The molar mass composition of the lipoprotein core at the end of infusion is given in Table II.

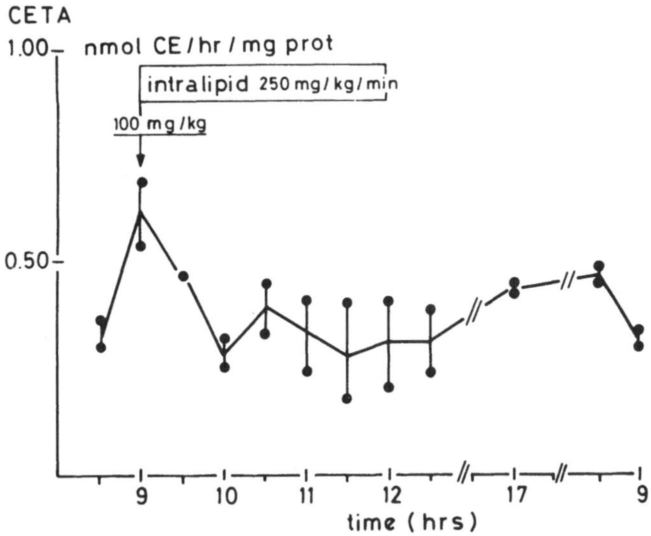

Fig. 4. Cholesterylester transfer activity (CETA), measured in delipidated plasma by assay of transfer of C-14 cholesterolester from pool serum HDL to pool serum unlabeled LDL. The dots and vertical lines represent duplicate measurements. The CETA does not change and is in the area of CETA present in normal plasma.

Table 2. Lipoprotein core composition in topfraction or "chylomicrons" (d 1.020) and bottom fraction or HDL (d 1.020 - 1.130) at the end of 3 hrs infusion of triglyceride rich particles in an abetalipoproteinemic patient.

	Triglyceride mmol/l (%)		Cholesterylester mmol/l (%)	
"Chylomicrons"	13.0	(92.1)	1.12	(7.9)
HDL	0.12	(7.9)	1.40	(92.1)

It is shown in Figure 1 that the density peak of the HDL particles shifted slightly to the left (lower density) after infusion, probably because the particles contained more triglyceride. The in vitro measured cholesterol ester transfer activity did not change appreciably during infusion (Fig. 4) and was within the range measured in normal subjects(8).
Triglyceride and cholesterol levels returned to almost pre infusion levels 5 hours after discontinuation of infusion and were basal again at 24 and 25 hours.

DISCUSSION

The study offers direct evidence that exchange of cholesterol ester and triglyceride between high density lipoprotein (HDL) and triglyceride rich particles occurs in vivo. The exchange and/or transfer of lipoprotein core constituents cholesterylester (CE) and triacylglycerol (TG) between lipoproteins has been demonstrated in vitro, mediated by lipid transfer protein(s) and possibly inhibited by a lipid transfer inhibitor(2). These proteins are present in the plasma of different species including man(9); a cholesterolester transfer protein has been characterized and its DNA sequence has been published(10). It is not known to what extent the transfer or exchange of CE and TG occurs in vivo, although it is assumed that CE enrichment of TG carrying particles, such as chylomicrons and Very Low Density Lipoproteins (VLDL) takes place during their catabolism in the vascular compartment. It has been noted that CE transfer in vitro is dependent on particle surface and core composition, that preferential transfer occurs to triglyceride rich particles(3,4). It is therefore of interest to know whether the presence of apolipoprotein B, the structural protein of triglyceride rich lipoproteins, is a necessary prerequisite for the transfer or exchange to occur. We were able to study both problems in a patient with abetalipoproteinemia, who was no apo B containing lipoproteins in the circulation. We found that after infusion of artificial TG rich particles, not containing apo B, CE accumulated in these particles while at the same time TG appeared in HDL, that did not contain TG before infusion. Since the artificial TG rich particles are very alike in size and composition to chylomicrons and since the period of infusion, i.e. three hours, is comparable to a normal postprandial period, the quantitated neutral lipid transfer may well be representative of the normal in vivo situation. This would mean that 25% of the cholesterolester present in plasma is transferred to triglyceride rich particles, making up 9% of their molar mass. Simultaneously 25% of the triglyceride in plasma is transferred from triglyceride rich particles to HDL, making up 9% of their molar mass. On a molar base this would mean that 1 mol of cholesterylester is exchanged against 2.1 mol of triacylglycerol. This transfer occurred in the absence of apolipoprotein B. HDL of the patient contained apo A, C and E and it is conceivable that apo C and E transferred to the infused triglyceride rich particles(11). These findings could be used in favour of the argument that lipid transfer proteins are rather unspecific mediators for neutral lipid transfer and that they facilitate equilibration of core neutral lipids between circulating lipoproteins.

In conclusion we have been able to demonstrate in vivo mass transfer of cholesterylester from HDL to triglyceride rich particles in exchange for mass triglyceride transfer from these particles to HDL in the absence of apo B. Since triglyceride rich particle remnants are cleared rapidly by the liver this would underline the role of the reverse cholesterol transport pathway in the removal of cholesterol from the plasma.

REFERENCES

1. P. Herbert, G. Assmann, A.M. Gotto, and D.S. Fredrickson, Familial lipoprotein deficiency, in: The Metabolic Basis of Inherited Disease. J.B.Stanbury et al. (eds), McGraw-Hill, New York, p. 589-621 (1982).
2. Y.C. Marcel, Lecithin: cholesterol acyl transferase and intravascular cholesterol transport, Adv.Lip.Res., 19:117 (1982).
3. S. Eisenberg, Preferential enrichment of large sized very low density lipoprotein populations with transferred cholesterylester, J. Lipid. Res., 26:487 (1985).
4. R.P.F. Dullaart, J.E.M. Groener, and D.W. Erkelens, Effect of composition of very low and low density lipoproteins on the rate of cholesterylester transfer from high density lipoproteins in man, studied in vitro, Eur.J.Clin.Invest., 17:241 (1987).
5. T.G. Redgrave, D.C.K. Roberts, and C.E. West, Separation of plasma lipoproteins by density gradient ultracentrifugation, Anal.Biochem., 65:42 1975.

6. R.P.F. Dullaart, B. Speelberg, H.J. Schuurman, et al., Epitopes of apolipoprotein B-100 and B-48 in both liver and intestine, J.Clin.Invest., 78:1397 (1986).
7. D.W. Erkelens, C. Chen, C.D. Mitchell, J.A. Glomset, Studies on the interaction between apolipoproteins A and C and triacylglycerol rich particles, Biophys.Biochem.Acta, 665:221 (1981).
8. J.E.M. Groener, A.E. van Rozen, D.W. Erkelens, Cholesterylester transfer activity, localisation and role in distribution of cholesterylester over lipoproteins in man, Atherosclerosis, 50:261 (1984).
9. Y.C. Ha, P.J. Barter, Differences in plasma cholesterylester transfer activity in sixteen vertebrate species, Comp.Biochem.Physiol., 71B: 265 (1982).
10. D. Drayna, A.S. Jarnagin, J. Mclean, W. Henzel, W. Kohr, C. Fielding, and R. Lawn, Cloning and sequencing at human cholesterylester transfer protein cDNA, Nature, 327:632 (1987).
11. D.W. Erkelens, J.D. Brunzell, E.L. Bierman, Availability of apolipotein C II in relation to the maximal removal capacity for an infused triglyceride emulsion in man, Metabolism, 28:495 (1979).

ENHANCED CHOLESTERYL ESTER TRANSFER ACTIVITY IN CYCLOPHOSPHAMIDE-TREATED RABBITS : RELATIONSHIP WITH LIPOLYTIC ENZYMES

N. Dousset, A.M. Julia, H. Chap and L. Douste-Blazy

INSERM U 101, Biochimie des Lipides, Hôpital Purpan
31059 Toulouse Cedex, France

INTRODUCTION

Hyperlipaemia occurs in male New Zealand rabbits after antimitotic treatment (cyclophosphamide) or after exposition to lethal levels of whole-body ionizing radiation (1,2). The treated rabbits were hypertriglyceridemic and hypercholesterolemic with an especially significant increase in VLDL cholesteryl ester (3). There was an accumulation of VLDL and a decrease in HDL (1). On the other hand, an inhibition of the post-heparin plasma lipoprotein lipase occurred after these treatments (4,5). Partial purification of lipoprotein lipase and of hepatic triacylglycerol lipase have shown that the absence of HDL observed in treated rabbits may be due to the LPL deficiency.

The purpose of this study was to study the plasma cholesteryl ester transfer activity in cyclophosphamide-treated rabbits and to investigate the possible relationship with lipolytic enzymes.

MATERIAL AND METHODS

- Animals

Male New-Zealand white rabbits weighing approximately 2-2.5 kg were used. All the animals were fasted 15 h before treatment. Treated rabbits received a maximal dose representing 1/2 LD_{50} (65 mg/kg) of cyclophosphamide . No food was given to the animals after the injection of cyclophosphamide into an ear vein. Control animals were also deprived of food over a period of time corresponding to that of treated animals.

Blood was drawn by intracardiac puncture into EDTA (final concentration : 1 mg/ml) 16 h after cyclophosphamide injection and under the same conditions for untreated rabbits.

- Lipoprotein preparations

Lipoprotein fractionation was performed in a Beckman L 8.70 ultracentrifuge using a 50 Ti rotor at 4° C. VLDL (d < 1.006 g/ml) were isolated by centrifugation at 105 000 x g for 18 h, after removing chylomicrons by centrifugation at 15 000 x g for 30 min. LDL (low density lipoproteins) and HDL (high density lipoproteins) were isolated at d < 1.063 g/ml and d < 1.21 g/ml, respectively, by centrifugation at 105 000 x g_{av}

for 20 h and 24 h. HDL and the plasma fraction of d > 1.21 g/ml which contained CETP were dialyzed against a 5 mM Tris-HCl buffer (pH 7.4) containing 0.5 mM EDTA and 0.15 M NaCl.

Subfractionation of VLDL was carried out by heparin-sepharose chromatography as described by Huff and Telford (6).

- Determination of cholesteryl ester transfer activity

Labelling of lipoproteins was as described by Marcel et al. (7). A stabilized (^{3}H)cholesterol (Amersham) emulsion in 5 mM Tris-EDTA, containing 0.05 % NaN_3 and 2 % fatty-free bovine serum albumin buffer was incubated with total plasma for 1 h at 37° C in the presence of DTNB (final concentration 0.5 mM). β-Mercaptoethanol (final concentration 15 mM) was added to stimulate LCAT activity. After 20 h of incubation, plasma labelled lipoproteins were separated by ultracentrifugation and dialysed against a buffer containing 5 mM Tris HCl, 0.5 mM EDTA, 150 mM NaCl, pH 7.4.

CETP activity was measured as the ability of protein fractions to promote the transfer of radiolabelled cholesteryl ester from HDL to VLDL at 37° C. Labelled HDL (33 µg cholesteryl ester) were incubated with unlabelled VLDL (33 µg cholesteryl ester) in the presence of CETP (3.5 mg protein) in a final volume of 1 ml saline solution containing 0.01 % NaN_3 for 3 h at 37° C. 0.5 mM DTNB was added to inhibit LCAT activity (8) Incubation was stopped by transfer to cruded ice and by immediately diluting with 0.15 M NaCl solution. VLDL were reisolated by centrifugation at 105 000 x g for 18 h at the density of 1.006. Lipids were analysed after extraction by thin-layer chromatography for their radioactive esterified and free cholesterol content.

- Radiation inactivation method

For radiation inactivation, aliquots (0.3 ml) of the plasma fraction of d > 1.21 g/ml either from control or treated rabbits were lyophilized in 1.5 ml Eppendorf microfuge tubes. The tubes were irradiated at room temperature (26 ± 2°C) in a ^{60}Co irradiator (Gammacell model 220, Atomic Energy of Canada, Ottawa) delivering about 2.5 Mrad per h (9).

The irradiator was calibrated with enzymes of known radiation sensitivities according to Beauregard and Potier (10). Three tubes were exposed to each radiation dose. Non-irradiated tubes served as controls. The logarithm of remaining enzyme activity was plotted vs radiation dose and a regression line was drawn through the points.

The empirical equation of Kepner and Macey (9) was used to relate the radiation dose (in Mrad) necessary to inactivate CETP to 37 % of its initial activity (D_{37}) to the Mr :

$$Mr = 6.4 \times 10^5 \times 1/D_{37}.$$

- Apolipoprotein preparation

Apolipoproteins were fractionated by chromatography on Ultrogel AcA 54 and by preparative electrofocusing according to the procedure previously described (5).

RESULTS AND DISCUSSION

1. Chemical composition of VLDL and HDL in control and treated rabbits

Cyclophosphamide injection induced in rabbits a pronounced hypertriglyceridemia and hypercholesterolemia accompanied by an accumulation of VLDL. Lipid composition of these VLDL and HDL in treated rabbits showed that HDL cholesterol was very low, compared to a high VLDL cholesterol. This increase of VLDL cholesterol was due to an increase of both esterified and free cholesterol. On the other hand, HDL esterified cholesterol markedly decreased while the amount of HDL free cholesterol was unchanged. Furthermore, there was a 3-fold increase in triacylglycerol level in VLDL from treated rabbits. These data are to be compared with the cholesteryl ester rich VLDL, as Roth et al. (11) have observed in cholesterol-fed rabbits. The HDL appeared enriched in triacylglycerol but poor in cholesteryl ester. The same HDL exist in patients with hypertriglyceridemia due to a lack of lipoprotein lipase activity (12).

The VLDL and HDL apoprotein levels were also increased after cyclophosphamide treatment. These lipoproteins are rich in threonine-poor apolipoproteins which have a low molecular weight and relatively high isoelectric points (1).

Furthermore, subfraction of VLDL by heparin-sepharose chromatography shows that the population of VLDL particles unbound to heparin on heparin-sepharose column was increased in treated rabbits. Similar findings are described for human VLDL from hyperlipoproteinemic subjects (6) and for survivors of cerebral infarction (13). In fact, Matsuda et al. (13) have recently reported the accumulation of partially catabolized particles of VLDL in these patients. They suggested that cholesterol ester may be excesssively transferred from HDL to VLDL during the disturbed catabolism of VLDL.

2. <u>Transfer of cholesteryl ester from rabbit HDL to VLDL</u>

- Since antimitotic drug provokes modifications on VLDL and HDL, concerning especially their lipid and protein composition, the increase of lipids seems to imply a remodelling of the VLDL core which is enriched in triacylglycerols and cholesteryl esters (3). A high level of triacylglycerols in the VLDL could be favorable to cholesteryl ester transfer from HDL with a reciprocal movement of triacylglycerols (14). These results are confirmed by experiments realized with triacylglycerol rich VLDL lipolyzed with LPL. On the contrary, transfer activity was enhanced with lipolyzed VLDL from control rabbits. LPL, by its lipolytic activity, accelerates the cholesteryl ester transfer. These results agree with Tall et al. (15) and Sammett and Tall (16) admitting a synergic effect between CETP and lipoprotein lipase. The enhanced CETP activity during lipolysis is related to the accumulation of products of lipolysis, especially fatty acids in the lipoproteins, augmenting the binding of CETP to these lipoproteins (16).

The decrease obtained in transfer activity with VLDL lipolyzed from treated rabbits may be explained by two means : it is possible that after lipolysis, triacylglycerol level was the same in VLDL, and then cholesteryl ester transfer occurred in the same way than in control rabbits. Conversely, the lipoprotein lipase may induce a removal of excess surface apoproteins, including threonine-poor apolipoproteins, which might be involved in the cyclophosphamide-induced increase in the transfer reaction. We have investigated the possible influence in vitro of these apoproteins on the CETP activity in treated rabbits and demonstrated that apolipoproteins did not induce any modification of CETP activity. If the threonine-poor apolipoproteins play a great part in the development of hypertriglyceridemia (4,5), these apolipoproteins cannot be involved in the increased CETP activity in cyclophosphamide treated rabbits.

- We have also examined the possibility that cyclophosphamide treatment resulted in an abnormal cholesteryl ester transfer protein. We have determined the molecular weight of CETP by radiation inactivation method. This method allows to determine in situ the Mr of enzymes without prior purification as reported (9). The radiation inactivation of rabbit CETP from the plasma fraction of d > 1.21 g/ml is shown in figure.

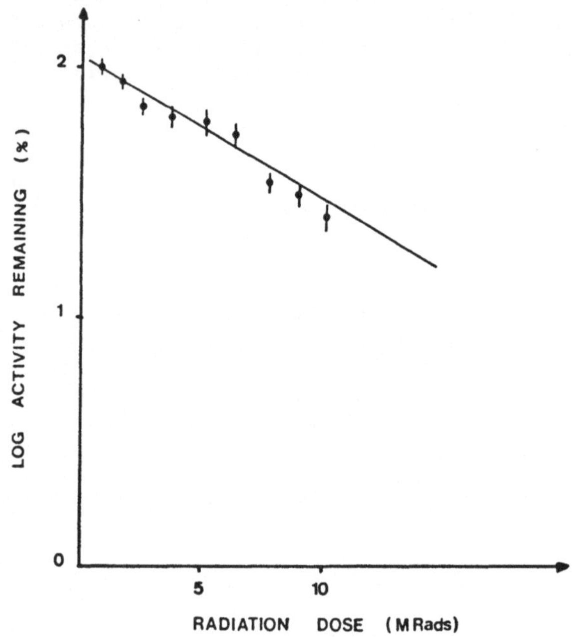

Fig.1. Radiation inactivation of CETP from control rabbits. The regression lines were fitted by the least-squares method.

We obtained similar results for cyclophosphamide-treated rabbits. The radiation dose leading to reduction to 37 % of initial activity (D_{37}) was obtained at 9.0 Mrad for control rabbits. This allowed an estimate of 71 000 ± 2 600 for control CETP and 72 000 ± 2 200 for cyclophosphamide rabbit CETP. Similar results were obtained for rabbits after a whole-body irradiation (17). These results agree with an unchanged activity of the protein. The functional unit of CETP is a monomer of about 70 000 daltons. With the radiation inactivation method, we observed a molecular weight similar to that of the enzyme subunit (M_r = 70 000 as determined by us (17) using SDS-polyacrylamide gel electrophoresis. These results confirm that CETP activity modification (3) in treated rabbits is not due to a change in the enzyme structure.

In conclusion, antimitotic agent does not alter the cholesteryl ester transfer protein molecule. Consequently, the transfer activity is modified only by an enrichment in triacylglycerol of cholesteryl ester acceptor lipoproteins (VLDL).

3. Relationship with lipolytic enzymes

On the other hand, an inhibition of the post-heparin plasma lipoprotein lipase occurred after the antimitotic treatment as after whole-body irradiation (4,5). Partial purification of lipoprotein lipase and of hepatic triacylglycerol lipase have shown that the absence of HDL observed in treated rabbits may be due to the LPL deficiency. A defect of enzyme secretion may be involved (unpublished). The deficiency of extrahepatic lipoprotein lipase explains the accumulation of partially catabolized particles of VLDL in these animals. As observed in treated rabbits, LPL inhibition by intravenous infusion of antibodies resulted in a marked increase in the largest VLDL in the cynomolgus monkey (18) and in a concomitant elevation of TG-rich lipoproteins and a decrease in HDL cholesterol due perhaps to a HDL cholesterol exchange for TG found in the increased pool of VLDL.

We have also observed that hepatic triacylglycerol lipase was normal in cyclophosphamide-treated rabbits. Therefore HDL are represented by a smaller population than the control ones. As for hypertriglyceridemic subjects (19), the small particle size of HDL is the possible consequence of the sequential actions in vivo of increased CET activity and normal H-TGL activity on HDL. Lipid exchange combined with hydrolysis of back-transferred triglyceride molecules by H-TGL can explain the conversion of HDL_2 to HDL_3, as described previously (20).

In summary, one may say, that several pathways do exist for explain the increased CETP activity in cyclophosphamide-treated rabbits : the most important probably is the inhibition of LPL resulting in an increase of larger VLDL. In relation with this disturbed catabolism of VLDL, the transfer of cholesteryl esters from HDL to VLDL by CETP is enhanced. H-TGL on the other hand plays also its role by hydrolyzing transferred triglycerides, favoring formation of small HDL particles.

Acknowledgements

The authors are grateful to Dr M. Potier for fruitful collaboration and to Mrs J. Manent for her excellent technical assistance. The radiation facility is supported by the Medical Research Council of Canada(MP).

References

1. A.M. Loudet, N. Dousset, M. Carton and L. Douste-Blazy, Effects of an antimitotic agent (cyclophosphamide) on plasma lipoproteins, Biochem. Pharm. 33:2961 (1984).

2. R. Feliste, N. Dousset, M. Carton and L. Douste-Blazy, Changes in plasma apolipoproteins following whole-body irradiation in rabbit, Radiat. Res. 87:602 (1981).

3. A.M. Loudet, N. Dousset, B. Perret, M. Ierides, M. Carton and L. Douste-Blazy, Triacylglycerol increase in plasma very low density lipoproteins in cyclophosphamide-treated rabbit : relationship with cholesteryl ester transfer activity, Biochim. Biophys. Acta, 794:444 (1985).

4. N. Dousset, R. Feliste, M. Carton and L. Douste-Blazy, Irradiation-induced free cholesterol accumulation in very-low density lipoproteins Role of lipoprotein lipase deficiency, Biochim. Biophys. Acta, 794: 444 (1984).

5. A.M. Loudet, N. Dousset, M. Carton and L. Douste-Blazy, Effect of an antimitotic agent (cyclophosphamide) on post-heparin plasma lipoprotein lipase activity in rabbit, Biochem. Pharm. 34:3597 (1985).

6. M.W. Huff and D.E. Telford, Characterization and metabolic fate of two very-low-density lipoprotein subfractions separated by heparin-sepharose chromatography, Biochim. Biophys. Acta, 796:251 (1984).

7. Y.L. Marcel, C. Vezina, B. Teng, and A. Sniderman, Transfer of cholesterol esters between high density lipoproteins and triglyceride rich lipoproteins controlled by a plasma protein factor, Atherosclerosis 35:127 (1980).

8. A.S. Garfinkel, E.S. Kempner, O. Ben-Zeec, J. Nikazy, S.J. James and M.C. Schotz.
Lipoprotein lipase : size of functional unit determined by radiation inactivation, J. Lipid Res. 24:775 (1983).

9. G.R. Kepner and R.I. Macey, Membrane enzyme systems. Molecular size determination by radiation inactivation, Biochim. Biophys. Acta, 163: 188 (1968).

10. G. Beauregard and M. Potier, Radiation inactivation of enzymes at low doses rates : identical molecular weights of rat liver cytosolic and lysosomal neuraminidases. Anal.Biochem., 122:379 (1982).

11. R.I. Roth, J.W. Gaubatz, A.M. Gotto and J.R. Patsch, Effect of cholesterol feeding on the distribution of plasma lipoproteins and on the metabolism of apolipoprotein E in the rabbit, J. Lipid Res., 24:1 (1983).

12. R.J. Deckelbaum, S. Eisenberg, Y. Oschry, M. Cooper and C. Blum, Abnormal high density lipoproteins of abetalipoproteinemia:relevance to normal HDL metabolism, J. Lipid Res. 23:1274 (1982).

13. M. Matsuda, T. Miyahara, A. murai, N. Fujimoto and M. Kameyama, Lipoprotein abnormalities in survivors of cerebral infarction with a special reference to apolipoproteins and triglyceride-rich lipoproteins, Atherosclerosis, 63:131 (1987).

14. S. Eisenberg, Preferential enrichment of large-sized very low density lipoprotein populations with transferred cholesteryl esters, J. Lipid Res, 26:487 (1985)

15. A.R. Tall, D. Sammett, G.M. Vita, R. Deckelbaum and T. Olevecrona, Lipoprotein lipase enhances the cholesteryl ester transfer protein mediated transfer of cholesteryl esters from high density lipoproteins to very-low-density lipoproteins, J. Biol. Chem., 259:9587 (1984).

16. D. Sammett and A.R. Tall, Mechanisms of enhancement of cholesteryl ester transfer protein activity by lipolysis, J. Biol. Chem. 260: 6687 (1985).

17. M. Ierides, N. Dousset, M. Potier, J. Manent, M. Carton and L. Douste-Blazy, Cholesteryl ester transfer protein. Size of the functional unit determined by radiation inactivation, FEBS Lett. 193;59(1985).

18. I.J. Goldberg, N.A. Le, H.N. Ginsberg, R.M. Krauss and F.T. Lindgren, Lipoprotein metabolism during acute inhibition of lipoprotein lipase in the Cynomolgus Monkey, J. Clin. Invest. 81:561 (1988).

19. G.J. Hopkins and P.J. Barter, Role of triglyceride rich lipoproteins and hepatic lipase in determining the particle size and composition of high density lipoproteins, J. Lipid Res. 27:1265 (1986).

20. R.J. Deckelbaum, S. Eisenberg, Y. Oschry, E. Granot, I. Sharon and G. Bengtsson-Olivecrona,, Conversion of human plasma high density lipoprotein-2 to high density lipoprotein-3. Roles of neutral lipid exchange and triglyceride lipases, J. Biol. Chem. 261:5201 (1986).

ROLE OF APOLIPOPROTEIN A IV IN THE INTERCONVERSION OF HDL SUBCLASSES

P. Gambert, L. Lagrost, A. Athias, S. Bastiras and C. Lallemant

Laboratoire de Biochimie des Lipoprotéines
Faculté de Médecine
Hopital du Bocage - CHRU Dijon - France

Apolipoprotein A-IV (apo A-IV) is a protein of molecular weight 46 000, present in lymph and plasma chylomicrons, very low density lipoproteins (VLDL) and high density lipoproteins (HDL) in various species including man (1, 2) and is presently considered as a real apolipoprotein. However, unlike most apolipoproteins, apo A-IV has a low affinity for plasma lipoproteins (3-5). It is a relatively hydrophilic protein with the weakest lipid affinity of any human apolipoprotein (6). It is marginally stable in aqueous solution compared to other apolipoproteins and its lipid binding properties are highly sensitive to the environment (5). It presents high self-association tendancies, mainly as dimer associations. All these special properties could be explained by the structural characteristics of the molecule which presents an association of a highly alpha-helical structure (35-54%) (5, 7) and a strongly hydrophobic domain.

Human apo A-IV is unusually distributed in plasma. After sequential ultracentrifugation 90% of apo A-IV is associated with the d$>$1.21 fraction and only 10% with the triglyceride rich lipoproteins in lipemic plasma ; in fasting plasma 98% of apo A-IV is found in the d$>$1.21 fraction (4). In contrast, after fractionation by gel chromatography 15 to 25% of apo A-IV is associated with HDL (8). This discrepancy suggests an artefactual redistribution of apo A-IV during ultracentrifugation.

It appears that apo A-IV dissociates quickly from the chylomicron surface after this particle enters the circulation (9) and is transfered first in the lipoprotein free fraction and then into the HDL (10). Thus, apo A-IV is distributed in three distinct metabolic pools, triglyceride-rich lipoproteins, HDL and lipoprotein free fraction, but only a small fraction of the bulk of apo A-IV in the lipoprotein-free fraction exchanges freely with apo A-IV from the two other pools (11).

Recent data suggest a role of apo A-IV in reverse cholesterol transport. It facilitates cellular cholesterol efflux (12), is a ligand responsible for binding and uptake of HDL by rat hepatocytes (13). It activates the enzyme lecithin cholesterol acyl transferase (LCAT). This activating effect depends on the saturation level of the hydrocarbon chain of the acyl donor and is even higher than that of apo A-I (14, 15). During LCAT activity, apo A-IV lipoprotein affinity increases (16-18) and the movements of apo A-IV in plasma closely parallel the formation and the interlipoprotein exchange of cholesteryl esters (17).

Both LCAT activity and apo A-IV seem to be related with the presence of small HDL in the plasma. Small HDL are abundant in LCAT deficiency states (19), are the best substrate for LCAT (20) and are suspected to have an important role in the reverse cholesterol transport (21). On the other hand apo A-IV has been localized in small particles (7.8 - 8.0 nm diameter) containing apo A-I and lipids (8) and in lipoproteins smaller than the typical HDL_3 (10). According to some authors apo A-IV would exist in a "minilipoprotein form", a lipid-protein complex comprising apo A-IV, apo A-I, an unknown 59 000 molecular weight protein and lipids (11).

All these obsvervations suggest a role of apo A-IV in HDL mediated lipid exchanges and transfers.

We undertook the study of the influence of apo A-IV on HDL alterations induced in vitro by protein preparations active on lipid transfer in the absence of lipoprotein lipase and LCAT activities. We (22, 23) and others (24, 25) have previously shown that in these incubation conditions a HDL conversion occurs, characterized mainly by a displacement of HDL particle diameters towards the large size range. The Lipid Transfer Protein (LTP) fraction was partially purified from fresh human serum according to the general procedure of Pattnaik et al (26). After 25-50% ammonium sulfate precipitation and ultracentrigugation at d = 1.21, the infranatant was applied on a Phenyl Sepharose column and the water eluted fraction was further purified on a Carboxymethyl Sepharose column.

Apo A-IV was extracted from human serum as previously described (27) and a further purification was achieved by using a preparative electrophoresis (28).

In a first set of experiments total serum lipoproteins were incubated with LTP in the presence or in the absence of apo A-IV. The gradient gel electrophoresis profile of controls kept at 4°C revealed a bimodal HDL distribution with subpopulations of average diameters 10.2 and 11.2 nm. Besides these two main fractions, a discrete population was detected in the small size range (7.8 nm). This profile was not modified by incubating lipoproteins in absence of LTP, whether apo A-IV was added or not. When lipoproteins were incubated in the presence of LTP and in the absence of apo A-IV, a major change in the particle size distribution was observed in the HDL range. There was a general displacement of HDL towards the large size range with a decrease of the 10.2 nm peak and a shift of the large size fraction from 11.2 to 13.0 nm. The minor 7.8 nm peak increased slightly. When lipoproteins were incubated wtih both LTP and apo A-IV, the same displacement of HDL particle diameters towards the large size range occured but, in addition, there appeared an abundant new small size lipoprotein fraction of mean apparent diameter 7.4 nm.

Immunodetection of apo A-IV after electrotransfer on nitrocellulose membranes showed two bands when apo A-IV was added to the reaction mixture. These two bands corresponded to the apparent molecular weight of monomeric (69 000) and dimeric (138 000) forms of apo A-IV in native conditions (5). Both were present in control and incubated mixtures, but the latter was much more abundant after incubation. Apo A-IV was not detected in lipoprotein fractions, neither in the main HDL fractions nor in the 7.4 nm fraction.

In a second set of experiments, HDL_3 subfraction was used instead of total lipoproteins. In these conditions, the addition of apo A-IV resulted in a marked enhancement of the HDL conversion process. A much greater proportion of the original HDL particles was converted into the small 7.4 nm particles. This effect of apo A-IV was dose dependant and specific. Other apolipoproteins (apo A-I, apo A-II, apo E) were unable to promote the formation of small HDL.

The results of the present study demonstrate a new property of apo A-IV. This protein was found to be necessary, in addition to LTP and serum lipoproteins, for the formation of a very small size lipoprotein fraction (mean diameter 7.4 nm). The general HDL size enlargement observed after incubation of serum lipoproteins with LTP, in the absence of apo A-IV, was very similar to that which has been previously described in incubated total serum (22, 23) and which, since then, has been associated to the action of a specific HDL-conversion factor (24, 25) or of plasma LTP (29). Introduction of apo A-IV into the reaction mixture did not noticeably modify these HDL alterations but was followed by the appearance of a lipoprotein fraction of mean diameter 7.4 nm which was not detected before incubation. The appearance of such a lipoprotein fraction has previously been observed in various conditions of incubation. Hopkins et al. (20) have found a 7.4 nm lipoprotein fraction after incubation of HDL_3 and VLDL (or Intralipid) with heat-inactivated lipoprotein-free plasma. Rye and Barter (25) reported the appearance of a lipoprotein of the same size after incubation of HDL_3 with a partially purified HDL-conversion factor in the presence of a LCAT inhibitor. Ellsworth et al. (29) observed a similar 7.4 nm lipoprotein after incubation of an HDL_3 subfraction and VLDL remnants with a lipid transfer complex deprived of LCAT activity. A common feature of these and our studies is the incubation of HDL, in the absence of LCAT activity, with a more or less purified plasma protein fraction able to promote a change in the HDL size distribution. The need for apo A-IV is reported here for the first time but the presence of this apolipoprotein in the reaction mixtures of the previous studies cannot be ruled out. In any case, in our experimental conditions apo A-IV was a necessary factor for the formation of the observed 7.4 nm lipoprotein fraction.

The way apo A-IV acts in the process leading to the formation of this very small lipoprotein remains to be elucidated. It is not via the known LCAT activating effect of apo A-IV since no increase in cholesterol ester content was detected in our incubation conditions. Immunoblotting experiments showed that apo A-IV was not a component of the newly formed lipoprotein fraction since it was not detected at the 7.4 nm level and was only present as mono- or dimeric free forms. Thus, we are led to suggest that apo A-IV may act as a cofactor by modifying the action of LTP, or another protein factor associated with it, in the complex process of lipid transfer and HDL conversion.

ACKNOWLEDGEMENTS

This work was supported by the Université de Bourgogne, the Conseil Régional de Bourgogne and the Groupement de Recherches sur l'Athérome et sa Prévention (G.R.A.P.). Stan Bastiras, a visiting investigator from University of Adelaide, South Australia, was the recipient of a fellowship from the Conseil Régional de Bourgogne. L. Lagrost was a recipient of a fellowship from the Groupement des Industries de Santé du Centre Est.

REFERENCES

1. J.B. Swaney, F. Braithwaite and H.A. Eder, Characterization of the apolipoproteins of rat plasma lipoproteins, Biochemistry, 16, 271-278 (1977).
2. M. Lefevre and P.S. Roheim, Metabolism of apolipoprotein A-IV, J. Lipid Res., 25, 1603-1610 (1984).
3. U. Beisiegel and G. Utermann, An apolipoprotein homologue of rat apolipoprotein A-IV in human plasma . Isolation and partial characterization, Eur. J. Biochem., 93, 601-608 (1979).
4. P.H.R. Green, R.M. Glickman, J.W. Riley and E. Quinet, Human apolipoprotein A-IV : intestinal origin and distribution in plasma, J. Clin. Invest., 65, 911-919 (1980).
5. R.B. Weinberg and M.S. Spector, The self-association of human apolipoprotein A-IV. Evidence for an In Vivo circulating dimeric form, J. Biol. Chem., 260, 14279-14286 (1985).
6. R.B. Weinberg, Differences in the hydrophobic properties of discrete alpha-helical domains of rat and human apolipoprotein A-IV, Biochim. Biophys. Acta, 918, 299-303 (1987).
7. E. Dvorin, N.W. Mantulin, M.F. Rohde, A.M.Jr. Gotto, M.J. Pownall and B.C. Sherrill, Conformational properties of human and rat apolipoprotein A-IV, J. Lipid Res., 26, 38-46 (1985).
8. C.L. Bisgaier, O.P. Sachdev, L. Megna and R.M. Glickman, Distribution of apolipoprqtein A-IV in human plasma, J. Lipid Res., 26, 11-25 (1985).
9. N.H. Fidge, The redistribution and metabolism of iodinated apolipoprotein A-IV in rats, Biochim. Biophys. Acta, 619, 129-141 (1980).
10. G. Ghiselli, S. Krishnan, Y. Beigel and A.M.Jr. Gotto, Plasma metabolism of apolipoprotein A-IV in humans, J. Lipid Res., 27, 813-827 (1986).
11. T. Ohta, N.H. Fidge and J. Nestel, Studies on the In Vivo and In Vitro distribution of apolipoprotein A-IV in human plasma and lymph, J. Clin. Invest., 76, 1252-1260 (1985).
12. Y. Stein, O. Stein, M. Lefevre and P.S. Roheim, The role of apolipoprotein A-IV in reverse cholesterol transport studied with cultured cells and liposomes derived from an ether analog of phosphatidylcholine, Biochim. Biophys. Acta, 878, 7-13 (1986).

13. E. Dvorin, N.L. Gorder, D.M. Benson and A.M.Jr. Gotto, Apolipoprotein A-IV : a determinant for binding and uptake of high density lipoproteins by rat hepatocytes, J. Biol. Chem., 261, 15714-15718 (1986).
14. A. Steinmetz and G. Utermann, Activation of Lecithin:Cholesterol Acyl Transferase by human apolipoprotein A-IV, J. Biol. Chem., 260, 2258-2264 (1985).
15. C.H. Chen and J.J. Albers, Activation of Lecithin:Cholesterol Acyl Transferase by apolipoprotein E-2, E-3, and A-IV isolated from human plasma, Biochim. Biophys. Acta, 836, 279-285 (1985).
16. J.G. Delamatre, G.A. Hoffmeier, A.G. Lacko and P. Roheim, Distribution of apolipoprotein A-IV between the lipoprotein and the lipoprotein-free fractions of rat plasma : possible role of Lecithin:Cholesterol Acyl Transferase, J. Lipid Res., 24, 1578-1585 (1983).
17. R.B. Weinberg and M.S. Spector, Lipoprotein affinity of human apolipoprotein A-IV during cholesterol esterification, Biochem. Biophys. Res. Commun., 135, 756-763 (1986).
18. C.L. Bisgaier, E.S. Lee, R.M. Glickman, A method to screen apolipoprotein polymorphisms in whole plasma : description of apolipoprotein A-IV variants in dyslipidemias and a reassessment of apolipoprotein A-I in Tangier disease, Biochim. Biophys. Acta., 918, 242-249 (1987).
19. J.A. Glomset, A.V. Nichols, K.R. Norum, W. King and T. Forte, Plasma lipoproteins in familial Lecithin:Cholesterol Acyl Transferase deficiency. Further studies of very low and low density lipoprotein abnormalities, J. Clin. Invest., 52, 1078-1092 (1973).
20. G. Hopkins, L.B.F. Chang and P.J. Barter, Role of lipid transfers in the formation of a subpopulation of small high density lipoproteins, J. Lipid Res., 26, 218-229 (1985).
21. P.J. Barter, G.J. Hopkins, O.V. Rajaram and K.A. Rye, Factors that induce changes in the particle size of high density lipoproteins in "Atherosclerosis VII", N.H. Fidge and P.J. Nestel, eds, Elsevier Science Publishers, Amsterdam (1986).
22. P. Gambert, C. Lallemant, A. Athias and P. Padieu, Alterations of HDL Cholesterol distribution induced by incubation of human serum, Biochim. Biophys. Acta, 713, 1-9 (1982).
23. P. Gambert, C. Lallemant and E. Louvrier, More on the enlargement of high density lipoproteins induced by incubation of human serum, Atherosclerosis, 53, 221-223 (1984).
24. K.A. Rye and P.J. Barter, Evidence of the existence of a high density lipoprotein transformation factor in pig and rabbit plasma, Biochim. Biophys. Acta, 795, 230-237 (1984).

25. K.A. Rye and P.J. Barter, Changes in the size and density of human high density lipoproteins promoted by a plasma conversion factor, Biochim. Biophys. Acta, 875, 429-438 (1986).

26. N.M. Pattnaik, A. Montes, L.B. Hughes and D.B. Zilversmit, Cholesteryl exchange protein in human plasma. Isolation and characterization, Biochim. Biophys. Acta, 530, 428-438 (1978).

27. R.B. Weinberg and A.M. Scanu, Isolation and characterization of human apolipoprotein A-IV from lipoprotein-depleted serum, J. Lipid Res., 24, 52-59 (1983).

28. S. Meunier, P. Gambert, J. Desgres and C. Lallemant, Preparative electrophoresis of human apolipoprotein E : an improved method, J. Lipid Res., 27, 1324-1327 (1986).

29. J.L. Ellsworth, M.L. Kashyap, R.L. Jackson and J.A.K. Harmony, Human plasma lipid transfer protein catalyzes the speciation of high density lipoproteins, Biochim. Biophys. Acta, 918, 260-266 (1987).

HDL RECEPTOR AND REVERSE CHOLESTEROL TRANSPORT IN ADIPOSE CELLS

Ronald **Barbaras**[o⊕], Pascal **Puchois**[+], Paul **Grimaldi**[o],
Ahmed **Barkia**[+], Jean-Charles **Fruchart**[+], and Gérard **Ailhaud**[o]

o Centre de Biochimie (CNRS), Parc Valrose, 06034 Nice
⊕ Laboratoires Fournier, 21121 Fontaine-les-Dijon
+ Institut Pasteur (SERLIA), 59019 Lille, France

Epidemiological studies have shown a relationship between low concentrations of high density lipoprotein (**HDL**) cholesterol and the incidence risk of cardiovascular diseases[1,2]. Recent pharmacological studies[3] have clearlv demonstrated the protective role of HDL in that respect. In the last decade much attention has been paid to the significance of apolipoprotein determination[4,5]. Thus, $ApoA_I$, the major protein of HDL, appears to be a better predictor of coronary artery disease (**CAD**) than HDL cholesterol[6,7]. Two main types of lipoprotein particles are identified within HDL : those which contain $ApoA_I$ and $ApoA_{II}$ (LpA_I:A_{II}) and those which contain $ApoA_I$ but not $ApoA_{II}$ (LpA_I). It has been shown quite recently that CAD subjects are characterized by a different distribution of $ApoA_I$ between LpA_I and LpA_I:A_{II}, the data supporting the view that LpA_I might represent the "antiatherogenic" fraction of HDL[8].

Taken together, the above observations raise the main question of the role of these various particles in cholesterol metabolism. In that respect ApoE-free HDL has been long known to bind to a variety of cells and to promote reverse cholesterol transport[9]. Among peripheral tissues, adipose tissue is recognized both in man and rodents for its ability to accumulate, store and, when needed, mobilize a large pool of unesterified cholesterol[10,11]. Thus adipose cells represent a cell type highly suitable to study reverse cholesterol transport. Unfortunately adipocytes isolated from adipose tissue loose their viability within a few hours, preventing the analysis of middle-term and long-term responses. During the last decade have been established in our laboratory preadipocyte cell lines from adipose tissue of genetically-obese ob/ob mice[12] and their lean counterpart[13]. The validity of these cellular models is supported by i) the biochemical properties of differentiated cells which are similar, if not indentical, to those of adipocytes isolated from fat tissue and ii) the ability of undifferentiated cells to differentiate <u>in vivo</u> within a few weeks into fully mature fat cells after their injection into athymic mice, under conditions where these cells could be unambiguously demonstrated not to be fat cells originating from the host animal[14]. Most of the studies, if not otherwise stated, were performed with Ob1771 cells, a subclone of Ob17 cells established from ob/ob mice[15,16].

CHARACTERIZATION OF LDL AND HDL BINDING AND CHOLESTEROL FLUX/EFFLUX IN OB1771 CELLS

The binding of human lipoproteins to mouse adipose cells and the study of their functional properties were made feasible owing to extensive homologies existing between rat, mouse and human $ApoA_I$ in various parts of the molecules[17,18]. In addition important homologies do exist between rat (and likely mouse) and human ApoB, including the consensus region of ApoB and ApoE which should be involved in the binding to the ApoB,E receptor[19-22]. The binding of ^{125}I-LDL was competitively inhibited by LDL > VLDL > total HDL ; human HDL and mouse LDL were equipotent in competition assays. Methylated LDL and ApoE-free HDL were not competitors. In contrast, the binding of ^{125}I-ApoE-free HDL was competitively inhibited by ApoE-free HDL > total HDL and that of ^{125}I-HDL_3 by mouse HDL. Thus mouse adipose cells possess distinct ApoB,E and ApoE-free HDL binding sites which can recognize heterologous or homologous lipoproteins. Further studies of ApoE-free HDL binding sites revealed that the binding of ^{125}I-HDL_3 was competitively inhibited by $ApoA_I$-containing (dimyristoylphosphatidylcholine) liposomes > mouse HDL > HDL_3. The binding of ^{125}I-$ApoA_I$ and ^{125}I-$ApoA_{II}$ liposomes was competitively inhibited by HDL_3, $ApoA_I$- and $ApoA_{II}$-containing liposomes, whereas dimyristoylphosphatidylcholine liposomes containing or not cholesterol did not interfere with the binding of labeled HDL_3 or apolipoprotein-containing liposomes. Thus differentiated Ob1771 cells have specific binding sites for ApoE-free HDL and $ApoA_I$ (or A_{II}) is the ligand for these binding sites.

During the course of these studies, it was observed that the endogenous cholesterol synthesis was nil[15] but the most striking observation was the fact that long-term exposure of adipose cells to LDL and HDL_3 did not affect the number of ApoB,E receptor sites and that of ApoE-free HDL receptor sites. In other words, the "buffering" capacity of adipose cells seems limited with respect to the regulation of cholesterol content. This lack of cholesterol homeostasis would explain the rather unique ability of adipose tissue <u>in vivo</u> to accumulate and mobilize a <u>large</u> pool of unesterified cholesterol[10,11]. Since differentiated Ob1771 cells were able to bind, internalize and degrade LDL[15], it appeared that adipose cells did not show an efficient cholesterol homeostasis <u>in vitro</u> and thus, as a first prediction, should accumulate cholesterol. The second prediction was that cholesterol-preloaded cells should mobilize cholesterol when exposed to appropriate lipoprotein particles. As shown by the results of Table 1, both predictions were fullfilled. It is of interest to note that, with respect to LDL, the apparent K_d value for binding and the EC_{50} value for cholesterol accumulation on one hand, and, with respect to HDL_3 or $ApoA_I$ -containing liposomes, the apparent K_d values for binding and the EC_{50} values for cholesterol mobilization (efflux) on the other hand, are within the same range of concentrations. These results suggest that specific binding to these distinct sites was a pre-requisite to cholesterol accumulation and subsequently to cholesterol mobilization. It is also of interest to note that cholesterol accumulation was taking place in the presence of LDL under the form of unesterified cholesterol only, in agreement with the fact that, at least in rat fat tissue, the majority (75-95%) of adipocyte cholesterol is unesterified and associated with central oil (triacylglycerol) droplet[10,11]

COMPARATIVE PROPERTIES OF LIPOPROTEIN PARTICLES CONTAINING APO AI AND/OR APO AII

Long-term exposure of adipose cells to LDL cholesterol as a function of LDL concentration led to an accumulation of cellular un-

Table 1. Binding Parameters of Lipoprotein Particles and Cholesterol Accumulation/Mobilization

LIGAND	BINDING (4°C)		CHOLESTEROL ACCUMULATION/MOBILIZATION (37°C)	
	Apparent Kd (μg/ml)	Bmax (sites/cell)	EC 50 (μg/ml)	[C] max (μg/ml)
LDL (accumulation)	31	15 000	30	50
HDL_3 (mobilization)	18	200 000	55	100
Apo AI/DMPC (mobilization)	30	180 000	42	90

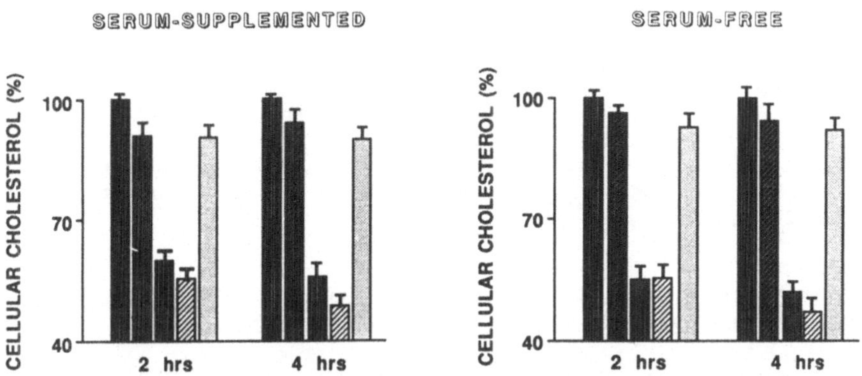

Fig.1. $ApoA_I$-dependent cholesterol efflux in cholesterol-preloaded Ob1771 cells

Differentiated Ob1771 cells were maintained in a medium containing either 10% lipoprotein-deficient fetal bovine serum or 1 g/l of serumalbumin (fatty acid poor) and exposed for 48 h at 37°C to 50 μg/ml of LDL. After careful washing, cells were maintained either in serum-supplemented or in serum-free medium as above, and exposed or not (defining time zero) to 100 μg/ml of HDL_3 or 100 μg/ml of $ApoA_I$ (or A_{II})-containing DMPC liposomes. Control experiments were performed in the presence of DMPC liposomes. 100% corresponds to the values obtained for cholesterol-preloaded cells in the absence of any subsequent addition, i.e. 33 and 36 μg of cholesterol per mg of cell protein in serum-supplemented and serum-free medium, respectively. Under these conditions, the cholesterol content of cholesterol-preloaded cells remained apparently unchanged, within experimental errors, between time zero and time 4 h. (■) control, (▨) DMPC, (▩) HDL_3, (▧) $ApoA_I$ DMPC, (▦) $ApoA_{II}$ DMPC.

esterified cholesterol. Promoting cholesterol accumulation by such a biological mean was satisfactory whereas cholesterol accumulation using different approaches led to irreproducible results. A net reverse cholesterol transport (cholesterol efflux) from cholesterol-preloaded cells was observed by exposure to HDL_3 or $ApoA_I$-containing liposomes, whereas $ApoA_{II}$-containing liposomes did not promote anv cholesterol efflux. As shown in Figure 1, similar results were obtained in serum-supplemented or serum-free medium, excluding a possible role of some serum components (including LCAT or CETP possibly present in fetal bovine serum) in mediating the effect of $ApoA_I$ and/or the lack of effect of $ApoA_{II}$. It should be pointed out that the rate of cholesterol efflux is rather high compared to that determined in other cell types when comparisons can be made. No cholesterol efflux could be observed in the absence of any addition or in the presence of DMPC liposomes. Thus it is clear that cholesterol efflux was dependent upon the presence of $ApoA_I$ and that $ApoA_{II}$ molecules inserted into liposomes showed no ability to promote cholesterol efflux, despite their ability to recognize cell-surface binding sites. In other words, $ApoA_{II}$ behaved as a typical "antagonist" of cholesterol efflux. Long-term exposure to LpA_I and $LpA_I{:}A_{II}$ particules isolated from the HDL fraction by immuno-affinity chromatography showed that LpA_I particles only were able to promote cholesterol efflux from cholesterol-preloaded, differentiated Ob1771 cells. Dose-response curves indicated that LpA_I particles were active within a physiological range of concentrations, whereas $LpA_I{:}A_{II}$ had no effect at all concentrations[23]. Recent and unpublished experiments have shown that LpA_I particles are indeed present in the human interstitial fluid surrounding peripheral cells, at concentrations compatible with a physiological role played by these particles. Last but not least, LpA_I and $LpA_I{:}A_{II}$ are equipotent in inhibiting the binding of $^{125}I\text{-}HDL_3$ to differentiated Ob1771 cells, excluding that the lack of effect of $LpA_I{:}A_{II}$ could be due to a lack of recognition by the HDL receptor sites.

RELATIONSHIP IN ADIPOSE CELLS BETWEEN THE PRESENCE OF RECEPTOR SITES FOR HDL AND THE PROMOTION OF REVERSE CHOLESTEROL TRANSPORT

In order to establish whether receptor sites for HDL were indeed required for the promotion of cholesterol efflux, use was made of Ob17 cells in which have been induced genetically defined alterations of the growth control mechanism bv transferring cloned oncogenes[24]. Ob17PY cells were obtained after transfer of the complete early region of polyoma virus whereas Ob17MT cells were obtained after transfer of a modified genome encoding only the middle T protein. The broad range of phenotypes thus generated has offered us unique opportunities to study reverse cholesterol transport in adipose cells as cells of the Ob17MT18 subclone had a 3-fold higher number of HDL receptor sites than cells of the parental Ob17 clone whereas growing Ob17PY cells did not have any detectable sites (see below).

As a pre-requisite to study the critical role, if any, of HDL receptor sites and to undertake their purification, conditions for their visualisation were searched and found using bivalent cross-linking reagent disuccimidyl suberate at 4°C in the presence of $ApoA_I$-containing liposomes and intact Ob1771 cells or derived crude membranes[25]. The existence of two specific cell-surface protein components of M_r 100,000 and 130,000 was demonstrated. It is possible that two different proteins of $M_r \sim 70{,}000$ and $\sim 100{,}000$ able to bind one molecule of $ApoA_I$ of M_r 28,000 are indeed present in adipose cells. Alternatively the possibility of either a single glycoprotein able to bind one molecule of $ApoA_I$ but having different degrees of glycosylation, or a single glycoprotein able to cross-link one or two molecules of $ApoA_I$, could be envisionned. The key observation in our study on the role of HDL receptor sites in the promotion of cholesterol efflux was that no binding of HDL_3, $ApoA_I$ or $ApoA_{II}$ was observed in growing Ob17PY cells and derived crude membranes, in contrast to growing or growth-arrested Ob1771 cells (see Table 1) or Ob17MT cells (not shown). After thymidine block, growth-arrested Ob17PY cells became able to recover in parallel binding activity for HDL_3, $ApoA_I$ and $ApoA_{II}$. The possibility that this recovery was an event common to various cell surface receptors is not very likely since ApoB,E and transferrin receptor sites were present in both growing and growth-arrested Ob17PY cells as well as in Ob1771 cells. The recovery of HDL receptor sites in growth-arrested Ob17PY cells was rapid (16 h) and prevented in actinomycin D- or cycloheximide-treated cells, adding further support to the conclusion that these sites are protein component(s). When experiments of reverse cholesterol transport were performed, the results showed that, after cholesterol accumulation taking place in the presence of LDL cholesterol, subsequent exposure to HDL_3 or $ApoA_I$ (but again not $ApoA_{II}$) under identical conditions to cholesterol efflux from Ob1771 cells and growth-arrested Ob17PY cells but not from growing Ob17PY cells. Thus it appears that the presence of high-affinity receptor sites for HDL in intact adipose cells is required for the promotion of reverse cholesterol transport.

CONCLUSIONS

In vivo adipose tissue seems to be, among peripheral tissues, a tissue lacking a tight control of cholesterol homeostasis. Cultured adipose cells *in vitro* appear to behave in a way similar to adipocytes *in vivo* and appear also to be cellular models quite suitable to study i) reverse cholesterol transport and the role of HDL receptor sites and ii) the functional properties of various and well-characterized particles of the human HDL fraction. Last but not least, the "all or none" modulation of HDL binding activity offers to our knowledge the unique opportunity to investigate the expression of functional HDL receptor protein and possibly that of the cognate mRNA.

ACKNOWLEDGEMENTS

The authors wish to thank Miss V. Boivin and Mrs. B. Barhanin for expert technical help, Dr. J. Barhanin for helpful advice in cross-link experiments and Mrs. G. Oillaux for expert secretarial assistance. This work was supported by the "Centre National de la Recherche Scientifique" (CNRS LP7300), by the "Fondation pour la Recherche Médicale Française" (Nice) and by Institut Pasteur (Lille).

REFERENCES

1. T. Gordon, W. P. Castelli, M. C. Hjortland, W. B. Kannel and T. R. Dawber, High density lipoprotein as a protective factor against coronary heart disease : the Frammingham Study, Am. J. Med. 62:707 (1977).
2. N. E. Miller, O. H. Forde, D. S. Thelle and O. D. Mjos, The Tromso Heart Study: high-density lipoprotein and coronary heart disease: a prospective case control study, Lancet 1:965 (1977).
3. M. Heikki Frick et al., Helsinki heart study: Primary-prevention trial with gemfibrozil in middle-aged men with dyslipidiemia: safety of treatment, changes in risk factors, and incidence of coronary heart disease, N. Engl. J. Med. 317:1237 (1987).

4. P. Alaupovic, The physiochemical and immunological heterogeneity of human plasma high density lipoproteins, in: "Clinical and Metabolic Aspects of High Density Lipoproteins", N. E. Miller and G. J. Miller, eds., Elsevier Science Publishers BV, Amsterdam (1985).
5. P. Alaupovic, W. J. McConathy, M. D. Curry and J. D. Fesmire, Characterization of dyslipoprtoeinemias by apolipoprotein profiles, in: "Lipoproteins and Coronary Atherosclerosis", G. Nodesa, C. Fragiacamo, R. Fumagolli and R. Paoletti, eds., Elsevier Biomedical Press, Amsterdam (1982).
6. P. Avogaro, G. Bittolo Bon, G. Cazzolato and G. B. Quinci, Are lipoproteins better discriminators than lipids for atherosclerosis?, Lancet 1:901 (1979).
7. G. Bon Bittolo, G. Cazzolato, M. Saccardi, G. M. Kostner and P. Avogaro, Total plasma apo E and high density lipoprotein apo E in survivors of myocardial infraction, Atherosclerosis 53:69 (1984).
8. P. Puchois, A. Kandoussi, P. Fievet, J. L. Fournier, M. Bertrand, E. Koren and J. C. Fruchart, Apolipoprotein A-I containing lipoproteins in coronary artery disease, Atherosclerosis 68:35 (1987).
9. A. R. Tall and D. M. Small, Body cholesterol removal: role of plasma high-density proteins, Adv. Lipid Res. 17:1 (1980).
10. R. K. Krause and A. D. Hartman, Adipose tissue and cholesterol metabolism, J. Lipid Res. 25:97 (1984).
11. A. Angel and B. Fong, Lipoprotein interactions and cholesterol metabolism in human fat cells, in: "The Adipocyte and Obesity: Cellular and Molecular Mechanisms", A. Angel, C. H. Hollenberg and D. A. K. Roncari, eds., Raven Press, New York (1983).
12. R. Négrel, P. Grimaldi and G. Ailhaud, Establishment of preadipocyte clonal line from epididymal fat pad ob/ob mouse that responds to insulin and to lipolytic hormones, Proc. Natl. Acad. Sci. USA 75:6054 (1978).
13. C. Forest, A. Doglio, L. Casteilla, D. Ricquier and G. Ailhaud, Expression of the mitochondrial uncoupling protein in brown adipocytes. Absence in brown preadipocytes and BFC-1 cells. Modulation by isoproterenol in adipocytes, Exp. Cell Res. 168:233 (1987).
14. D. Gaillard, P. Poli and R. Négrel, Characterization of ouabain-resistant mutants of the preadipocyte Ob17 clonal line. Adipose conversion in vitro and in vivo, Exp. Cell Res. 156:513 (1985).
15. R. Barbaras, P. Grimaldi, R. Négrel and G. Ailhaud, Binding of lipoproteins and regulation of cholesterol synthesis in cultured mouse adipose cells, Biochim. Biophys. Acta 845:492 (1985).
16. R. Barbaras, P. Grimaldi, R. Négrel and G. Ailhaud, Characterization of high-density lipoprotein binding and cholesterol efflux in cultured mouse adipose cells, Biochim. Biophys. Acta 888:143 (1986).
17. P. Forgez, M. J. Chapman, S. C. Rall Jr. and M. C. Camus, The lipid transport system in the mouse, Mus musculus: isolation and characterization of apolipoproteins B, A-I, A-II, and C-III, J. Lipid Res. 25:954 (1984).
18. C. G. Miller, T. D. Lee, R. C. LeBoeuf and J. E. Shively, Primary culture of apolipoprotein A-II from inbred mouse strain BALB/c, J. Lipid Res. 28:311 (1987).
19. T. L. Innerarity, E. J. Friedlander, S. C. Rall, K. H. Weisgraber and R. W. Mahley, The receptor-binding domain of human apolipoprotein E, J. Biol. Chem. 258:12341 (1983).

20. T. J. Knott, R. J. Pease, L. M. Powell, S. C. Wallis, S. C. Rall Jr., T. L. Innerarity, B. Blackhart, W. H. Taylor, Y. Marcel, R. W. Mahley, B. Levy-Wilson and J. Scott, Complete protein sequence and identification of structural domains of human apolipoprotein B, Nature 323:734 (1986).
21. C. Y. Yang, S. H. Chen, S. H. Gianturco, W. A. Bradley, J. T. Sparrow, M. Tanimura, W. H. Li, D. A. Sparrow, H. DeLoof, M. Rosseneu, F.S. Lee, Z. W. Gu, A. M. Gotto Jr. and L. Chan, Sequence, structure, receptor-binding domains and internal repeats of human apolipoprotein B-100, Nature 323:738 (1986).
22. A. J. Lusis, R. West, M. Mehrabian, M. A. Reuben, R. C. LeBoeuf, J. S. Kaptein, D. F. Johnson, V. N. Schumaker, M. P. Yuhasz, M. C. Schotz and J. Elovson, Cloning and expression of apolipoprotein B, the major protein of low and very low density lipoproteins, Proc. Natl. Acad. Sci. USA 82:4597 (1985).
23. R. Barbaras, P. Puchois, J. C. Fruchart and G. Ailhaud, Cholesterol efflux from cultured adipose cells is mediated by LpA_I particles and not by $LpA_I:A_{II}$ particles, Biochem. Biophys. Res. Commun. 142:63 (1987).
24. P. Grimaldi, D. Czerucka, M. Rassoulzadegan, F. Cuzin and G. Ailhaud, Ob17 cells transformed by middle-T-only gene of polyoma virus differentiate in vitro and in vivo into adipose cells, Proc. Natl. Acad. Sci. USA 81:5440 (1984).
25. R. Barbaras, P. Puchois, P. Grimaldi, A. Barkia, J. C. Fruchart and G. Ailhaud, Relationship in adipose cells between the presence of receptor sites for high density lipoproteins and the promotion of reverse cholesterol transport, Biochem. Biophys. Res. Commun. 149:545 (1987).

MOLECULAR ANALYSIS OF ATHEROGENIC LIPOPROTEIN PARTICLES IN ADEQUATELLY CONTROLLED TYPE I DIABETES MELLITUS

Catherine Fievet [1], Luc Méjean [2], Pierre Drouin [3], and
Jean Charles Fruchart [1]

[1] INSERM U 279 et SERLIA - Institut Pasteur - Lille (F)
[2] INSERM U308 - Nancy (F)
[3] Hôpital Jeanne d'Arc - Dammartin les Toul (F)

INTRODUCTION

The high frequency of accelerated atherosclerosis in diabetic patients has resulted in numerous studies of lipoprotein lipids and lipoprotein levels in which comparisons have been made with normal subjects [1,2,3] .However, such lipoprotein abnormalities , according to epidemiological surveys, are minor and so, do not explain a such development of atherosclerosis in these patients. It is now well appreciate that atherosgenesis is not only related to the concentrations of major lipids and lipoproteins in plasma, but that it may also be associated with subtle abnormalities of the structure of certain lipoproteins or with altered metabolism of apolipoproteins. Using new methodologies which permit a quantitative analysis of discrete lipoprotein particles different from their apolipoprotein composition and epitope expression [4,5], we discuss the clinical significance of this approach to determine the role played by subtle abnormalities in the composition of apolipoprotein B - containing particles in generating atherogenic risk in well-controlled type I insulin - dependent diabetic patients.

MATERIALS AND METHODS

Patients

Fifety - four male patients with type I diabetes were investigated.Another group of 54 healthy adults served as controls and were sex - , age - and body mass index - matched with the patients. The mean level of glycosylated hoemoglobin of the patients was 6.7 % and so they were considered in a good metobolic state.Another group of 54 healthy adults served as controls and were sex - , age - and body mass index - matched with the patients. The mean level of glycosylated hoemoglobin of the patients was 6.7 % and so they were considered in a good metabolic state. Table 1 summarizes the clinical characteristics of the subjects.

Table 1. Clinical characteristics of non diabetic and diabetic subjects

	Age (years)	Body mass Index [a]	Hb A 1C (%)
non diabetic subjects (n = 54)	42 ±15	24.5±34	not determined
diabetic patients (n = 54)	42±15	24.6±3.5	6.7±1.3

[a] body mass index = $\frac{\text{weight in Kg}}{(\text{height in mm})2}$

Assays

Blood was drawn following a 12 - h overnight fast. Serum samples, obtained by low - speed centrifugation, were kept at + 4°C and always used within 48 h.

Total triglycerides (TG), cholesterol (CH) and phospholipids (PL) were determined by enzymatic methods [6, 7, 8]. Using a selective precipitation test with concanavalin A, we could separate two subpopulations of lipoproteins, one which contained apolipoprotein - B (LpB) and one which contained no apolipoprotein - B (Lp nonB) [9]. CH and PL were determined in these two fractions, corresponding to LpB - CH or - PL, and to Lp nonB - CH or - PL. Apolipoprotein - B (Apo B) was evaluated in total sera and in three define particles Lp BL3, Lp BL5, Lp BL7. These particles shared specific epitopes to three well characterized monoclonal antibodies (BL3, BL5, BL7) produced by the Research Center Clin-Midy (Montpellier - France) [10, 11]. BL3 and BL7 recognize sequence epitopes whereas BL5 recognizes one conformational epitope (the mapping of these epitopes were made by Dr. Marcel and Milne - Montréal - Canada) (results not published). The measurements of Apo B were made using a non competitive enzyme - linked immunosorbent assay [11]. Statistical analysis was performed by standard methods. The significance of the mean differences between the groups was estimated by a two-tailed t-test. Spearman rank correlation coefficient r was used to study the degree of association between two variables

RESULTS AND DISCUSSION

Figure 1 represented the results obtained about lipids in total serum, lipids in lipoprotein subfractions and Apo B determinations in the both groups. There was no significant difference in total TG in patients and in controls. Total CH and PL concentrations were significantly reduced in patients when compared to controls and these falls occured only in Lp B particles. The content of CH and PL in Lp non B particles was slighty increased in diabetic group but not significantly. The ratio LpB-PL/LpB-CH was thus significantly lower in diabetics and so, abnormalities exist in lipidic composition of their Lp B particles. The serum Apo B profile of diabetic patients was expressed as percent of the levels in controls. A significant decrease in total Apo B was observed in patients, this drop being reliable to only Lp BL5 particles ; no statistical differences were obtained for Lp BL3 and Lp BL7 particles between the both groups.

No significant correlation was found between glycemic - diabetic control (as assessed by Hb A 1 C measurements) and any biochemical parameters we assessed. A correlation study between the variations in accessibility of the different epitopes recognized by the monoclonal antibodies BL3, BL5 and BL7, and the variations of the ratio Lp B-PL/Lp B-CH showed a significant correlation only with Lp BL5 particles in diabetic group. This is very interesting since that BL5 recognizes one conformational epitope and that lipids we measured are rather located at the surface of the lipoproteins particles. So, in diabetic patients, the excess atherosclerotic vascular disease may be related in part to an abnormal composition of the surface of Apo B - containing particles.

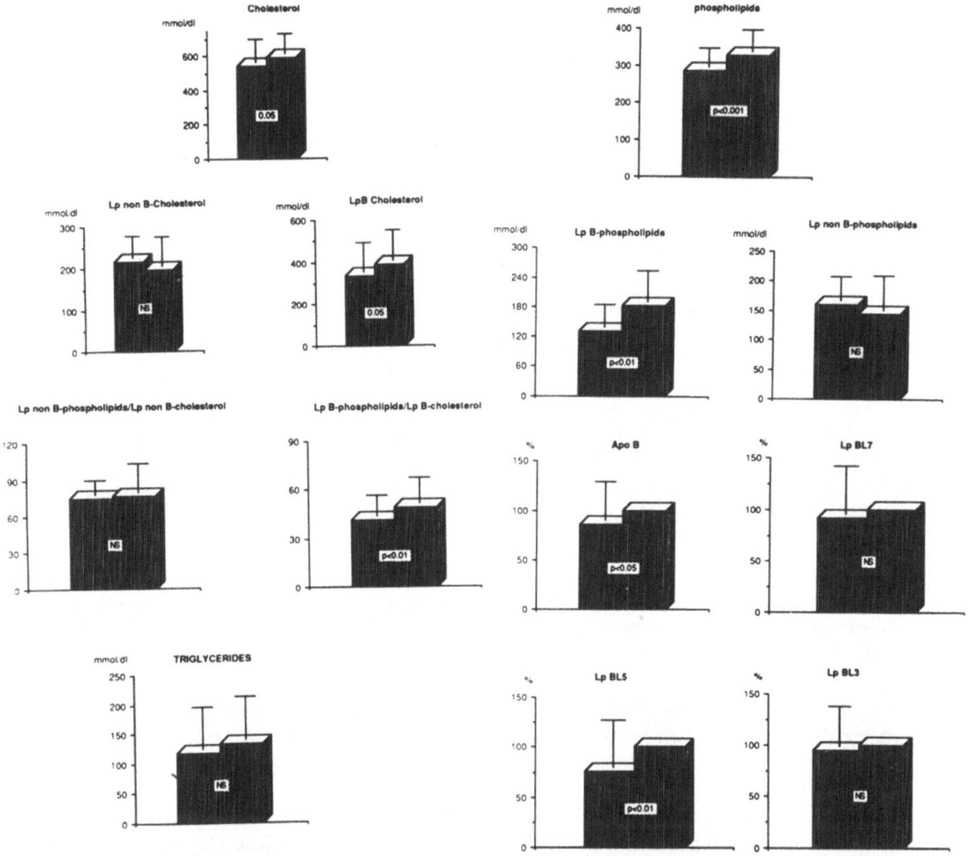

Fig.1. Mean (± SD) fasting sera total lipids, lipids in Lp B and Lp non B particles, Apo B and Lp BL3, Lp BL5 and LpBL7 particles in non diabetic controls ▨ and patients with type I diabetes ■

REFERENCES

1. J. A.Colwell, Atherosclerosis in diabetes mellitus, in : "Diabetes Annual/3", K.G.M.M. Alberti, L. P. Krall, eds., Elsevier Science Publishers BV, Amsterdam, New-York, Oxford, 325 (1987).

2. R. B. Goldberg, Lipid disorders in diabetes, Diabetes Care 4 : 561 (1981).

3. E. A. Nikkilä, Plasma lipid and lipoprotein abnormalities in diabetes, in : "Diabetes and Heart Disease", R.J. Jarret Ed, Elsevier Science Published BV, Amsterdam, New-York, Oxford, 133 (1984).

4. P. Alaupovic, Conceptual development of the classification systems of plasma lipoproteins, Protides Biol. Fluids Proc. Colloq. 19 : 9 (1972).

5. J. C. Fruchart, Polyclonal, oligoclonal and monoclonal antibody mapping of lipoprotein particles, in : Atherosclerosis VII, Fidge NH, J.P. Nestel, Eds, Elsevier Science Publishers BV, 287 (1986).

6. J. C. Fruchart, P. Duthilleul, A. Daunizeau, P. Comyn, Dosage du cholestérol total à l'aide d'une méthode enzymatique utilisant un monoréactif, Pharma. Biol. 24 : 227 (1980).

7. J. Ziegenhorn, K. Baril, R. Deeg,Improved kinetic method for automated determination of serum triglycerides, Clin. Chem. 26 : 973 (1980).

8. M. Takayama, S. Itoh, T. Nagasaki, I. Tanimizu, A new enzymatic method for determination of serum choline containing phospholipids, Clin. Chim. Acta 79 : 93 (1977).

9. W. Mac Conathy, P. Alaupovic, Studies on the interaction of concanavalin A with the major density classes of human plasma lipoproteins : evidence for the specific binfing of lipoprotein B in its associated and free forms, FEBS Lett 41 : 174 (1974).

10. S. Salmon, S. Goldstein, D. Pastier, I. Theron, M. Berthellier, M. Ayrault-Jarrier, M. Dubarry, R. Rebourcet, B. Pau, Monoclonal antibodies to low density lipoprotein used for the study of low- and very-low-density lipoprotein, in "Elisa" and immunoprecipitation technics. Biochim. Biophys. Res. Commun 125 : 704 (1984).

11. I. Luyeye, C. Fievet, J.C. Dupont, C. Durieux, N. Slimane, J.F. Lecocq, C. Demarquilly, J.C. Fruchart, Human apolipoprotein B. Evidence for its immunochemical heterogeneity using monoclonal antibodies and an immunoenzymometric assay, Clin. Biochem.in press (1987).

QUANTITATIVE ABNORMALITIES OF LIPOPROTEIN PARTICLES IN CHRONIC HEMODIALYSIS PATIENTS

H.J. Parra[1], C. Cachera[1,2,3], K. Equagoo[3], M. Dracon[3], J.C. Fruchart[1,2], and A. Tacquet[3]

1 - SERLIA et Inserm U 279, Institut Pasteur, Lille (France)
2 - UER de Pharmacie de Lille II, Lille (France)
3 - Laboratoire de Recherches Néphrologiques, Hôpital Calmette, Lille (France)

INTRODUCTION

Complex and characteristic abnormalities have been described in chronic renal failure (CRF) patients treated by hemodialysis. Hypertriglyceridemia is the most usual abnormality associated with an increased level of the triglyceride content of very low density lipoproteins (VLDL) which represent a mixture of particles differing in density, size, charge distribution and apolipoprotein (apo) composition[1]. In normolipidemic subjects these particles contain various combinations of apolipoproteins B, C and E with the major lipoprotein species LpB:C:E characterized by the presence of all three apolipoproteins[2]. Recent studies tend to establish whether CRF is characterized by a specific apolipoprotein profile[3,4] by determining the levels of the different apolipoproteins in this disease.

In a previous report, we have shown that patients with CRF exhibit a consistent increase in apo B- and apo C-III-containing lipoproteins[5]. Measurement of apo C-III in lipoproteins with and without apolipoprotein B were assessed in lipoprotein fractions prepared by selective precipitation with concanavalin A[6].

Recently we have developed a procedure for the direct quantification of the populations of apo B-containing lipoprotein particles by an enzyme linked differential antibody immunosorbent assay[7,8].

Having the tools for the analysis of these particles we have studied 76 hemodialysis patients and 76 matched controls.

In addition to usual lipids - cholesterol, triglycerides, phospholipids and HDL-cholesterol - we have measured apolipoproteins A-I, A-II, B, C-III, E and lipoprotein particles LpC-III:B and LpE:B.

METHODS

Patients

76 patients (25 women, 51 men) with chronic renal failure and receiving hemodialysis (CRF group) were investigated. The patients had a mean age of 48 $\pm$ 16 years. The renal diagnosis was chronic glomerulonephritis in 44 patients, polycystic kidney disease in 11 patients, chronic intertitial nephritis in 11 patients and other or unknow diagnosis in 10 patients. Another group of 76 healthy volunteers

selected in a center for preventive medicine (Institut Pasteur, Lille), sex and age matched with the patients was used as controls.

Biochemical analysis

Blood was drawn following a 12 H overnight fast at least two days after the previous dialysis.

All biochemical study was made on fresh sera stored à +4°C for no longer than three days. Preservatives were added in samples to prevent degradation of lipoproteins[3].

Total cholesterol (CH), triglycerides (TG) and phospholipids were determined by enzymatic methods[9,10,11]. HDL cholesterol (HDL-CH) was determined in the supernate after precipitation with sodium phosphotungstate-$MgCl_2$[12].

Apolipoprotein analyses were carried out by non competitive enzyme linked immunosorbant assays according to previously described procedures for apo A-I, A-II and B[13], apo C-III[14] and apo E[15].

Lipoprotein particles LpC-III:B and LpE:B were determined by an enzyme linked differential antibody immunosorbant assay. To directly determine associated apo B, we coated microtiter plates with antibody to apo C-III or apo E, blocked the non specific binding sites, and incubated plate with plasma, immobilizing the lipoproteins containing apo C-III or apo E. The unbound constituents of plasma were washed away, peroxidase-labeled antibody to apo B was added and the resulting colour measured[8].

Mean, standard deviation were calculated by conventional methods. Significance of differences was estimated by the student test. The level of significance was taken as $p < 0,05$.

RESULTS

The lipid profile (Table 1) of patients with CRF receiving hemodialysis is characterized by markedly increased triglycerides, decreased total cholesterol and HDL cholesterol while phospholipids are not modified in comparison with the control group.

Table 1. Serum Cholesterol (CH), Triglycerides (TG), Phospholipids (PL) and HDL-Cholesterol (HDL-CH) in Patients With Chronic Renal Failure Treated by Hemodialysis (CRF) and in Controls (Values Are Given as Mean ± SD)

		Control Group	CRF Group
Age	years	47.5 ± 16	48 ± 16
CH	g/l	2.31 ± 0.33	2.14 ± 0.55[a]
TG	g/l	1.01 ± 0.30	1.95 ± 1.05[c]
PL	g/l	2.54 ± 0.21	2.52 ± 0.53
HDL-CH	g/l	0.57 ± 0.19	0.44 ± 0.25[b]

Significance of difference between patients and controls
[a] $p < 0.05$
[b] $p < 0,001$
[c] $p < 0.000001$

Table 2. Serum Apolipoproteins (Apo) and Lipoprotein Particles LpC-III:B, LpE:B in Patients With Chronic Renal Failure Treated by Hemodialysis (CRF) and in Controls (Values Are Given as Mean ± SD)

		Control Group	CRF Group
Apo A-I	g/l	1.35 ± 0.38	1.08 ± 0.26[b]
Apo A-II	g/l	0.236 ± 0.041	0.179 ± 0.055[d]
Apo B	g/l	1.05 ± 0.21	1.11 ± 0.32
Apo C-III	g/l	0.055 ± 0.023	0.129 ± 0.07[d]
Apo E	g/l	0.043 ± 0.015	0.06 ± 0.03[a]
LpC-III:B	g/l	0.229 ± 0.084	0.389 ± 0.18[c]
LpE:B	g/l	0.296 ± 0.135	0.525 ± 0.283[c]

Significance of difference between patients and controls
[a] $p < 0.0001$
[b] $p < 0,00001$
[c] $p < 0.000001$
[d] $p < 0.000001$

Table 2 shows the apolipoprotein profile and lipoprotein particles LpB:C-III and LpB:E in the two groups.

Apolipoprotein A-I and specially apolipoprotein A-II are significantly lower in patients than in controls ; Apo B levels are not modified. In contrast, patients with CRF have a highly significant elevation of apo C-III and apo E. The most characteristic feature of the lipoprotein profile in CRF patients is the significant elevation of LpC-III:B ($p < 1.10^{-9}$) and LpE:B particles ($p < 1.10^{-8}$).

DISCUSSION

As reported in earlier studies[5,16], hypertriglyceridemia is of frequent occurence in patients with chronic renal failure (CRF) treated by hemodialysis. Metabolic disorder in lipids and lipoproteins represents an integral part of pathologic changes in this disease.

The greatly elevated levels of apo C-III and decreased levels of apo A-I and apo A-II in CRF patients were in agreement with our previous study[5] and with the recent results of Attman et al[4]. However, their findings of reduced level of apo E could not be confirmed here ; this discrepancy can be explained by the dialysis treatment.

In this study, we have made use of newer methods for the quantitation of LpC-III:B and LpE:B lipoprotein particles. Our results show a significantly increase of the apo B-containing particles in CRF patients, whereas levels of apo B have been found to be similar to those of controls. This caracteristic profile, reflecting the degree and extent of accumulating remnant particles appears to be a potential predictor of coronary artery disease in patients with chronic renal failure. It is now evident that these subjects commonly have premature and accelerated coronary disease. More adequate control of plasma lipoprotein levels might be desirable for these patients. As a guide to the efficacy of lipoprotein regulation, quantitation of lipoprotein particles can be of great help.

REFERENCES

1. P. Puchois, P. Alaupovic, and J. C. Fruchart, Mise au point sur les classifications des lipoprotéines plasmatiques, Ann. Biol. Clin. 43:831 (1985).

2. P. Alaupovic, C. S. Wang, W. J. McConathy, D. Weiser, and D. Downs, Lipolytic degradation of human very low density lipoproteins by human milk lipoprotein lipase : the identification of lipoprotein B as the main lipoprotein degradation product, Arch. Biochem. Biophys. 244:226 (1986).

3. P. O. Attman, P. Alaupovic, C. Knight, C.S. Wang, and H. Bass, Apolipoprotein pattern in patients with chronic renal failure, Contr. Nephrol. 41:328 (1984).

4. P. O. Attman, P. Alaupovic, and A. Gustafson, Serum apolipoprotein profile of patients with chronic renal failure, Kidney Int. 32:368 (1987).

5. D. Parsy, M. Dracon, C. Cachera, H. J. Parra, G. Vanhoutte, A. Tacquet, and J. C. Fruchart, Lipoprotein abnormalities in chronic haemodialysis patients, Nephrology Dialysis and Transplantation (in press), 1988.

6. M. J. McConathy, and P. Alaupovic, Studies on the interaction of concanavalin A with major density classes of human plasma lipoproteins. Evidence of the specific binding of lipoprotein B in its associated and free forms, FEBS Let. 41:174 (1974).

7. E. Koren, P. Puchois, P. Alaupovic, J. Fesmire, A. Kandoussi, and J. C. Fruchart, Quantification of two different types of apolipoprotein A-I containing lipoprotein particles in plasma by enzyme-linked differential-antibody immunosorbent assay, Clin. Chem. 33:38 (1987).

8. A. Kandoussi, P. Puchois, D. Parsy, and J. C. Fruchart, Quantitative determination of different apo B containing lipoproteins by an enzyme linked immunosorbent assay : lipoprotein containing apo B and C-III, apo B and E. J. Lipid Res., Submitted for publication.

9. J. C. Fruchart, P. Duthilleul, A. Daunizeau, and P. Comyn, Dosage du cholestérol total à l'aide d'une méthode enzymatique utilisant un monoréactif, Pharm. Biol. 24:227 (1980).

10. J. Ziegenhorn, K. Baril, and R. Deeg, Improved kinetic method for automated determination of serum triglycerides, Clin. Chem. 26:973 (1980).

11. M. Takayama, S. Itoh, and T. Nagasaki, A new enzymatic method for determination of serum cholin containing phospholipids, Clin. Chim. Acta 79:93 (1977).

12. T. H. Grove, Effet of nagent pH on determination of high density lipoprotein cholesterol by precipitation with sodium phosphotungstate-magnesium, Clin. Chem. 25:560 (1979).

13. J. C. Fruchart, C. Fievet, and P. Puchois, Apo-lipoproteins, in: "Methods of Enzymatic Analysis" H. U. Bergmeyer, ed., Verlag Chemie (1985).

14. D. Parsy, V. Clavey, C. Fievet, I. Kora, P. Duriez, and J. C. Fruchart, Quantification of apolipoprotein C-III in serum by a non competitive immunoenzymometric assay, Clin. Chem. 31:1632 (1985).

15. M. Koffigan, I. Kora, V. Clavey, J. M. Bard, J. Chapman, and J. C. Fruchart, Quantification of human apolipoprotein E in plasma and lipoprotein subfractions by a non-competitive enzyme immunoassay, Clin. Chim. Acta 163:245 (1987).

16. R. Hahn, K. Oette, H. Mondorf, K. Finke, H.G. Sieberth, Analysis of cardiovascular risk factors in chronic hemodialysis patients with special attention to the hyperlipoproteinemias, Atherosclerosis 48:279 (1983).

LIPOPROTEIN PARTICLES IN HYPERTRIGLYCERIDEMIC STATES

P. Alaupovic, M. Tavella, J.M. Bard, C.S. Wang, P.O. Attman, E. Koren, C. Corder, C. Knight-Gibson and D. Downs

Lipoprotein and Atherosclerosis Research Program
Oklahoma Medical Research Foundation
Oklahoma City, Oklahoma, USA 73104

INTRODUCTION

The compositional and metabolic heterogeneity of operationally defined plasma lipoprotein classes (1-3) has necessitated the introduction of a classification system that utilizes apolipoproteins as specific markers for identifying and distinguishing discrete lipoprotein particles (1,4). In this system, lipoprotein particles are characterized and defined by their apolipoprotein composition (1,4). Studies on the quantification and distribution of apolipoproteins (4,5) have shown that apolipoprotein (Apo)B and ApoA (A-I + A-II) form two major groups of plasma lipoproteins. These two major lipoprotein groups may be separated (6) by immunoprecipitation or immunoaffinity chromatography of whole plasma (6). The use of these procedures results in the isolation of ApoA-containing lipoproteins free of ApoB. The fractionation of ApoA-containing lipoproteins into two major discrete lipoprotein particles LP-A-I and LP-A-I:A-II by immunoaffinity chromatography on an immunosorber with polyclonal antibodies to ApoA-II has already been described by Cheung and Albers (7). To identify discrete lipoprotein particles of the ApoB group of lipoproteins, we have developed a procedure based on sequential immunoprecipitation of ApoB-containing lipoproteins with polyclonal antisera to apolipoproteins B, E, C-III and, if necessary, C-II and C-I (6,8). The fractionation of very low density (VLDL, d < 1.006 g/ml) and two subfractions of low density (LDL_1, d = 1.006-1.019 g/ml; LDL_2, d = 1.019-1.063 g/ml lipoproteins from normolipidemic subjects by sequential immunoprecipitation showed that each of these density classes consists of a mixture of distinct lipoprotein particles including cholesterol ester-rich LP-B and triglyceride- rich LP-B:C-I:C-II:C-III:E (LP-B:C:E) and LP-B:C-I:C-II:C-III (LP-B:C) particles (8). The LP-B:C:E family of particles in some normolipidemic and hypercholesterolemic subjects also contained varying amounts of LP-B:E particles. In addition, small amounts of LP-B:C-I:E, LP-B:C-II, LP-C-III and LP-E particles were detected in some but not all subjects or density classes. Each of the major ApoB-containing families of particles was shown to represent a polydisperse system of particles heterogeneous with respect to size, hydrated density, and lipid/protein ratio, but homogeneous with respect to the qualitative apolipoprotein composition.

Measurement of lipoprotein particle profiles of normal, hypercholesterolemic and hypertriglyceridemic subjects has shown that each consists of characteristic concentration patterns of qualitatively identical LP-B, LP-B:C:E and LP-B:C particles (8). Hypercholesterolemic states are characterized by elevated concentrations of cholesterol ester-rich LP-B particles, and the hypertriglyceridemic states by increased levels of triglyceride-rich LP-B:C:E and LP-B:C particles when compared with the corresponding profile of normolipidemic subjects.

Results of our recent study (9) on the possible pathogenetic mechanisms responsible for the impaired metabolism of triglyceride-rich lipoproteins in Tangier disease have shown that, in comparison to normal controls, Tangier patients had significantly decreased levels of plasma post-heparin lipoprotein lipase activity, increased levels of hepatic triglyceride lipase activity, and a lower reactivity of VLDL with human milk lipoprotein lipase. It has also been established that the protein composition of VLDL particles is characterized by a significant increase in the percent content of ApoA-II, suggesting a possible association between the abnormal apolipoprotein composition and an abnormally low reactivity of triglyceride-rich lipoproteins. To establish the lipoprotein form of VLDL-ApoA-II and its possible association with low reactivity, the VLDL from Tangier patients were fractionated on an immunosorber with monoclonal antibodies to ApoA-II ("Pan ApoA-II" monoclonal antibody) (10). The retained fraction was identified as a triglyceride-rich LP-A-II:B: C-I:C-II:C-III:D:E ("LP-A-II:B" complex) particle, while the unretained fraction consisted of a mixture of LP-B, LP-B:C:E and LP-B:C particles. In Tangier patients the LP-A-II:B complex accounted for 80-90% of the total ApoB content of VLDL. Moreover, it was shown by the measurement of the pseudo- first-order rate constant (k_1) of triglyceride hydrolysis that LP-A-II:B complex was responsible for the low reactivity of Tangier VLDL with human milk lipoprotein lipase.

The purpose of this study was to determine the profiles of discrete ApoB-containing lipoprotein particles in some selected primary and secondary hypertriglyceridemic states including the possible occurrence of LP-A-II:B complex.

METHODS

Study Subjects

Normolipidemic male and female subjects, 30-60 years of age, consisted of employees of the Oklahoma Medical Research Foundation. They were classified as normolipidemics according to the recommended criteria of the Lipid Research Clinics of the National Institutes of Health (11). All subjects were healthy, asymptomatic Caucasians with no history of familial hyperlipoproteinemia or diabetes. Their weight/height index was between 0.9 and 1.1, and their alcohol consumption did not exceed 50 g/week. Patients with primary hyperlipoproteinemias, phenotype IV and phenotype V, were selected from those attending the Lipid Research Clinic of the Oklahoma Medical Research Foundation. They were classified according to the criteria of the Lipid Research Clinics of the National Institutes of Health (11). None of the patients were on lipid-lowering drugs or diets expected to affect plasma lipids for at least four weeks prior to the blood collection. Patients with chronic renal failure were previously described (12). Twelve patients with type II diabetes mellitus were randomly selected from among Oklahoma Indian diabetics attending the Indian Hospital in Claremore, Oklahoma (13). None of the patients were on lipid-lowering drugs or diets expected to affect plasma lipids for at least four weeks prior to the blood collection. However, patients with type II diabetes mellitus were treated either with hypoglycemic oral agents or with low doses of insulin. All normolipidemic and hyperlipoproteinemic subjects provided informed consent. Blood samples were drawn from all subjects after an overnight fast of 12 hours. Blood samples were drawn into tubes that contained EDTA, and the plasma samples were collected by low-speed centrifugation.

Isolation of Lipoprotein Particles

The VLDL, LDL_1 and LDL_2 were isolated by sequential ultracentrifugation as previously described (14). Major ApoB-containing lipoprotein particles were isolated by sequential immunoprecipitation according to a previously described procedure (8,15). Concentrations of the major ApoB-containing lipoprotein particles were expressed in terms of their apolipoprotein concentrations (mg/dl).

Monoclonal antibodies to human plasma ApoA-II were prepared by a previously described procedure (16). The coupling of "pan" ApoA-II monoclonal antibody to Affi-gel 10 was carried out as previously described (17). The 50 ml glass column (1.2 x 49 cm, K-50 Pharmacia, Uppsala, Sweden) was first packed with 25 ml of Sephadex G-25 followed by 10 ml of ApoA-II antibody-coupled Affi-gel 10 and another protective layer of 5 ml of Sephadex G-25. The VLDL samples were applied to the column and incubated in the Affi-gel layer for 12 hours at room temperature. The unretained fraction, free of ApoA-II, was eluted with 0.05 M Tris-HCl buffer, pH 7.4, containing 0.5 M NaCl and 1.5 mg/ml EDTA. After the absorbance at 280 nm returned to baseline, the column was washed with 150 ml of the above described buffer, and the retained fraction, containing all LP-A-II:B complex, was eluted with 5 ml of 3 M NaSCN, pH 7.4. The bottom layer of Sephadex G-25 allowed an immediate separation of the retained lipoprotein fraction from the dissociating agent, which resulted in two distinct peaks at 280 nm, the first one consisting of lipoprotein and the second on of NaSCN. Column chromatography was carried out at room temperature and was monitored by measuring absorption at 280 nm.

Determination of k_1-values

The preparation of purified human milk lipoprotein lipase and the measurement of k_1-values of retained and unretained VLDL fractions separated by immunoaffinity chromatography on an immunosorber with "pan" monoclonal antibody to ApoA-II were previously described (9).

Lipid and apolipoprotein analyses

Neutral lipids (cholesterol esters, free cholesterol and triglycerides) were quantified by the gas-liquid chromatographic procedure of Kuksis et al. (18). The quantification of apolipoproteins A-II, B, C-II, C-II, C-III, D and E was carried out by electroimmunoassays developed in this laboratory (8,9,12,15).

RESULTS

All four hyperlipoproteinemic states were characterized by significantly higher levels of plasma triglycerides than normals (Table 1). Only patients with type IV and type V disease had higher plasma cholesterol than normal controls. The most characteristic feature of apolipoprotein profiles of all four hyperlipoproteinemias was a significantly lower concentration of ApoA-I and significantly higher levels of ApoC-III in comparison with normolipidemic subjects. Comparatively higher levels of ApoB were only observed in patients with type IV hyperlipoproteinemia and in patients with chronic renal failure. Both the patients with type IV and type V disease had a significantly higher concentration of ApoE than normals. Patients with primary hyperlipoproteinemias had significantly higher levels of plasma triglycerides than patients with secondary hyperlipoproteinemias and, as an expected consequence, higher concentrations of apolipoproteins C-III and E (Table 1). One could predict on the basis of lipid and apolipoprotein profiles that the former group of hypertriglyceridemics would have higher concentrations of

Table 1. Plasma Lipids and Apolipoproteins of Patients with Primary and Secondary Hyperlipoproteinemias

Dyslipoproteinemia	TC	TG	Apolipoproteins			
			A-I	B	C-III	E
			mg/dl			
Hyperlipoproteinemia Type IV (n = 5)	267[a,c] (37)	847[d] (410)	129[b] (22)	144[c] (30)	33.0[d] (8.1)	22.0[d] (2.6)
Hyperlipoproteinemia Type V (n = 6)	305[d] (77)	1407[d] (487)	124[b] (13)	113 (16)	41.3[d] (11.3)	21.2c (8.7)
Chronic Renal Failure (n = 10)	232 (73)	201[c] (214)	105[d] (38)	123 (51)	19.3[d] (7.7)	10.9 (4.9)
Diabetis Mellitus Type II (n = 12)	194 (43)	262[c] (210)	107[c] (28)	126[b] (32)	16.0[c] (9.0)	13.0 (4.0)
Normals (n = 19)	199 (31)	69 (26)	162 (32)	101 (22)	9.6 (3.0)	10.0 (5.0)

[a]Mean (SD)

[b]Significantly different from normals at $p < 0.05$.

[c]Significantly different from normals at $p < 0.01$.

[d]Significantly idfferent from normals at $p < 0.001$.

triglyceride-rich LP-B:C and LP-B:C:E particles and lower concentrations of cholesterol ester-rich LP-B particles than the latter group; and that both groups would be characterized by lower levels of ApoA-containing lipoprotein particles than normolipidemic subjects. A study on the distribution of ApoC-III and ApoE between high density lipoproteins (HDL, d = 1.063-1.21 g/ml) and VLDL + LDL showed that all four hypertriglyceridemic groups had higher percentages of both apolipoproteins in very low and low density lipoproteins than normal subjects. There were some characteristic differences between patient groups. In type V disease 90% of total ApoC-III content was present in VLDL, whereas in type IV disease, type II diabetes and chronic renal failure these percentages were 64, 43 and 23, respectively. The percent distribution of ApoE was similar to that of ApoC-III. For comparison, normal subjects had only 10% of total ApoC-III and 11% of total ApoE in VLDL. In contrast to patients with primary hyperlipoproteinemias, patients with chronic renal failure and patients with type II diabetes had higher percentages of ApoC-III and ApoE in LDL_1 and LDL_2 suggesting also a different distribution of triglyceride-rich lipoproteins among VLDL and LDL. To some extent, all these predictions were substantiated by determination of lipoprotein particle profiles.

Each of the four hypertriglyceridemic groups of patients had a characteristic profile of ApoB-containing particles (Tables 2 and 3). In general, patients with primary hypertriglyceridemias had higher total concentrations of LP-B:C and LP-B:C:E particles and lower total concentrations of LP-B particles than patients with secondary hypertriglyceridemias. However, patients with type IV and type V disease and patients with type II diabetes

Table 2. Concentrations of Major ApoB-Containing Lipoprotein Particles in Density Classes of Patients with Primary Hyperlipoproteinemias

Lipoprotein Particles	VLDL	LDL_1	LDL_2
		mg/dl	
Hypertriglyceridemia Type IV (n = 5)			
LP-B	0.25[a] (0.1-1.0)	4.4 (0.5-9.5)	31.7 (7.6-45.3)
LP-B:C:E	31.0 (13.6-37.9)	4.7 (2.8-6.7)	13.9 (1.3-25.2)
LP-B:C	43.9 (22.2-90.8)	5.9 (1.6-15.9)	9.1 (1.6-16.0)
Hypertriglyceridemia Type V (n = 6)			
LP-B	9.3 (1.0-17.4)	5.0 (1.6-9.4)	24.0 (15.3-35.3)
LP-B:C:E	22.3 (14.2-30.3)	2.0 (1.1-3.3)	7.2 (3.6-11.6)
LP-B:C	28.2 (14.5-53.6)	1.2 (0.2-2.3)	3.1 (1.1-5.2)

[a]Mean (range of values)

had 70-80% of LP-B:C and LP-B:C:E particles in VLDL, whereas patients with chronic renal disease had only 25-30% of these particles in VLDL. This finding suggests that the accumulation of triglyceride-rich particles in chronic renal failure may be due to a different underlying metabolic defect than in the other three hypertriglyceridemic states.

The characteristic feature of the lipoprotein particle profile in patients with type IV disease is that the concentration of LP-B:C particles in VLDL is higher than that of LP-B:C:E particles. The levels of LP-B are proportionately higher in LDL_2 than VLDL or LDL_1 (Table 2). In contrast, patients with type V hyperlipoproteinemia have almost equal concentrations of LP-B:C and LP-B:C:E particles in VLDL. Whereas their total concentration of LP-B particles does not differ from that of type IV patients, they have a relatively high proportion of these particles in VLDL (24%) and low in LDL_2 (62%).

Patients with chronic renal failure a have slightly higher concentration of LP-B:C:E particles than LP-B:C particles in VLDL. However, as mentioned earlier, both of these lipoproteins are present in greater proportions in LDL_1 and LDL_2 than in VLDL. They also have a relatively high concentration of LP-B particles in VLDL (Table 3). Diabetic patients have higher levels of LP-B:C than LP-B:C:E particles in VLDL. Another characteristic feature of their lipoprotein particle profile is the high concentration of LP-B and low concentrations of LP-B:C and LP-B:C:E particles in LDL_2.

Table 3. Concentrations of Major ApoB-Containing Lipoprotein Particles in Density Classes of Patients with Secondary Hyperlipoproteinemias

Lipoprotein Particles	VLDL	LDL_1	LDL_2
		mg/dl	
Chronic Renal Failure (n=10)			
LP-B	4.4[a] (0.1-17.7)	8.8 (1.4-26.9)	51.1 (26.0-88.0)
LP-B:C:E	9.4 (1.7-26.8)	5.4 (1.4-17.0)	12.9 (0.1-51.7)
LP-B:C	7.7 (0.2-23.0)	6.4 (0.1-23.2)	15.6 (1.8-39.6)
Diabetes Mellitus, type II (n=12)			
LP-B	1.9 (0.1-7.8)	5.9 (2.4-13.1)	57.6 (30-106)
LP-B:C:E	9.1 (3.6-23)	1.3 (0.1-5.0)	2.8 (0.1-9.3)
LP-B:C	11.3 (0.1-79)	0.9 (0.1-2.6)	2.3 (0.1-11.6)
Normals (n=19)			
LP-B	1.3 (0.2-3.1)	2.9 (0.1-8.0)	48.0 (29-79)
LP-B:C:E	3.4 (0.8-7.3)	2.1 (0.5-7.5)	7.8 (0.9-23.8)
LP-B:C	1.7 (0.1-5.0)	0.7 (0.1-1.4)	5.0 (0.6-16.1)

[a]Mean (range of values)

Although the concentrations of LP-A-I and LP-A-I:A-II particles were not determined in this study, it was established that the ApoA-I/ApoA-II ratios in all hypertriglyceridemic groups were significantly lower (1.8-2.05) than in normals (2.4), suggesting a possible reduction in apparently antiatherogenic LP-A-I particles (19).

Results of preliminary studies on the presence in VLDL of a newly identified LP-A-II:B complex, characteristic of VLDL in Tangier disease (10), have shown that this complex triglyceride-rich lipoprotein also occurs in VLDL of normal and hyperlipoproteinemic subjects (Table 4). Patients with type II diabetes have lower and patients with type V disease or chronic renal failure have higher percentage of LP-A-II:B complex than normals. The lipid composition of LP-A-II:B complex isolated from VLDL of patients with type V hyperlipoproteinemia was characterized by a high relative content of triglyceride (55-60%) and lower contents of cholesterol esters (10-11%),

Table 4. Percent Distribution of LP-A-II:B:C:D:E (LP-A-II:B Complex) Particles in VLDL from Patients with Primary and Secondary Hypertriglyceridemia[a]

Dyslipoproteinemia	ApoA-II Column	
	Retained Fraction (LP-A-II:B Complex)	Unretained Fraction (Free of LP-A-II:B Complex)
	%	
Hyperlipoproteinemia, type V	45.0	55.0
Chronic Renal Failure	44.0	56.0
Diabetes Mellitus, type II	27.0	73.0
Normals	36.0	64.0

[a]Distribution of lipoprotein particles in retained and unretained fractions is based on the content of ApoB.

free cholesterol (5-6%) and phospholipids (14-16%). The lipid composition of LP-B:C:E and LP-B:C particles isolated from VLDL of type V patients was very similar, if not identical, to that of LP-A-II:B complex. Both the retained (LP-A-II:B complex) and unretained (LP-B:C:E + LP-B:C) fractions from anti-ApoA-II column were characterized by measurement of k_1-values. The mean k_1-value of LP-A-II:B complex from four separate VLDL samples was found to be highly significantly lower than the k_1-values of corresponding LP-B:C:E + LP-B:C particles (0.0148 ± 0.002 vs 0.0300 ± 0.002 min^{-1}; $p < 0.0001$). This finding shows clearly that discrete lipoprotein particles isolated from the same density class have not only different apolipoprotein composition but also different biological properties.

CONCLUSIONS

1. Patients with primary (type IV and type V hyperlipoproteinemias) and secondary (type II diabetes and chronic renal failure) hypertriglyceridemias have distinct concentration profiles of LP-B, LP-B:C:E and LP-B:C particles. Patients with primary hypertriglyceridemias have higher concentreations of triglyceride-rich LP-B:C and LP-B:C:E particles, while patients with secondary hypertriglyceridemias have higher levels of cholesterol ester-rich LP-B particles.

2. Fractionation of triglyceride-rich lipoproteins on an immunosorber with antibodies to ApoA-II revealed that ApoB-containing lipoprotein particles consist of two subpopulations, one of which was identified as LP-B:C:E and the other as LP-A-II:B:C:D:E (LP-A-II:B complex).

3. The percent content of LP-A-II:B complex in VLDL was highest in patients with chronic renal disease and lowest in patients with type II diabetes.

4. Chemically distinct triglyceride-rich lipoprotein families (LP-A-II:B complex and LP-B:C:E particles) also have distinct metabolic properties.

5. The possible significance of discrete lipoprotein particles as markers of abnormal synthetic or catabolic processes as well as their relative atherogenic potential remain to be explored and determined in future studies.

ACKNOWLEDGEMENTS

We wish to thank Mr. J. Fesmire and Mr. R. Whitmer for their excellent technical assistance and Ms. J. Pilcher for typing the manuscript.

This study was supported in part by Grant HL23181 from the U.S. Public Health Service and by the resources of the Oklahoma Medical Research Foundation.

REFERENCES

1. P. Alaupovic, The role of apolipoproteins in lipid transport processes, La Ricerca Clin. Lab. 12:3 (1982).

2. C. J. Packard and J. Shepherd, Models and mechanisms in very low density lipoprotein metabolism, Eur. J. Clin. Invest. 15:51 (1985).

3. K. Lippel, S. Gianturco, A. Fogelman, P. Nestel, S. M. Grundy, W. Fisher, A. Chait, J. Albers, and P.S. Roheim, Lipoprotein heterogeneity workshop, Arteriosclcerosis 7:315 (1987).

4. P. Alaupovic, The concepts, classification systems, and nomenclature of human plasma lipoproteins, in: "CRC Handbook of Electrophoresis. Lipoproteins: basic principles and concepts", L. A. Lewis and J. J. Opplt, eds., CRC Press, Inc., Boca Raton, Florida, Vol. I (1980).

5. P. Alaupovic, The physicochemical and immunological heterogeneity of human plasma high-density lipoproteins, in: "Clinical and Metabolic Aspects of High-density Lipoproteins", N. E. Miller and G. J. Miller, eds., Elsevier Science Publishers B.V., Amsterdam (1984).

6. P. Alaupovic, E. Koren, W. J. McConathy, M. Tavella, C. Knight-Gibson, and J. D. Fesmire, Immunochemical methods of isolating and characterizing ApoB-containing lipoproteins, in: "Proceedings of the Workshop on Lipoprotein Heterogeneity", K. Lippel, ed., National Institutes of Health, NIH Publication No. 87-2646, Bethesda, MD (1987).

7. M. Cheung and J. J. Albers, Characterization of lipoprotein particles isolated by immunoaffinity chromatography. Particles containing A-I and A-II and particles containing A-I but not A-II, J. Biol. Chem. 259:12201 (1984).

8. P. Alaupovic, M. Tavella, and J. Fesmire, Separation and identification of ApoB-containing lipoprotein particles in normolipidemic subjects and patients with hyperlipoproteinemias, Adv. Exp. Med. Biol. 210:7 (1987).

9. C.-S. Wang, P. Alaupovic, R. E. Gregg, and H. B. Brewer, Jr., Studies on the mechanism of hypertriglyceridemia in Tangier disease. Determination of plasma lipolytic activities, k_1 values

and apolipoprotein composition of the major lipoprotein density classes, Biochim. Biophys. Acta 920:9 (1987).

10. P. Alaupovic, C. Knight-Gibson, C.-S. Wang, D. Downs, E. Koren, H. B. Brewer, Jr., and R. Gregg, Possible association of a newly recognized ApoA-II-containing lipoprotein and hypertriglyceridemia in Tangier disease, Arteriosclerosis 7:536a (1987).

11. Lipid Research Clinics Laboratory Manual 1, DHEW No. (NIH)75-628, National Heart and Lung Institute, Bethesda, MD (1974).

12. P. O. Attman, P. Alaupovic, and A. Gustafson, Serum apolipoprotein profile of patients with chronic renal failure, Kidney International 32:368 (1987).

13. J. M. Bard, P. Alaupovic, and D. Shafer, Lipoprotein abnormalities common to all patients with type II diabetes mellitus, Arteriosclerosis 7:536a (1987).

14. P. Alaupovic, D. M. Lee, and W. J. McConathy, Studies on the composition and structure of plasma lipoproteins. Distribution of lipoprotein families in major density classes of normal human plasma lipoproteins, Biochim. Biophy. Acta 260:689 (1972).

15. P. Alaupovic, C.-S. Wang, W. J. McConathy, D. Weiser, and D. Downs, Lipolytic degradation of human very low density lipoproteins by human milk lipoprotein lipase: the identification of lipoprotein B as the main lipoprotein degradation product, Arch. Biochem. 244:226 (1986).

16. E. Koren, W. J. McConathy, J. Keeling, R. Schwiedessen, C. Knight-Gibson, G. Wen, and Z. Reiner, The use of "pan" monoclonal antibody for quantification of apolipoprotein A-II, Arteriosclerosis 6:521a (1986).

17. N. Dashti, P. Alaupovic, C. Knight-Gibson, and E. Koren, Identification and partial characterization of discrete ApoB-containing lipoprotein particles produced by human hepatoma cell line HepG2, Biochemistry 26:4837 (1987).

18. A. Kuksis, J. J. Myher, L. Marai, and K. Gehler, Determination of plasma lipid profiles by automated gas chromatography and computerized data analyses, J. Chromatogr. Sci. 13:423 (1975).

19. P. Puchois, A. Kandoussi, P. Fievet, J. L. Fourrier, M. Bertrand, E. Koren and J. C. Fruchart, Apolipoprotein A-I containing lipoproteins in coronary artery disease, Atherosclerosis 68:35 (1987).

IN VIVO METABOLISM OF APOLIPOPROTEINS C-II AND C-III IN NORMAL AND HYPERTRIGLYCERIDEMIC SUBJECTS

C.L. Malmendier, J-F. Lontie, D. Dubois, C. Delcroix, T. Magot*, and L. De Roy†

Fondation de Recherche sur l'Athérosclérose, Brussels
*Laboratoire de Physiologie de la Nutrition, Université de Paris-Sud, Orsay, France, †Hôpital Militaire, Brussels, Belgium

INTRODUCTION

Two papers on apo C-II and apo C-III metabolism have been published in 1981 and 1982 (1,2), only one comparing normal and hyperlipoproteinemic subjects (1). The kinetic analysis of ^{125}I labeled VLDL was followed in plasma for 48 h only and there were no urine data. The essential limitation of these early studies was their short duration due to the low specific activity of each individual C peptide obtained after whole VLDL-apoprotein labeling (apo B, C, E) necessitating tedious and error-prone separation procedures.

To overcome these difficulties, more recent experiments extended up to 15 days were performed in normal subjects using labeled free apo C-I, apo C-II and C-III (3,4,5) injection. As immediately after their introduction into plasma the apoproteins associated to the different lipoproteins in proportion to their relative mass (5,6), the use of in vitro labeled homologous lipoproteins seemed even more physiological. This new methodology was applied to kinetic studies described in the present paper comparing the kinetic parameters of normal and hypertriglyceridemic patients.

MATERIAL AND METHODS

Patients

Ten normal subjects (8 male and 2 female) and four male hypertriglyceridemic patients with plasma triglyceride concentrations exceeding 300 mg/dl participated to the study. The clinical data are given in Table 1. Two patients (N° 11 and 14) were 134 and 137% of the ideal body weight. A detailed genetic classification of the type of hyperlipoproteinemia according to Goldstein et al. (7) was not carried out. The subjects were all studied as outpatients. None were on a special diet or lipid lowering drugs one month at least before the test. All patients gave informed consent to the project. Potassium iodide (500

Table 1. Clinical Data.

N°	Sex	Age	Weight	Height	Cholest.	TG	HDL-C	LDL-C	A-I	A-II	B	C-II	C-III
Subject		Yrs	kg	cm	mg / dl								
Normal													
1	M	28	63.5	168	210	96	40	152	156	46	93	2.7	6.6
2	F	23.5	68	171	183	110	42	119	178	26	77	4.1	9.8
3	M	28	79	172	168	63	45	110	150	30	49	2.4	3.7
4	M	29.5	71	179	234	103	59	134	221	52	113	3.7	11.4
5	M	23.5	73	170	177	139	36	102	157	47	80	3.9	7.8
6	M	23.5	72	185	149	70	43	92	163	39	59	1.9	7.5
7	M	23	71	177	160	82	43	101	145	34	64	1.2	6.1
8	M	23	72	180	160	58	50	98	147	37	60	3.2	7.4
9	M	34	82	172	213	116	49	141	165	46	94	2.6	11.4
10	F	26	70	169	204	68	52	138	191	52	82	2.3	7.8
MEAN ± S D					186±27	91±25	46±6	119±20	167±22	41±9	77±19	2.8±0.9	8.0±2.2
Hypertriglyceridemic													
11	M	48	94	178	247	310	30	-*	145	40	93	5.4	21.1
12	M	39	84.5	177	353	616	31.5	-	158	51	161	10.4	35.2
13	M	59	76	174	303	580	28.5	-	147	39	122	9.2	26.2
14	M	55	92	174	203	544	22.5	-	127	30	70	6.8	23.0
MEAN ± S D					277±65	513±138	28±3.9	-	144±13	40±9	111±39	8.0±2.2	26.4±6.2

* It was not possible to estimate LDL-cholesterol levels by the Friedewald's formula.

mg/day) was given in two divided doses 3 days prior to and extending throughout the study in order to prevent iodine uptake by the thyroid.

Study protocol

In normal subjects, 20 to 60 µCi of ^{125}I labeled free apo C-II or HDL-apo C-III were injected intravenously. Although these tracers were valid in normal subjects, we decided to test and adopt a new, even more physiological, labeling technique to study pathological patients. In hypertriglyceridemic patients, blood was withdrawn 4 days before the test and 4ml of plasma was incubated for 30 min at 37°C with labeled ^{125}I apo C-II and ^{131}I apo C-III (0.2 mg each). Immediately after incubation the labeled plasma was ultracentrifuged at d 1.25 g/ml, desalted on PD10 column (Pharmacia) and dialyzed quickly with several changes against 0.15 M NaCl containing 0.01% EDTA and merthiolate, pH 7.4. The homologous labeled lipoproteins (50 µCi in 4 ml) were reinjected in each patient intravenously. Following injection, 12 ml of blood was obtained at 10, 30, 60 min, 2, 4, 8 hours and daily at 8.00 a.m. (fasting) for 15 days. Urines were collected from "0 time" to 4 h, 4 to 8, and 8 to 24 h and for every 24-hour period thereafter, in bottles containing a preservative (8).

Radioactivity of ^{125}I and ^{131}I in total plasma, urine, and gel chromatography (FPLC) fractions was assessed using a Beckman gamma 5500 counting system.

Analytical methods

Lipid and lipoprotein analysis

Plasma triglyceride (TG), total cholesterol (TC), HDL- and LDL-cholesterol (HDL-C and LDL-C) determinations were performed in duplicate by the methods of the Lipid Research Clinics (9). LDL-C was calculated by the formula of Friedewald et al. (10) in normals. The formula might not be applied in hypertriglyceridemia.

Apolipoproteins

Apolipoproteins A-I, A-II and B were analyzed by ELISA (11, 12) in plasma. Apolipoproteins C-II and C-III were quantitated by sandwich ELISA in the plasma and in the lipoprotein fractions by modification of (13) and (14).

Column chromatography

Total plasma lipoproteins were isolated from 4 ml plasma by ultracentrifugation for 20 h at 12°C at a density of 1.225 g/ml. The supernatant (in 2 ml) was chromatographed without prior dialysis on Superose 6 prep grad column using a FPLC system (Pharmacia) as described in (15). The absorbance was monitored continuously at 280 nm. All tubes were assayed in a Beckman gamma 5500 counting system. Thereafter all fractions separated as VLDL, IDL+LDL, HDL were pooled and apolipoproteins C-II and C-III determined in each lipoprotein.

Table 2. Kinetic Parameters of Apolipoproteins C-II and C-III Metabolism.

		Apoprotein C-II				Apoprotein C-III			
Subject N°	Plasma Vol/ml	Conc mg/dl	FCR pools/day	T days	SR mg/kg.day	Conc mg/dl	FCR pools/day	T days	SR mg/kg.day
Normal									
1	2622	2.7	0.490	2.62	0.55				
2	2808	4.1	0.510	2.91	0.87				
3	3262	2.4	0.554	2.87	0.57				
4	2932	3.7	0.358	3.27	0.55				
5	3015	3.9	0.492	2.81	0.79				
6	2974					7.5	0.725	2.48	2.28
7	2840					6.1	0.987	1.94	2.36
8	2974					7.4	0.735	2.63	2.19
9	3280					11.4	0.713	2.44	2.54
10	2891					7.8	0.742	2.47	2.45
Mean ± SD		3.4±0.7	0.480±0.073	2.90±0.24	0.67±0.15	8.0±2.0	0.780±0.116	2.39±0.26	2.36±0.14
Hypertriglyceridemic									
11	3880	5.4	0.515	2.22	1.15	21.1	0.773	2.15	6.73
12	3490	10.4	0.340	3.25	1.46	35.2	0.569	3.00	8.27
13	3140	9.2	0.355	4.43	1.35	26.2	0.500	5.41	5.41
14	3800	6.8	0.279	5.01	0.78	23.0	0.413	5.18	3.92
Mean ± SD		7.9±2.2	0.372±0.100	3.74±1.23	1.19±0.30	26.4±6.2	0.564±0.153	3.94±1.61	6.08±1.86

Calculation of the kinetic parameters of apo C-II and apo C-III metabolism

The plasma radioactivity decay curves were plotted on semi-logarithmic paper against time. The kinetic parameters were calculated by the two-pool Matthews model (16). Mean residence time (T) and fractional catabolic rate (FCR) were calculated from the area under the curve according to (17). Under steady conditions synthetic rate (SR) was calculated as plasma pool of apo C-II or C-III times the FCR and normalized by kg body weight. Plasma volume was estimated by the radioactivity dilution technique and also as 4.13% body weight. The FCR was measured independently by relating the daily urinary excretion rate of radioactivity to the radioactivity in plasma (U/P ratio).

RESULTS

Plasma lipids and apolipoproteins (Table 1)

The four hypertriglyceridemic patients had plasma triglycerides varying from 310 to 616 mg/dl, two of them (N° 12 and 13) had increased total cholesterol, only one (N° 12) had a high apo B concentration. All HDL-cholesterol were low. Apo C-II and C-III values were 2 to 4 times higher than normal values.

Plasma radioactivity decay curves

Fig. 1 shows the plasma decay curves of the 4 hyperTG patients and the mean curve of normal subjects, and the mean U/P ratio curves. Contrasting with relatively small variations in the normal plasma curves, there was a large dispersion in the hyperTG curves. Mean U/P ratio curves of both apo C-II and C-III were markedly lower in hyperTG than in normals.

Lipoprotein fractions (Fig. 2)

The hyperTG patients profiles show much higher VLDL and lower HDL, compared to normal, and two patients had elevated LDL. Most apo C-II and C-III radioactivity in hyperTG was located in VLDL (48 and 41% respectively) and IDL+LDL (24 and 29%), only 26 and 24% being recovered in HDL. In normals the corresponding percentages of apo C-II and C-III were in VLDL (9 and 9%) in IDL+LDL (23 and 18%) and in HDL (67 and 74%).

Kinetic parameters (Table 2)

Assuming that apo C-II and C-III constitute an homogeneous pool, kinetic parameters were calculated according to Matthews' model analysis. For apo C-II, FCR in normal subjects varied from 0.358 to 0.554 pools/day and residence times from 2.62 to 3.27 days. In hyperTG, FCR ranged from normal (0.515) to very low values (0.279) and residence times were slightly higher than normal in 2 patients. Synthetic rates were also higher than normal in 3 patients. For apo C-III, FCR were lower and residence times higher than normal, except for one patient (N° 11). The synthetic rates were 2 times -or more- higher than normal.

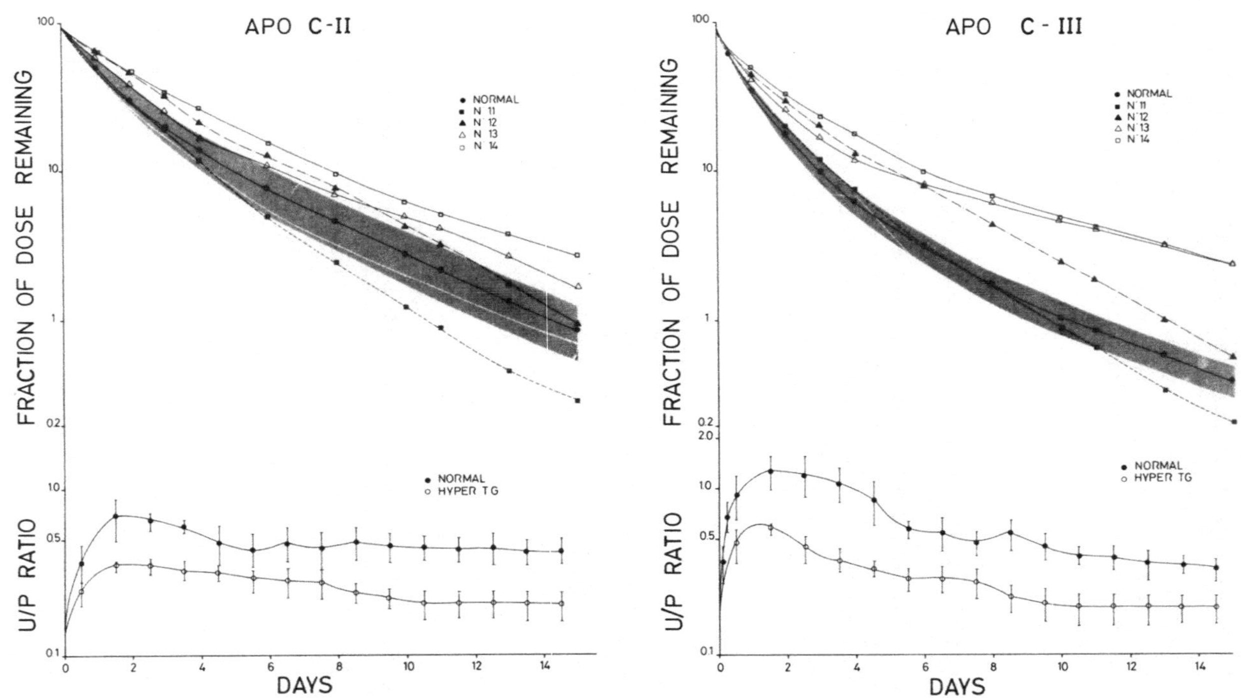

Fig. 1. Apo C-II and C-III plasma decay curves and mean U/P ratios. Shaded areas indicate I SD on either side of the mean plasma curve.

DISCUSSION

Reflexions on apo C kinetics

a) The duration of the experiment must be related to the half-life of carrier lipoproteins. As we stressed already in previous papers (4, 18), it is essential to extend the experimental time up to 15 days in order to follow the catabolism of the HDL particles, main reservoir of apo C (4).

b) The problem of steady-state conditions has never been mentioned in studies on hypertriglyceridemic patients. It is obvious that these patients are more prone to dramatic changes in their triglyceride plasma levels during the day. The blood samplings must therefore be withdrawn at the same hour before the morning breakfast, after a 12 hour fast, in order to minimize the effects of the meals.

c) FCR using a multicompartmental model is calculated from the area under the plasma specific activity curve. It may also be calculated from the U/P ratio. However this FCR is needful only for the metabolism of any apo C as a whole without giving information on their fate on individual lipoproteins. Due to the very fast exchanges of apo C between lipoproteins, they represent one pool considered as homogeneous.

d) There may be several models derived from the plasma data fitted to a sum of three exponentials. It is difficult to choose between them without the urine excretion data. The simultaneous fitting of both data allow the restriction of model number. The shape of the U/P curve may be very informative on the potential existence of two or more populations of particles carrying apo C's.

Comparison between apo C-II and C-III metabolism

In both normals and hyperTG, FCR (calculated from the plasma curves or from U/P ratios) were higher for apo C-III than for apo C-II. U/P ratio curve of apo C-III displays, better than that of apo C-II, the biphasic shape of the urinary excretion suggesting the existence of different catabolic routes from the plasma compartment. Residence times differed less than FCR. Production rates were 4 to 6 times higher for apo C-III than for apo C-II.

Apo C-II and C-III metabolism in hypertriglyceridemic compared to normal subjects

Hypertriglyceridemia affected less FCR than production rate but this effect was more marked for apo C-III than apo C-II. Patient 11 with normal cholesterol and moderately elevated triglycerides, showed FCR and residence time in the normal range for both apo C-II and apo C-III, only the production rate being 2 or 3 times increased. Patient 14 with normal cholesterol but high TG, had reduced catabolic rates and prolonged residence times but the synthesis was only slightly increased in apo C-III. Patients 12 and 13, having in common an increased cholesterol and the highest triglyceride levels, showed FCR slightly lower than the lowest normal value, prolonged residence times but also largely increased synthetic rates. Although the number of patients was limited, it thus appears that two or even three very different patterns are observable in the hypertriglyceridemics. The increase in cholesterol was due to high LDL chol-

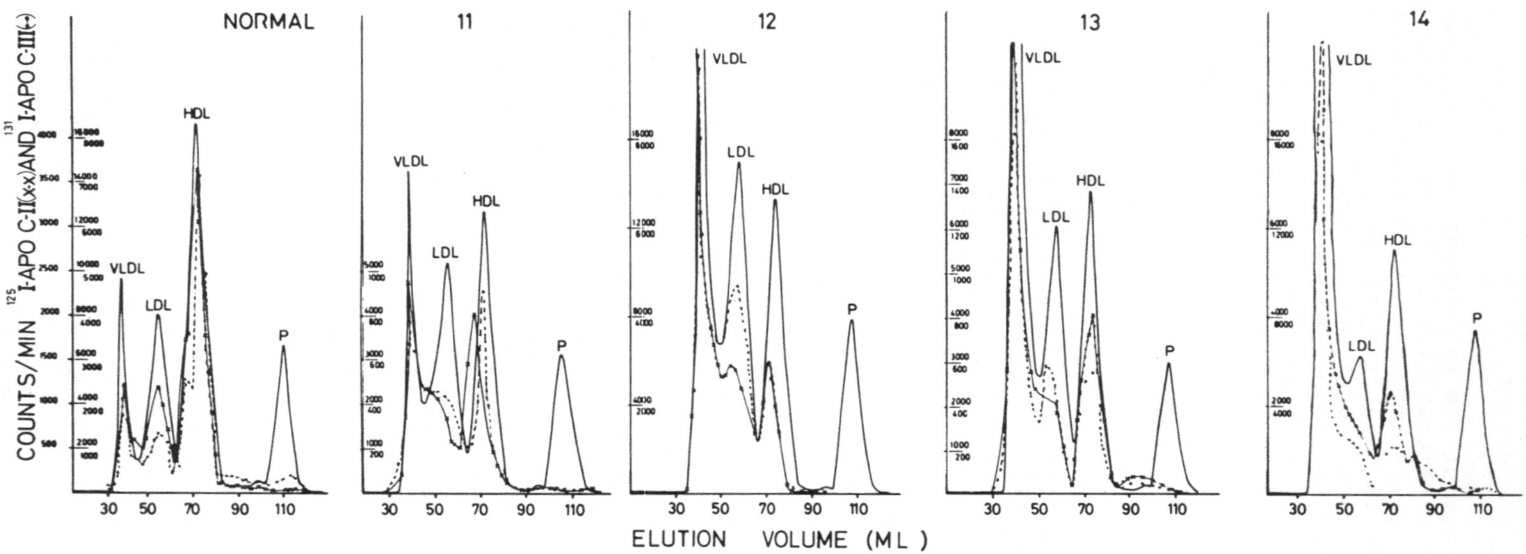

Fig. 2. FPLC chromatography on Superose 6 prep grad of lipoproteins isolated from 4 ml of plasma for a typical normal subject and for the 4 hypertriglyceridemic patients. Protein (280 nm) and apo C-II and C-III radioactivity profiles.

esterol in N° 12 and VLDL cholesterol in N° 13, in agreement with their respective apo B levels.

Mechanism of hypertriglyceridemia

One of the major factors responsible for hypertriglyceridemia is the defective lipolysis either due to an absence or reduction of LPL (Types I and V) (19), a functional deficiency or absence of apo C-II (20, 21) or a reduction in LPL activity in relation to a lower apo C-II/C-III ratio (22). Apo C-II normally circulates in much higher concentration than is necessary to adequately activate lipoprotein lipase (23). As it appears that apo C-II level, when sufficient, is not critical, the role of apo C-III may become crucial to explain the lipolytic defect.

Our patients showed: no absence of LPL activity because the lactescence after a normal meal disappeared progressively but much slower than in normal subjects (lipoprotein lipase determinations were not performed), and no absence but an increase in apo C-II level. No major change in apo C-II/C-III ratio was observed. The major observation is the increase in absolute terms of apo C-III plasma concentration, more related to a constant increase in synthesis than to an inconstant reduction in FCR.

The following sequence of events, at least in the 4 patients studied, might be: increased apo C-III synthesis, acting as an inhibitor of lipolysis (24) and of triglyceride-rich remnants uptake by liver (25), causing an hypertriglyceridemia; the latter will require an activation of lipoprotein lipase activity that will induce an increased synthesis of the co-factor, apo C-II. Thus apo C-II synthesis increase may be secondary to that of apo C-III, primary determinant.

REFERENCES

1. M. W. Huff, N. H. Fidge, P. J. Nestel, T. Billington, and B. Watson, Metabolism of C-apolipoproteins: kinetics of C-II, $C\text{-}III_1$ and $C\text{-}III_2$, and VLDL-apolipoprotein B in normal and hyperlipoproteinemic subjects, J. Lipid Res., 22:1235 (1981).
2. M. W. Huff, and P. J. Nestel, Metabolism of apolipoproteins CII, $CIII_1$ and $CIII_2$ and VLDL-B in human subjects consuming high carbohydrate diets, Metabolism, 31:493 (1982).
3. C. L. Malmendier, J-F. Lontie, G. A. Grutman, and C. Delcroix, Metabolism of apolipoprotein C-I in normolipoproteinemic human subjects, Atherosclerosis, 62:167 (1986)
4. C. L. Malmendier, J-F. Lontie, G. Grutman, and C. Delcroix, Metabolism of apolipoprotein C: kinetic studies in human subjects: a critical review, In: Lipoproteins and Atherosclerosis, C. L. Malmendier,and P. Alaupovic, eds, Plenum Press, New York (1986).
5. C. L. Malmendier, J-F. Lontie, G. A. Grutman, and C. Delcroix, Metabolism of apolipoprotein C-III in normolipemic human subjects, Atherosclerosis, 69:51 (1988).
6. M. L. Kashyap, L. S. Srivastava, B. A. Hynd, G. Perisutti, D. W. Brady, P. Gartside, and C. J. Glueck, The role of high density lipoprotein apolipoprotein CII in triglyceride metabolism, Lipids, 13:933 (1978).
7. J. L. Goldstein, W. R. Hazzard, H. G. Schrott, E. L. Bierman, and A. G. Motulsky, Hyperlipidemia in coronary heart disease. I. Lipid levels in 500 survivors of myocardial infarction, J. Clin. Invest., 52:1533 (1973).

8. J. L. Steinfeld, R. R. Paton, A. L. Flick, R. A. Milch, F. E. Beach, and D. L. Tabern, Distribution and degradation of human serum albumin labeled with I^{131} by different techniques, Ann. N. Y. Acad. Sci, 70:109 (1957).
9. Lipid Research Clinics Program. Manual of Laboratory Operations. Lipid and Lipoprotein Analysis (DHEW Publication N° (NIH) 75-628), National Institutes of Health, Bethesda, MD, 1974.
10. W. T. Friedewald, R. I. Levy, and D. S. Fredrickson, Estimation of the concentration of low-density lipoprotein cholesterol in plasma, without use of the preparative ultracentrifuge, Clin. Chem., 18:499, (1972).
11. D. Y. Dubois, F. Cantraine, and C. L. Malmendier, Comparison of different sandwich enzyme immunoassays for the quantitation of human apolipoproteins A-I and A-II, J. Immunol. Methods, 96:115 (1987).
12. J-C. Fruchart, C. Fievet, and P. Puchois, Apolipoproteins, in : Methods of Enzymatic Analysis, Vol. VIII, H. U. Bergmeyer, ed., Verlag Chemie, Weinheim (1985).
13. J. Bury, G. Michiels, and M. Rosseneu, Human apolipoprotein C-II quantitation by sandwich enzyme-linked immunosorbent assay, J. Clin. Chem. Clin. Biochem., 24:457 (1986).
14. J. Bury, and M. Rosseneu, Enzyme linked immunosorbent assay for human apolipoprotein C-III, J. Clin. Chem. Clin. Biochem., 23:63 (1985).
15. Y. C. Ha, and P. J. Barter, Rapid separation of plasma lipoproteins by gel permeation chromatography on agarose gel Superose 6B, J. Chromatogr., 341:154 (1985).
16. C. M. E. Matthews, The theory of tracer experiments with iodine 131-labeled plasma proteins, Phys. Med. Biol., 2:36 (1957).
17. A. Rescigno, and E. Gurpide, Estimation of average times of residence, recycle and interconversion of blood-borne coumpounds using tracer methods, J. Clin. Endocrinol. Metab., 36:263 (1973).
18. C. L. Malmendier, and J-F. Lontie, Turnover studies of apolipoproteins C: a first critical appraisal, in: Proceed. 2nd Europ. Workshop on Lipid Metabolism, Munich (1988). Springer Verlag, Heidelberg, in press.
19. D. S. Fredrickson, J. L. Goldstein, and M. S. Brown, The familial hyperlipoproteinemias, chapter 30, in: The Metabolic Basis of Inherited Disease, J. B. Stanbury, J. B. Wyngaarden, and D. S. Fredrickson, eds., McGraw-Hill, New York (1978).
20. W. C. Breckenridge, J. A. Little, G. Steiner, A. Chow, and M. Poapst, Hypertriglyceridemia associated with deficiency of apolipoprotein C-II, N. Engl. J. Med., 298:1265 (1978).
21. W. C. Breckenridge, Apolipoprotein C-II deficiencies: in vivo models for assessing the significance of defective lipolysis on lipoprotein metabolism, in: Eicosanoids, Apolipoproteins, Lipoprotein Particles and Hypertriglyceridemia, C. L. Malmendier, and P. Alaupovic, eds., Plenum Press (1988). This volume.
22. L. A. Carlson, and D. Ballantyne, Changing relative proportions of apolipoproteins CII and CIII of very low density lipoproteins in hypertriglyceridaemia, Atherosclerosis, 23:563 (1976).
23. N. E. Miller, S. N. Rao, P. Alaupovic, N. Noble, J. Slack, J. D. Brunzell, and B. Lewis, Familial apolipoprotein CII deficiency: plasma lipoproteins and apolipoproteins in heterozygous and homozygous subjects and the effects of plasma infusion, Eur. J. Clin. Invest., 11:69 (1981).

24. W. V. Brown, and M. L. Baginsky, Inhibition of lipoprotein lipase by an apoprotein of human very low density lipoprotein, Biochem. Biophys. Res. Comm., 46:375 (1972).
25. E. Windler, and R. J. Havel, Inhibitory effects of C apolipoproteins from rats and humans on the uptake of triglyceride-rich lipoproteins and their remnants by the perfused rat liver, J. Lipid. Res., 26:556 (1985).

PATHOGENESIS OF HYPERTRIGLYCERIDEMIA: IMPLICATIONS FOR CORONARY HEART DISEASE AND THERAPY

Gloria Lena Vega and
Scott M. Grundy

University of Texas Southwestern Medical Center at Dallas
5323 Harry Hines Boulevard
Dallas, Texas 75235-9052

INTRODUCTION

The role of hypertriglyceridemia in the causation of coronary heart disease (CHD) is a subject of continuing controversy. Some investigators hold that hypertriglyceridemia constitutes a major risk factor for CHD[1-4], but others[5] claim the opposite, namely, that elevated triglyceride levels are of little significance for CHD risk. The very fact that so many experienced investigators hold such divergent views attests to the complexity of the issue. The crux of the problem appears to be that a high proportion of patients with CHD have hypertriglyceridemia, and yet high triglycerides per se are not a major causative factor; instead, elevated plasma triglycerides frequently appear to be associated with other atherogenic factors that have not been fully defined. This paper will examine the complexity of the hypertriglyceridemic state, and it will consider the following questions: (a) what are the basic pathways of triglyceride metabolism in humans? (b) what are the causes of hypertriglyceridemia? (c) what is the relation of hypertriglyceridemia to CHD?, and (d) what constitutes a rational approach to the treatment of hypertriglyceridemia?

PATHWAYS OF TRIGLYCERIDE METABOLISM

Transport of Exogenous Triglycerides

Dietary triglycerides are hydrolyzed in the intestine into monoglycerides and fatty acids, both of which are taken up by the intestinal mucosa and are resynthesized into triglycerides. These triglycerides are incorporated into lipoproteins called chylomicrons. The major "structural" apolipoprotein of chylomicrons is apo B-48. In addition, chylomicrons contain apolipoproteins A-I, A-II, and A-IV. Chylomicrons are secreted into intestinal lymph and enter the systemic circulation through the thoracic duct. When these particles enter peripheral capillaries, they come in contact with the enzyme, lipoprotein lipase, and through the action of this enzyme, triglycerides are hydrolyzed to free fatty acids. When hydrolysis of triglycerides is almost complete, a residual lipoprotein, called a chylomicron remnant, is released into the circulation and is cleared by the liver. During lipolysis, chylomicron surface lipids--unesterified cholesterol and phospholipids--appear to be partly transferred to high density lipoproteins (HDL).

When chylomicrons enter the plasma, they seemingly acquire apolipoproteins C-II, C-III, and E by transfer from HDL. Apo C-II is required for activation of lipoprotein lipase, and during lipolysis, this and other C apolipoproteins are released into the circulation and are either transferred to other triglyceride-rich lipoproteins or temporarily sequestered with HDL. Apo E remains with chylomicron remnants, and it may play a role in hepatic uptake of chylomicron remnants. Whether there is a unique "apo E receptor" on liver cells has not been confirmed, preliminary reports to the contrary not withstanding. Further research will be required to elucidate the mechanism of hepatic uptake of chylomicron remnants.

Transport of Endogenous Triglycerides

The liver like the intestine secretes a triglyceride-rich lipoprotein, which is called very low density lipoprotein (VLDL). Actually, VLDL normally circulating in plasma probably are not identical to newly secreted VLDL, which we can call nascent lipoproteins. These nascent particles contain apo B-100 as their major apolipoprotein. In normal humans, triglycerides appear to be almost the sole lipid in nonpolar core of nascent VLDL; cholesterol esters seem to be present in only small amounts. As nascent VLDL circulate in plasma, two types of changes occur that transform them into VLDL found in circulating plasma. First, they may acquire more of the apo C's and apo E's for the surface coat from HDL, and at the same time, they receive cholesterol ester, again from HDL. And second, they interact with lipoprotein lipase where a portion of their triglycerides is removed.

The VLDL circulating in plasma continue to undergo these two modifications leading to progressively smaller particles that are increasingly enriched in cholesterol ester. The VLDL particles found in plasma consist of a continuum of lipoproteins containing varying quantities of triglycerides and cholesterol ester; those having more triglycerides are designated native VLDL, whereas those enriched in cholesterol are called VLDL remnants. This distinction obviously is arbitrary, and in fact, most of the circulating VLDL particles seemingly have lost some triglycerides and have acquired cholesterol ester, and thus probably deserve the name "remnants".

While these changes are occurring, VLDL at any stage of degradation are susceptible to uptake by the liver. One route of uptake is through receptors for low density lipoproteins (LDL)[6]. Although these receptors were named LDL receptors because they were first shown to recognize LDL, they can in fact remove particles containing apo B-100. Indeed, VLDL in the early stages of delipidation seemingly have a greater affinity for LDL receptors than LDL itself. This is because the VLDL contain apo E which also can bind to LDL receptors[7]. Apparently, particles having the largest number of apo E molecules, which may be the larger sizes of VLDL, will have the greatest affinity for LDL receptors. This effect should favor the direct removal of VLDL in the beginning of the delipidation chain[8]. The affinity of these particles for LDL receptors diminishes as particles move progressively down the delipidation chain. It is possible that larger VLDL are removed by pathways other than the LDL receptors, and if chylomicron-remnant receptors actually exist, larger VLDL may be cleared through this pathway as well. Recent studies indicate that large-sized VLDL are cleared rapidly from the circulation[9,10].

Early isotope-kinetic studies suggested that most VLDL are degraded to LDL. However, recent investigations indicate that a significant fraction of VLDL is cleared directly from the circulation. In fact, the latest data suggest that only about 20 to 40% of nascent lipoproteins entering plasma are degraded to LDL before being removed from the plasma[11]. When circulating VLDL are isolated, radiolabeled, and reinjected into plasma, a higher fraction of VLDL-apo B (i.e. 40 to 60%) is transformed to LDL-apo B[12,13]. However, it is likely that circulating VLDL are essentially VLDL remnants and have already lost some of their affinity for LDL receptors (or chylomicron-remnant receptors); as a result, they are more likely to be converted to LDL. Those VLDL that have the greatest affinity for receptors are quickly removed from the circulation and consequently are not accurately traced in isotope-kinetic studies[8,11,14].

From data of kinetics of lipoproteins containing apo B-100, some have concluded that LDL is secreted directly into plasma[11]. If this conclusion is taken literally, it would mean that the liver synthesizes LDL that are identical to those found in plasma, i.e. "mature" LDL are secreted into plasma. From what is known about LDL metabolism, it seems doubtful that direct production of LDL is possible. Most of the cholesterol esters in LDL have been synthesized in plasma through the enzyme, lecithin-cholesterol acyltransferase (LCAT). A more likely mechanism for "direct input" of LDL is that the liver secretes a nascent lipoprotein that is converted to LDL independently of the delipidation chain for VLDL[8,11,14]. Two possibilities can be considered. First, large, nascent VLDL could rapidly undergo lipolysis to form LDL. Or second, the liver could secrete smaller VLDL that are quickly converted to LDL. Current techniques cannot trace nascent lipoproteins in humans; consequently these two possibilities cannot be distinguished.

Normally, LDL contains only small quantities of triglycerides; thus LDL does not play a major role in triglyceride transport. Actually, LDL-triglycerides belong to two pools: (a) a rapidly catabolized component, possibly a substrate for lipoprotein lipase or hepatic triglyceride lipase, and (b) a component that remains untouched by lipases[15]. The latter is retained with LDL until it is removed from the circulation. The major removal pathway for LDL is through LDL receptors; however smaller quantities are cleared by nonreceptor pathways. At least two-thirds of circulating LDL are removed by the liver, whereas the remainder is cleared by extrahepatic tissues[6].

Only small amounts of triglyceride passes through HDL[16]; even so, HDL play an important role in plasma triglyceride metabolism. For example, HDL contain a pool of apo C's and apo E's, and these can be transferred to chylomicrons or VLDL to promote their catabolism. In addition, HDL serve as an acceptor for surface coat unesterified cholesterol and phospholipids that are released from triglyceride-rich lipoproteins during lipolysis. Finally, HDL-cholesterol esters can be transferred to the core of triglyceride-rich lipoproteins. The interaction between HDL and VLDL (or chylomicrons) affects the distributions of cholesterol and triglycerides among the major subfractions of HDL, specifically, HDL_3, HDL_{2a}, and HDL_{2b}. When plasma triglycerides are high, HDL_{2b} is relatively elevated at the expense of HDL_{2a}.

CAUSES OF HYPERTRIGLYCERIDEMIA

Three factors theoretically can contribute to the development of hypertriglyceridemia; these are: (a) overproduction of VLDL-triglycerides, (b) defective lipolysis of VLDL-triglycerides, and (c) defective uptake of

VLDL remnants. A defect in lipolysis of triglyceride-rich lipoproteins extends to chylomicrons as well as VLDL, and patients with lipolytic defects frequently have retention of both chylomicrons and VLDL in plasma. After partial lipolysis of VLDL, a defect in direct removal of VLDL remnants may contribute to their accumulation in plasma. Seemingly, many patients with hypertriglyceridemia have two or more defects. Indeed, many patients with one of these abnormalities apparently do not have hypertriglyceridemia because other determinants of triglyceride levels are normal or increased.

An increased input of VLDL-triglycerides can occur in either of two ways. First, VLDL particles may be enriched with triglycerides, possibly by increased hepatic synthesis of triglycerides. With this condition, the number of VLDL particles secreted by the liver may not be increased; rather, they are merely enriched in triglycerides. And second, the number of VLDL particles secreted by the liver may be increased, whereas each particle contains a normal amount of triglyceride. Differentiation between these two mechanisms may be important because overproduction of VLDL particles may impart a greater risk for CHD than mere enrichment of VLDL particles with triglycerides.

In the discussion to follow, the hypertriglyceridemias will be divided into (a) those in which the defects are limited to triglyceride metabolism and (b) those in which triglyceride defects are combined with other abnormalities of lipoprotein metabolism. The former category will be called primary (or secondary) hypertriglyceridemia, while the latter will be designated primary (or secondary) mixed hyperlipidemias.

Primary Hypertriglyceridemia

This condition can be defined as hypertriglyceridemia resulting from defect(s) limited to metabolism of plasma triglycerides. Concomitant abnormalities in the metabolism of lipoproteins are not present. Two degrees of severity--moderate (borderline) and marked (definite) hypertriglyceridemia--can be defined. The former represents a plasma triglyceride of 250 to 500 mg/dl, whereas the latter is a triglyceride level over 500 mg/dl. The terms "borderline" and "definite" were applied to hypertriglyceridemia by the National Institutes of Health Consensus Conference on Triglycerides[17]. However, the term "borderline" may understate the importance of triglyceride levels of 250 to 500 mg/dl, and thus we prefer the term "moderate". Two major mechanisms responsible for primary hypertriglyceridemia are overproduction and defective lipolysis of VLDL-triglycerides. Defects in removal of VLDL remnants also can be present, but since patients with this defect frequently have elevations of both cholesterol and triglycerides, remnant-removal defects will be discussed under mixed hyperlipidemias.

Overproduction of VLDL-triglycerides. Patients with primary overproduction of VLDL-triglycerides (without overproduction of VLDL-apo B) have been identified by Chait et al[18]; this abnormality was noted to occur in families, and thus was designated familial hypertriglyceridemia. When the input of VLDL-triglycerides is increased, the plasma triglyceride level of course will rise. The extent of rise will depend not only on the severity of overproduction but also on the rate of lipolysis of VLDL-triglycerides. Two patterns of lipolytic response to overproduction of VLDL-triglycerides have been postulated. The first pattern is that proposed by Farquhar, Reaven et al[19,20]; it is based on the assumption that in vivo kinetics of VLDL-triglycerides are similar to in vitro enzyme kinetics. The basic hypothesis is that the rise in triglyceride concentrations due to an increased influx of VLDL-triglycerides will saturate lipoprotein lipase; any further increases in production rates will cause a marked increase in triglyceride concentrations. Subsequently, however, several lines of evidence cast doubt on the assumption that triglyceride kinetics are strictly analogous to in vitro enzyme kinetics[21]. Even so, an element of "saturation" of the lipolytic system probably exists. The pattern of response in vivo more likely can be explained by hyperbolic curve relating production rates to concentrations, rather than the typical in vitro curve of enzyme saturation kinetics[21].

The mechanisms responsible for primary overproduction of VLDL-triglycerides have not been determined. One mechanism may be a peripheral resistance to insulin; this resistance could divert glucose or three-carbon fragments to the liver, which could compete with fatty acid oxidation in the liver; if so, more fatty acids would be available for incorporation into VLDL-triglycerides. Another mechanism could be a deregulation of lipid synthesis in the liver; many patients with familial hypertriglyceridemia have increased synthesis of both cholesterol and triglycerides[22], and thus could have oversynthesis of triglycerides as well. This response might be independent of insulin action.

Defective lipolysis of VLDL-TG. Another cause of primary hypertriglyceridemia can be a decrease in capacity for lipolysis of triglyceride-rich lipoproteins. Nikkila and Kekki[23] have proposed that lipolytic capacities are variable from one person to another. This variability in lipolytic capacity can be inferred from the relation between production rates and concentrations in a group of normal subjects. The scatter of points within the normal range suggests heterogeneity in clearance capacity[21]. If these points constitute so many points on saturation curves, there would be a variety of saturation curves indicative of polymorphism of lipolytic capacities. If this family of saturation curves truly exists within the general population, then some patients may develop hypertriglyceridemia even with production rates in the normal range. In fact, studies from our laboratory[21,24,25] show this to be the case; measurements of VLDL-triglyceride turnover in a large group of hypertriglyceridemic patients reveal a subgroup who had elevated VLDL-triglycerides and yet normal production rates. Another study from our institution[26] revealed one family in which all affected members had reduced fractional catabolic rates for VLDL-triglycerides. The findings in this family suggest that a reduced lipolytic capacity also can be responsible for familial hypertriglyceridemia.

An obvious cause of defective lipolysis is a congenital deficiency of lipoprotein lipase. However, complete absence of lipoprotein lipase is extremely rare. A reduced activity of lipoprotein lipase on the other hand might have many causes; potential defects include (a) reduced synthesis of lipase, (b) polymorphism in the primary structure of lipoprotein lipase, such that some variants will interact poorly with VLDL-triglycerides, (c) retarded migration of lipase to the endothelial surface of capillary, (d) an abnormality in the milieu in which lipase resides that interferes with its interaction with triglycerides, and (e) defects in the structure and composition of VLDL that interferes with its interaction with the lipase. To date, none of these defects have been identified, but all are theoretical possibilities that are worthy of investigation.

Secondary Hypertriglycerides

Excess dietary carbohydrates. Several investigations have shown that excess intakes of carbohydrates will increase the production of VLDL-triglycerides by the liver. Although the mechanism by which this occurs is not apparent, the increased synthesis of VLDL-triglycerides seemingly occurs without an overproduction of VLDL particles[27]. In some patients with primary endogenous hypertriglyceridemia, ingestion of excess carbohydrates can produce a striking increase in triglyceride levels.

Excess alcohol intake. Like carbohydrates, an excessive intake of ethanol can stimulate the hepatic synthesis of triglycerides, which "spills over" into plasma as VLDL-triglycerides[28]. Again, the number of VLDL particles entering plasma probably is not increased. Excess ingestion of ethanol clearly can worsen hypertriglyceridemia in individuals with an underlying primary form of hypertriglyceridemia. Furthermore, excess alcohol in the blood stream may interfere with activities of lipoprotein lipase and hepatic triglyceride lipase.

Noninsulin dependent diabetes mellitus (NIDDM). Two metabolic abnormalities may produce a rise in the plasma triglyceride concentrations in patients with NIDDM. A relative deficiency of peripheral insulin activity apparently reduces the synthesis of lipoprotein lipase, and at the same time, it indirectly stimulates the synthesis of VLDL-triglycerides[29,30]. The latter may be due to (a) increased levels of free fatty acids, (b) increased plasma glucose levels, (c) increased influx of three-carbon fragments into the liver, (d) hyperinsulinemia, and (e) hyperglucagonemia. All of these defects can be reversed in the earlier stages of NIDDM by reducing action of the peripheral action of insulin either by weight reduction or specific hypoglycemic therapy.

Renal failure. Yet another secondary cause of hypertriglyceridemia is chronic renal failure. A significant fraction of patients with chronic renal failure have elevated triglyceride concentrations. The mechanisms of high triglyceride levels were evaluated in a study of VLDL-triglyceride kinetics carried out in our laboratory[31]. Most patients with renal failure had increased concentrations of VLDL-triglyceride, but at the same time they had relatively normal production rates. Thus, the major abnormality causing hypertriglyceridemia in these patients was a defect in lipolysis of VLDL-triglycerides. The cause of reduced lipolytic capacity in chronic renal failure has not been determined, but it is possible that there might be a circulating inhibitor of lipoprotein lipase.

Combined Defects in Triglyceride Metabolism

When only a single defect in the metabolism of triglycerides is present, an affected patient usually manifests only mild hypertriglyceridemia, or if other metabolic processes are efficient, triglyceride levels may remain in the high-normal range. For example, many individuals ingesting excess carbohydrates or alcohol have overproduction of VLDL-triglycerides, and yet are able to maintain normal triglyceride levels. Presumably these patients have a highly effective lipolytic capacity. Other individuals may have a reduced lipolytic capacity and still not develop hypertriglyceridemia because of relatively low production rates for VLDL-triglyceride. We previously proposed that such individuals have a "latent" lipolytic defect[21].

Thus, many patients with definite hypertriglyceridemia probably have combined defects in triglyceride metabolism. If there is a family of "saturation" curves indicating differences in lipolytic capacities among individuals in the general population, the final level of triglycerides for any individual will depend both on the production rate and that person's lipolytic capacity. Certainly it may be possible to develop hypertriglyceridemia by either "pure" overproduction of VLDL-triglyceride or a "pure" lipolytic defect, but perhaps most hypertriglyceridemic patients have a combined overproduction and a relatively low lipolytic capacity. For example, Kesaniemi and Grundy[25] reported that almost all patients with marked hypertriglyceridemia [type 5 hyperlipoproteinemia (HLP)] have both of these defects in triglyceride metabolism.

METABOLIC CONSEQUENCES OF HYPERTRIGLYCERIDEMIA

The presence of hypertriglyceridemia in a patient appears to impart abnormalities in all of the lipoprotein fractions. Some of these changes may be secondary to physical chemical effects related to an excess of triglyceride-rich lipoproteins, whereas the underlying metabolic defect may be responsible for others. The changes that occur in each lipoprotein fraction can be reviewed briefly.

Chylomicrons

Since chylomicrons and VLDL compete for the same lipolytic system, an increase in circulating VLDL particles will interfere in the lipolysis of chylomicron-triglycerides[32]. This will delay the overall clearance of chylomicrons and cause them to be progressively rich in cholesterol esters. Thus, endogenous hypertriglyceridemia causes prolonged chylomicronemia and increases in partially-catabolized chylomicrons (remnants) following ingestion of a fat-containing meal.

VLDL and VLDL remnants

Hypertriglyceridemia usually is associated with a prolonged residence time of VLDL in the circulation. This gives rise to several changes in VLDL composition. For example, the particles become progressively enriched in cholesterol esters. In addition, as shown by Gianturco et al[33,34], "hypertriglyceridemic" VLDL are modified in a way to change their recognition by cells. Compared to normal VLDL, "hypertriglyceridemic" VLDL are more readily taken up by LDL receptors of fibroblasts grown in tissue culture. They likewise can be recognized as modified lipoproteins by macrophases. The former effect may be related to changes in the confirmation of apo E on the surface of the VLDL, whereas the latter may be due to partial degradation of the apo B-100 molecule.

Intermediate density lipoproteins (IDL)

An increase in concentration of VLDL almost always is accompanied by increased levels of IDL. Both IDL-cholesterol and IDL-apo B levels are raised. Presumably, the same metabolic defects that are responsible for the evaluation of VLDL extend into the IDL fraction which may account for the increase in the latter fraction.

Low density lipoproteins (LDL)

Hypertriglyceridemia patients usually have multiple abnormalities of LDL metabolism[35,36,37]. One defect is an increased fractional catabolic rate (FCR) for LDL. Many patients also have increased conversion of VLDL to LDL, possibly due a decreased direct removal of VLDL remnants. In addition, the LDL fraction is more heterogeneous in particle size and composition than normal[38]. One subfraction of LDL is abnormally enriched in triglycerides, whereas another is relatively depleted in cholesterol esters[37]. Particles in the latter subfraction tend to be abnormally small and dense.

High density lipoproteins (HDL)

Hypertriglyceridemia produces several alterations of HDL metabolism. Most hypertriglyceridemic patients have reduced concentrations of

HDL-cholesterol; this reduction appears to be due in part to exchange of triglycerides in VLDL with cholesterol esters of HDL. Apo A-I levels frequently are decreased in patients with elevated triglycerides, although the decrease usually is less than for HDL-cholesterol. The HDL_2 fraction almost always is decreased, whereas HDL_3 and HDL_{2b} may be normal.

Implications of lipoprotein changes for atherogenesis

Although patients with hypertriglyceridemia frequently have premature CHD, the mechanisms for enhanced atherogenesis are not understood. The presence of excess triglyceride in plasma may not be atherogenic per se. Triglycerides do not accumulate in atherosclerotic plaques, and high triglycerides are not an "independent" risk factor for CHD in epidemiological studies[5]. On the other hand, elevated triglyceride concentrations may be a marker for increased coronary risk. Furthermore, abnormal lipoproteins that are the metabolic consequence of hypertriglyceridemia may be atherogenic. Possible atherogenic factors could be: (a) periodic increases in cholesterol-rich, chylomicron remnants, (b) modified VLDL, (c) increased concentrations of IDL, (d) the presence of abnormally small and dense LDL, and (e) reduced concentrations of HDL-cholesterol and other modifications of HDL metabolism. All of these metabolic consequences are potentially atherogenic, but further studies will be required to prove the atherogenecity of any or all of these changes.

CONCOMITANT METABOLIC DEFECTS

Several abnormalities in the metabolism of lipoproteins may occur concomitantly with defects in triglyceride metabolism. Patients with these defects will have not only hypertriglyceridemia and its metabolic consequences but will have in addition other lipoprotein abnormalities that reflect the concomitant defect. Three major categories of the latter defects are (a) overproduction of VLDL particles, (b) defects in remnant removal, (c) reduced activity of LDL receptors, and (d) defects in HDL metabolism. When these concomitant defects are present the risk for CHD almost certainly is increased; indeed, some investigators believe that only these defects, and not hypertriglyceridemia, is responsible for the increased risk. Even if this is not true, their presence almost certainly heightens the risk for coronary disease in patients with elevated triglycerides. These concomitant defects will be considered briefly.

Overproduction of VLDL Particles

This disorder is characterized by an increased input of VLDL particles into the circulation. Isotope kinetic studies are characterized by an increased production of both VLDL-apo B and VLDL-triglycerides[16,39,40]. Affected patients may manifest several patterns of hyperlipidemia including (a) hypertriglyceridemia, (b) hypercholesterolemia, and (c) mixed hyperlipidemia (increased levels of both triglycerides and cholesterol)[41,42]. Although some patients with overproduction of VLDL particles may not have persistent hypertriglyceridemia, we have observed that they frequently have transitory hypertriglyceridemia[43]. The several causes of this disorder can be considered.

Familial combined hyperlipidemia (FCHL). In families having FCHL, affected members present a variety of lipoprotein patterns[41,42]. Isotope-kinetic studies indicate that most affected patients have overproduction of VLDL-apo B[18,39,40] and many have an associated overproduction of VLDL-triglyceride[24]. The mechanisms for multiple lipoprotein phenotypes in families with FCHL has not been determined with certainty. However, if a high prevalence of "latent" catabolic defects for different lipoprotein species exist in the general population a concomitant overproduction of VLDL particles may lead to an increase in one or another of these species[44].

The mechanisms responsible for overproduction of VLDL particles in FCHL have not been determined. Several possibilities however can be considered. First, the synthesis of apo B-100 by the liver could be increased; this defect should be lead to increased input of nascent lipoproteins. Second, the absolute synthesis of apo B could be normal, but nascent lipoproteins could be smaller than normal and would preferentially move through the chain-delipidation pathway into LDL; this pattern would give the isotope kinetic picture of overproduction of VLDL-apo B. Finally, there could be a defect in direct removal of nascent lipoproteins, so that an increased proportion would enter the delipidation chain. This latter abnormality might have several causes: abnormal nascent lipoproteins, sluggish lipolysis, or reduced activity of LDL receptors. None of these mechanisms for overproduction of VLDL particles have been proven or disproven, but all theoretically could give the kinetic pattern of overproduction of VLDL-apo B.

Obesity. The effects of obesity on lipoprotein metabolism resemble in many ways those of primary overproduction of VLDL particles. Obese patients have an excessive input of VLDL-apo B[12,13] and VLDL-triglyceride[45]; the former[12,13] indicates that the number of VLDL particles entering plasma is increased. The mechanism for VLDL overproduction have not been determined, but most likely there is an increased synthesis of both apo B and triglycerides in the liver. The degree of increase in plasma triglycerides in obese patients will depend on a given individual's lipolytic capacity. Likewise, the overall lipoprotein pattern in an obese patient will depend on his/her capacity to catabolize lipoproteins at the various sites in the delipidation cascade. Certainly many obese patients do not have hyperlipidemia in spite of overproduction of VLDL particles because they have efficient mechanisms to catabolize lipoproteins[45].

Nephrotic syndrome. Another cause of secondary mixed hyperlipidemia is the nephrotic syndrome. This disorder may give rise to multiple defects of lipoproteins. Studies in experimental animals[46] have shown that the nephrotic state causes excessive synthesis of VLDL particles. This response may be secondary to a reduced level of plasma albumin. In our studies[14] on four patients with the nephrotic syndrome, overproduction of LDL-apo B was noted, which most likely was indicative of an excessive production of apo B-containing lipoproteins. This excessive production of lipoproteins probably explains the hypercholesterolemia of the nephrotic syndrome. Beyond this, however, many individuals with the nephrotic syndrome have hypertriglyceridemia. In our nephrotic patients[14], the decay of VLDL-apo B was markedly delayed, and the shape of the curve suggested sluggish delipidation. Thus, the hypertriglyceridemia of the nephrotic syndrome appears to be secondary to delayed lipolysis of VLDL-triglycerides, and not mainly to overproduction of VLDL particles. The overproduction of apo B was manifest to a greater extent in the LDL fraction than in VLDL.

Defects in Remnant Removal

VLDL remnants are removed mainly by LDL receptors. At least three defects appear to interfere with direct hepatic uptake of VLDL remnants. The most striking example of remnant removal defect is the presence of an abnormal form of apo E, namely, apo E2[47]. The latter, apo E2, has a poor affinity for LDL receptors, and in E2/E2 homozygotes, VLDL remnants accumulate in plasma. Second, in patients with a reduced activity of LDL receptors, remnant uptake by the liver may be defective. Finally, patients with lipolytic defects may have a delayed uptake of VLDL remnants. If the latter in fact is true, its mechanism is unknown; possibly there is a relative deficiency of apo E, i.e. there may be insufficient apo E to supply the excessive number of VLDL particles in plasma.

Patients who have only one of these three defects often do not have marked accumulation of VLDL remnants. This is noted in patients with either the E2/E2 phenotype or deficiencies of LDL receptors. However, when two abnormalities [(e.g. (a) overproduction of VLDL + remnant removal defect or (b) two different remnant clearance defects)] are present in the same individual, then distinct hyperlipidemia will result[44]. Delayed removal of VLDL remnants leads to progressive accumulation of cholesterol ester, transforming normal VLDL remnants into beta-VLDL[48]. When this occurs, patients will develop mixed hyperlipidemia, i.e. increased triglycerides and cholesterol.

Reduced Activity of LDL Receptors

The LDL-receptor mediated clearance of LDL is a major pathway for clearance of triglyceride-rich lipoproteins as well as for LDL[6]. Thus, in hypertriglyceridemic patients who have a concomitant reduction in activity of LDL receptors, accumulation of cholesterol-rich VLDL and IDL should be accentuated. Likewise, the level of LDL-cholesterol may be raised to produce type 2b hyperlipoproteinemia, i.e. elevations of both VLDL and LDL. Since methods are not available to assess total LDL-receptor activity in humans, it has been impossible to prove that individual hypertriglyceridemic patients have a concomitant reduced activity of LDL receptors. Still, because of the high frequency of the two disorders in the general public, their coexistence in the same individual must not be rare.

Abnormalities in HDL Metabolism

Since hypertriglyceridemia per se can produce profound alterations in the metabolism of HDL, it is difficult to be certain that an underlying defect in HDL metabolism occurs together with a primary disorder of triglyceride metabolism. However, it has been noted that normalization of triglyceride levels in some hypertriglyceridemic patients does not raise HDL-cholesterol levels to normal. Whether this phenomenon reflects an underlying abnormality of HDL metabolism remains to be determined.

TREATMENT OF HYPERTRIGLYCERIDEMIA

Dietary Therapy

Diet modification has a definite role to play in treatment of hypertriglyceridemia. The first step is to modify dietary factors that alter the metabolism of triglycerides; thus, it is reasonable to eliminate excess ingestion of carbohydrates and alcohol. Certainly weight reduction in obese people, especially those with NIDDM, is indicated. In patients with marked hypertriglyceridemia and chylomicronemia, who very likely have a severe lipolytic defect for triglyceride-rich lipoproteins, restriction of dietary fat to 10 to 15% of total calories will reduce the burden chylomicrons, and will reduce the risk for acute pancreatitis.

The desirable level of fat intake for patients with moderate hypertriglyceridemia is open to question. Low-fat diets have become a standard recommendation for treatment of hyperlipidemia[49,50]. Current recommendations generally call for total fat intakes of 30% or less. Low-fat diets of necessity have an increased carbohydrate intake, which may raise the triglyceride levels, and they certainly will not reduce triglycerides. Further, very low intakes of fat will raise triglyceride levels and reduce HDL-cholesterol concentrations[51]. Therefore the potential benefit of diets relatively high in fat is worthy of consideration. Our studies have shown that monounsaturated fatty acids will reduce LDL-cholesterol levels as much as high-carbohydrate diets[51]; moreover, as compared to dietary carbohydrates, monounsaturates will lower plasma triglycerides and raise HDL-cholesterol levels. For this reason, we propose that for patients with hypertriglyceridemia who are not obese, a higher fat intake, e.g. 40 to 45% of total calories, may be preferable to low-fat diets; in addition, the increase in fats should be of the monounsaturated variety, and not saturated.

Nicotinic Acid

A powerful triglyceride-lowering drug is nicotinic acid. This drug appears to act primarily in the liver to devease hepatic synthesis of VLDL particles. This mechanism can account for the action of nicotinic acid to reduce both VLDL and LDL levels. Studies in our laboratory have shown that nicotinic acid decreases the production of VLDL-triglycerides[52]; also Langer et al[53] have reported that the drug reduces the production of LDL-apo B. Thus, for hypertriglyceridemic patients who have a concomitant overproduction of VLDL particles, as in those with FCHL, nicotinic acid appears to be the drug of choice. At the clinical level, nicotinic acid should be considered for those patients who have primary mixed hyperlipidemia, i.e. patients with increased plasma levels of both cholesterol and triglycerides. For most patients with moderate mixed hyperlipidemia, nicotinic acid should "normalize" both plasma cholesterol and triglycerides. Nicotinic acid certainly will reduce the triglyceride levels in patients with primary hypertriglyceridemia who have no concomitant defects in lipoprotein metabolism; however, because of the common occurrence of side effects with this drug, the question can be asked whether it is necessary to resort to nicotinic acid for patients of this type.

Fibric Acids

Several fibric acids--clofibrate, gemfibrozil, bezafibrate, and fenofibrate--are available for treatment of hyperlipidemia. These drugs have the potential to reduce the plasma triglyceride levels similarly to nicotinic acid. The mechanisms whereby the fibric acids lower the plasma triglycerides have been studied extensively, but remain to be determined with certainty. Two mechanisms have been proposed. In laboratory animals, the fibric acids appear to inhibit the synthesis of triglycerides in the liver[54]; if this same action pertains in humans, it should reduce hepatic secretion rates of VLDL-triglycerides. In contrast to nicotinic acid, the fibric acids appear to reduce hepatic secretion of VLDL-triglyceride more than of VLDL-apo B; this suggests that the fibrates decrease the triglyceride content of VLDL particles, without necessarily reducing the overall number of VLDL particles secreted into plasma. It also has been proposed that these drugs promote the lipolysis of VLDL-triglycerides, presumably by enhancing the activity of lipoprotein lipase[55]. Whether this increased activity is due to a direct effect of the fibric acids on the synthesis of lipoprotein lipase, or is secondary to reduced utilization of available lipase by a decreased influx of VLDL-triglycerides, has not been determined with certainty.

We might address first the usefulness of fibric acids for the treatment of primary hypertriglyceridemia. These drugs are valuable for marked hypertriglyceridemia with chylomicronemia in patients at increased risk for acute pancreatitis. In many patients of this type, the fibric acids will lower triglycerides to a range that greatly reduces the risk for pancreatitis. In patients with primary moderate hypertriglyceridemia, the fibric acids usually lower triglyceride levels to the normal range. Simultaneously, there usually is a rise in HDL-cholesterol levels, often to the normal range. Further, the LDL-cholesterol level concentrations will rise because readjustments the catabolism and composition of LDL to normal[36]. In patients with primary forms of hypertriglyceridemia, however, the LDL-cholesterol does not increase to above 160 mg/dl. Thus, in such patients, fibric acids will essentially correct the abnormalities in triglyceride metabolism and will normalize the lipoprotein profile.

The situation for primary mixed hypertriglyceridemia is more complicated. Whereas the fibric acids will correct the defects in triglyceride metabolism, other lipoprotein abnormalities will remain. The fibric acids are best for treatment of remnant removal defects, i.e. in familial dyslipoproteinemia; in this disorder, levels of beta-VLDL are markedly reduced. This response suggests that defects of triglyceride metabolism contribute in a major way to the development of type 3 HLP. In other forms of mixed hyperlipidemia, fibric-acid therapy will raise the LDL-cholesterol levels by the mechanisms described above, and in these patients, levels rise to abnormally high levels, i.e. over 160 mg/dl. Thus, the question can be asked whether fibric acids provide any benefit in patients with primary mixed hyperlipidemia. In this condition, their utility might be increased by combination with a cholesterol-lowering drug[56].

<u>HMG CoA Reductase Inhibitors</u>

Several drugs in this category are becoming available for treatment of hyperlipidemia. The drug studied in our laboratory is lovastatin. The reductase inhibitors generally are considered for treatment of hypercholesterolemia--or elevated LDL. However, in our view, these drugs may be useful for treatment of some forms of hypertriglyceridemia. Several potential uses can be examined.

One use of HMG CoA reductase inhibitors may be for treatment of familial dysbetalipoproteinemia. East et al[57], in a preliminary report, showed that lovastatin will reduce levels of beta-VLDL in one patient with type 3 HLP. Subsequently, Vega et al[8] examined the use of lovastatin in three more patients with dysbetalipoproteinemia. Again, lovastatin was effective in reducing levels of beta-VLDL. Presumably, this action is a result of induction of an increased number of LDL receptors, which can bind beta-VLDL. The data of this study[8] suggest that lovastatin is most effective in patients who do not have major concomitant defects in triglyceride metabolism. Those who have such defects may respond better to drugs that mainly affect triglyceride metabolism, such as gemfibrozil.

Another potential use of reductase inhibitors is for patients with nephrotic syndrome. Not only does lovastatin lower the LDL in this disorder, but it also reduces VLDL levels in those with associated hypertriglyceridemia[14]. As indicated before[14], these patients have a sluggish lipolysis of VLDL-triglycerides. When nephrotic patients are treated with lovastatin, the clearance of VLDL is accelerated; this probably is due to enhanced removal of VLDL remnants via an increased number of LDL receptors.

Still another usage of HMG CoA reductase may be in patients with NIDDM. Garg and Grundy[58] recently reported that lovastatin effectively lowers total cholesterol levels in men with NIDDM; this lowering extends to VLDL-cholesterol (and VLDL-triglycerides) as well as to LDL-cholesterol. Since increased levels of VLDL have been implicated as a risk factor for CHD in NIDDM, the VLDL-lowering action by lovastatin may provide a beneficial result.

Finally, in primary mixed hyperlipidemia of other forms (e.g. FCHL), HMG CoA reductase may have a role. For example, East et al[56] recently reported that the combination of gemfibrozil and lovastatin often will normalize the lipoprotein profile in patients with FCHL. Although gemfibrozil may be useful for correcting a reduced HDL-cholesterol in these patients, future studies may show that lovastatin alone may be sufficient in many patients with primary mixed hyperlipidemia.

REFERENCES

1. M. J. Albrink and E. B. Man, Serum lipids, hypertension and coronary artery disease, Am J Med 31:4 (1961).
2. A. M. Gotto, G. A. Gorry, J. R. Thompson, J. S. Cole, R. Trost, D. Yeshurun and M. E. DeBakey, Relationship between plasma lipid concentrations and coronary artery disease in 496 patients, Circulation 56:875 (1977).
3. L. A. Carlson, L. E. Bottinger and P-E. Ahfeldt, Risk factors for myocardial infarction in the Stockholm prospective study: A 14-year followup focusing on the role of plasma triglyceride and cholesterol, Acta Med Scand 206:351 (1979).
4. L. A. Carlson and L. E. Bottinger, Ischemic heart disease in relation to fasting values of plasma triglycerides and cholesterol, Lancet 1:865 (1972).
5. S. B. Hulley, R. H. Rosenman, R. D. Bawol and R. J. Brand, Epidemiology as a guide to clinical decisions: The association between triglyceride and coronary heart disease, N Eng J Med 302:1383 (1980).
6. M. S. Brown and J. L. Goldstein, A receptor-mediated pathway for cholesterol homeostasis, Science 232:34 (1986).
7. R. W. Mahley and T. L. Innerarity, Lipoprotein receptors and cholesterol homeostasis, Biochim Biophys Acta 737:197 (1983).
8. G. L. Vega, C. East and S. M. Grundy, Lovastatin therapy in familial dysbetalipoproteinemia: Effects on kinetics of apolipoprotein B, Atherosclerosis (in press).
9. C. J. Packard, A. Munro, A. R. Lorimer, A. M. Gotto Jr. and J. Shepherd, Metabolism of apolipoprotein B in large triglyceride-rich VLDL of normal and lipoprotein lipase-deficient humans, J Clin Invest 74:2178 (1984).
10. A. F. H. Stalenhoef, J. J. Malloy, J. P. Kane and R. J. Havel, Metabolism of apolipoprotein B-48 and B-100 of triglyceride-rich lipoproteins in normal and lipoprotein lipase-deficient humans, Proc Natl Acad Sci USA 81:1839 (1984).
11. W. F. Beltz, Y. A. Kesaniemi, B. V. Howard and S. M. Grundy, Development of an integrated model for analysis of the kinetics of apolipoprotein B in plasma lipoproteins VLDL, IDL, and LDL, J Clin Invest 76:575 (1985).
12. Y. A. Kesaniemi, W. F. Beltz and S. M. Grundy, Comparisons of metabolism of apolipoprotein B in normal subjects, obese patients, and patients with coronary heart disease, J Clin Invest 76:586 (1985).

13. G. Egusa, W. F. Beltz, S. M. Grundy and B. V. Howard, The influence of obesity on the metabolism of apolipoprotein B in man, J Clin Invest 76:596 (1985).
14. G. L. Vega and S. M. Grundy, Lovastatin therapy in nephrotic hyperlipidemia: Effects on lipoprotein metabolism, Kidney International (in press).
15. C. L. Malmendier and M. Berman, Endogenously labeled low density lipoprotein triglyceride and apoprotein B kinetics, J Lipid Res 19:978 (1978).
16. M. Berman, W.F. Beltz, R. Riemke, A. Sedaghat, and S.M. Grundy, HDL triglyceride kinetics and exchanges with VLDL in vivo. in "Lipoprotein Kinetics and Modeling", M. Berman, S.M. Grundy, and B.V. Howard, eds. Academic Press, New York pp. 299-305 (1982).
17. S. M. Grundy (chairman), E. Barrett-Connor, E. L. Bierman, T. B. Clarkson, W. R. Harlan, W. R. Hazzard, D. B. Hunninghake, D. T. Mason, G. O'Keefe, H. Rifkin, A. A. Spector, M. Winston and P. D. Wood, The Consensus Development Conference on the Treatment of Hypertriglyceridemia, National Institutes of Health. Published J Amer Med Assn 215:1196 (1984) and Arteriosclerosis 4:296 (1984).
18. A. Chait, J. J. Albers and J. D. Brunzell, Very low density lipoprotein overproduction in genetic forms of hypertriglyceridemia, Eur J Clin Invest 10:161 (1980).
19. J. W. Farquhar, R. D. Gross, R. M. Watner and G. M. Raven, Validation of liver and plasma triglyceride in man, J Lipid Res 6:119 (1965).
20. G. M. Reaven, D. B. Hill, R. C. Gross and J. W. Farquhar, Kinetics of triglyceride turnover of very low density lipoprotein of human plasma, J Clin Invest 44:1826 (1965).
21. S. M. Grundy and G. L. Vega, Are plasma triglyceride concentrations explained by saturation kinetics?, in "Lipoprotein Kinetics and Modeling," M. Berman, S. M. Grundy and B. V. Howard, eds., Academic Press, New York pp. 272-286 (1982).
22. S. M. Grundy, A. Chait and J. D. Brunzell, Familial combined hyperlipidemia workshop, Arteriosclerosis 7:203 (1987).
23. E. E. Nikkila and M. Kekki, Polymorphism of plasma triglyceride kinetics in normal human adult subjects, Acta Med Scand 190:49 (1971).
24. U. Beil, S. B. Grundy, J. R. Crouse and L. Zech, Triglyceride and cholesterol metabolism in primary hypertriglycerides, Arteriosclerosis 2§:44 (1982).
25. Y. A. Kesaniemi and S. M. Grundy, Dual defect in metabolism of very low density lipoprotein triglycerides in patients with type 5 hyperlipoproteinemia, J Amer Med Assn 251:2542 (1984).
26. F. L. Dunn, P. Raskin, D. W. Bilheimer and S. M. Grundy, The effect of diabetic control on very low-density lipoprotein - Triglyceride metabolism in patients with type II diabetes and marked hypertriglyceridemia, Metabolism 33:117 (1984).
27. J. Melish, N. A. Le, H. Ginsberg, D. Steinberg and W. V. Brown, Dissociation of apoprotein B and triglyceride production in very-low-density lipoproteins, Am J Physiol 239:E354 (1980).
28. J. R. Crouse and S. M. Grundy, Effects of alcohol on plasma lipoproteins and triglyceride metabolism in man, J Lipid Res 25:486 (1984).
29. J. J. Abrams, H. Ginsberg and S. M. Grundy, Metabolism of cholesterol and plasma triglycerides in nonketotic diabetes mellitus, Diabetes 31:903 (1982).
30. M. R. Taskinen, W. F. Beltz, I. Harper, R. M. Fields, G. Schonfeld, S. M. Grundy and B. V. Howard, Effects of NIDDM on very-low-density lipoprotein triglyceride and apolipoprotein B metabolism: Studies before and after sulfurylurea therapy, Diabetes 35:1268 (1986).

31. M. Sanfelippo, S. M. Grundy and L. Henderson, Transport of very low density lipoprotein triglyceride (VLDL-TG): Comparison of hemodialysis and hemofiltration, Kidney International 16:868 (1979).
32. S. M. Grundy and H. Y. I. Mok, Chylomicron clearance in normal and hyperlipidemic man, Metabolism 25:1227 (1976).
33. S. Gianturco, A. M. Gotto Jr., R. L. Jackson, J. R. Patsch, H. D. Sybers, O. D. Taunton, D. L. Yeshurun and L. C. Smith, Control of 3-hydroxy-3-methylglutaryl-CoA reductase activity in cultured human fibroblasts by very low density lipoproteins of subjects with hypertriglyceridemia, J Clin Invest 61:320 (1978).
34. S. H. Gianturco, S. G. Eskin, L. T. Navarro, C. J. Lahart, L. C. Smith and A. M. Gotto Jr., Abnormal effects of hypertriglyceridemic very-low-density lipoproteins on 3-hydroxy-3-methylglutaryl-CoA reductase activity and viability of cultured bovine aortic endothelial cells, Biochim Biophys Acta 618:143 (1980).
35. J. Shepherd, M. J. Caslake, A. R. Lorimar, B. D. Vallance and C. J. Packard, Fenofibrate reduces low density lipoprotein catabolism in hypertriglyceridemic subjects, Arteriosclerosis 5:162 (1985).
36. G. L. Vega and S. M. Grundy, Gemfibrozil therapy in primary hypertriglyceridemia associated with coronary heart disease, J Amer Med Assn 253:2398 (1985).
37. G. L. Vega and S. M. Grundy, Kinetic heterogeneity of low density lipoproteins in primary hypertriglyceridemia, Arteriosclerosis 6:395 (1986).
38. S. Eisenberg, D. Gavish, Y. Oschry, M. Fainaru and R. J. Deckelbaum, Abnormalities in very low, low, and high density lipoproteins in hypertriglyceridemia: Reversal toward normal with bezafibrate treatment, J Clin Invest 74:470 (1984).
39. E. D. Janus, A. M. Nicole, P. R. Turner, P. Magill and B. Lewis, Kinetic basis of the primary hyperlipidemias: Studies of apolipoprotein B turnover in genetically defined subjects, Euro J Clin Invest 10:161 (1980).
40. A. H. Kissebah, S. Alfarzi and P. W. Adamo, Integrated regulation for very low density lipoprotein triglyceride and apolipoprotein-B kinetics in man: Normolipidemic subjects, familial hypertriglyceridemia, and familial combined hyperlipidemia, Metabolism 30:856 (1981).
41. J. L. Goldstein, W. R. Hazzard, H. G. Schrott, E. L. Bierman and A. G. Motulsky, Hyperlipidemia in coronary heart disease. I. Lipid levels in 500 survivors of myocardial infarction, J Clin Invest 52:1533 (1973).
42. J. L. Goldstein, H. G. Schrott, W. R. Hazzard, E. L. Bierman and A. G. Motulsky, Hyperlipidemia in coronary heart disease. II. Genetic analysis of lipid levels in 176 families and delineation of a new inherited disorder, combined hyperlipidemia, J Clin Invest 52:1544 (1973).
43. G. L. Vega and S. M. Grundy, Kinetic heterogeneity of low density lipoproteins in primary hypertriglyceridemia, Arteriosclerosis 6: 395 (1986).
44. S. M. Grundy, Pathogenesis of hyperlipoproteinemia, J Lipid Res 25: 1611 (1984).
45. S. M. Grundy, H. Y. I. Mok, L. A. Zech, D. Steinberg and M. Berman, Transport of very low density lipoprotein-triglycerides in varying degrees of obesity and hypertriglyceridemia, J Clin Invest 63:1274 (1979).
46. E. Shafrir and T. Brenner, Lipoprotein lipid and protein synthesis in experimental nephrosis and plasmapheresis. I. Studies in the rat in vivo, Lipids 14:695 (1979).

47. R. J. Havel, Familial dysbetalipoproteinemia. New aspects of pathogenesis and diagnosis, Med Clin N Am 69:441 (1982).
48. J. P. Kane, G. Chen, R. L. Hamilton, D. H. Hardman, M. J. Mallory and R. J. Havel, Remnants of lipoproteins of intestinal and hepatic origin in familial dysbetalipoproteinemia, Arteriosclerosis 3:47 (1983).
49. S. M. Grundy (ed), A. M. Gotto Jr., E. L. Bierman et al., The Nutrition Committee, and Council on Arteriosclerosis of the American Heart Association. Recommendations for Treatment of Hyperlipidemia in Adults, Circulation 69:1067A (1984) and Arteriosclerosis 4:445A (1984).
50. Report of the National Cholesterol Education Program Expert Panel on Detection, Evaluation and Treatment of High Blood Cholesterol in Adults, Arch Intern Med 148:36 (1988).
51. S. M. Grundy, Comparison of monounsaturated fatty acids and carbohydrates for plasma cholesterol lowering, N Engl J Med 314: 745 (1986).
52. S. M. Grundy, H. Y. I. Mok, L. Zech and M. Berman, Influence of nicotinic acid on metabolism of cholesterol and triglycerides in man, J Lipid Res 22:24 (1981).
53. T. Langer and R. I. Levy, The effect of nicotinic acid on turnover of low density lipoproteins in Type II hyperlipoproteinemia, in "Metabolic Effects of Nicotinic Acid and Its Derivatives", K. F. Gey and L. A. Carlson, eds., Huber, Bern, New York (1971).
54. M. E. Maragandakis and H. Hankin, On the mode of action of lipid lowering agents. V. Kinetics of the inhibition in vitro of rat acetyl CoA carboxylase, J Biochem 246:348 (1971).
55. J. Boberg, M. Boberg, R. Gross, S. M. Grundy, J. Augustin and V. Brown, The effect of treatment with clofibrate on hepatic triglyceride and lipoprotein lipase activities of post heparin plasma in male patients with hyperlipoproteinemia, Atherosclerosis 27:499 (1977).
56. C. East, D. W. Bilheimer and S. M. Grundy, Combination drug therapy for treatment of familial combined hyperlipidemia, types IIB and IV, Ann Intern Med (in press).
57. C. A. East, S. M. Grundy and D. W. Bilheimer, Preliminary Report: Treatment of type 3 hyperlipoproteinemia with mevinolin, Metabolism 35:97 (1986).
58. A. Garg and S. M. Grundy, Lovastatin for lowering cholesterol levels in non-insulin-dependent diabetes mellitus, N Engl J Med 318:81 (1988).

INCREASED VLDL

M.J. Halpern and M.F. Mesquita

Center for Lipid Research and Dept. Biochemistry
Faculty of Medical Sciences
Campo Santana, 130- 1100 Lisbon, Portugal

Increased VLDL is observed in hyperlipemic dyslipoproteinemia (type IV, V and IIb) and in latent dyslipoproteinemia (latent type IV). In this paper we shall discuss type IV and latent type IV and we shall do some remarks on type V.

TYPE IV

Prevalence in Portugal

Is the most prevalent type in the Portuguese population. In table I we sumarize the results of our first screening in rural and urban population (1).

In table II we show the distribution of hyperlipemia observed in our laboratory in the last three months of 1987.

Serra e Silva (2) in a screening did in Coimbra (table III) showed the same results. It is worthwhile to note that the tracking in medical doctors showed a different distribution.

We observed this type in about 10% of 250 newborns (3, 4). Studies done in the families showed a genetic background. This finding raises a new question - the prevalence of type IV, hyperlipemia usually regarded as a characteristic of the adult.

Biochemical characterization

We observed increased apo B, decreased apo $CIII_1$ and increased apo $CIII_2$ (5).

We think that the increased apo B can be considered as a marker of combined familial hyperlipemia.

Relationship with hyperuricemia

We described a frequent association between hypertriglyceridemia and hyperuricemia (6, 7, 8).

If we compare the distribution of hyperlipemia in several leves of serum uric acid and it is easy to conclude the high distribution of the hypertriglyceridemic types (IV and IIb) in the higher levels of uric acid.

It is possible to obtain an uric and decrease with a hypocaloric and hyperproteic diet.

Relationship with fructose

Fructose loads showed us that the most part of type IV were glucide--sensitive (8).

Relationship with alcohol

To study alcohol sensitive we proposed a load fast with 0,4 l of wine with determination at o', 15', 30' 60' and 120'. In alcohol sensitive patients we observed an increase of triglycerides after 60'. That continue after take 120' sample (9).

Symptoms

The more frequent symptoms observed in these patients are dizziness, buzzing, post-prandial somnolence and asthenia. It is often observed an association with type IV, obesity, hyperuricemia, polyglobulia, decreased HDL, hyperinsulinism (plurymetabolic Syndrome).

LATENT TYPE IV

Definition

It is an increase of VLDL (as shown by electrophoresis or ultracentrigugation) with normal triglycerides (10, 11). After a three hours fasting we observe a triglyceride increase in this type but not in normal people (12).

Biochemical characterization (table VII)

The biochemical characterization shall be presented by M.F. Mesquita et all (5).

The existence of increased apo B-VLDL afford us to think in the possibility to afiliate this type on combined familial hyperlipemia.

TYPE V

It is often associated with pancreatitis.

As a therapeutic approach we think that it is important to avoid the transformation of chylomicra in remnants with a fibrate therapy. We propose a stepwise treatment - inicial treatment with short-chain fatty acids or fish oil for chylomicra disappearance and only after the fibrate therapy.

Table 1. Hyperlipemia in urban and rural populations

Pattern	Urban (nº 139)	Rural (nº 152)
Normal	71,9%	59,9%
IIa	2,2%	4,0%
IIb	2,2%	5,3%
IV	23,7%	30,9%

Table II. % Hyperlipemia observed in the laboratory

Pattern	%
IV	69,8
IIa	15,9
IIb	13,7
V	0,6

Table III.

Populations	% Cholesterol	increased Triglycerides
General	4,6	20,7
Doctors	15,8	11,1

Table IV. Triglycerides and uric acid

Pattern	Uric acid			
	< 5mg%	5.6	5.7	> 7
Normal	57,3	27,6	15,7	12
IIa	5,1	8,0	4,3	5,3
IIb	10,8	13,8	18,6	18,7
III	0,1	0	0	0
IV	26,7	40,5	60,0	64,0
V	0	1,1	1,4	0
n	157	87	70	75

Table V. Dyslipoproteinemia in CHD

Pattern	After	
	3 months	6 months
Normal	35,5 %	29,6
IIa	15,3	10,3
IIb	3,4	7,3
IV	38,4	40,6
Latent IV	7,3	11,1
Hyper VLDL my (IV+lat.IV)	45,7	52,3
n	177	145

Table VI. Dyslipoproteinemia in CVD versus age (%)

Age	45	45-54	55-64	65
Normal	0	0	0	40
IIb	0	20	0	0
HDL_2	83,2	0	8,4	0
IV	0	40,0	66,5	40
Lat. IV	16,8	40,0	25,1	20
IV + lat IV	16,8	80,0	91,6	60

Table VII. Dyslipoproteinemia in PAD versus age (%)

	45	45-54	55-64	65
Normal	0	0	18,5	8,3
IIb	0	37,5	11,2	8,3
HDL_2	33,3	12,5	22,2	25,0
IV	33,3	50,0 a)	22,2	49,1
Lat. IV	33,3	0	25,9	8,3
IV + Lat.IV	66,6	50,0	48,1	57,4

VLDL AND ATHEROGENESIS

This point is until now controversial. However some results shows that triglycerides can be considered a risk factor.

1. Increased triglycerides in M.I. Survivers (13)

2. In the families of M.I. patients the hypertriglyceridemia is prevalent (14, 15)

3. Discriminative analyses did in some epidemiological studies showed a good correlation between a CHD and triglycerides. (16, 17, 18, 19)

4. Hanefeld (20) Waldius (21) showed that when triglycreides decrease the coronary mortality fall.

5. Genetically hipertensive rats has normal cholesterol but increased triglycerides.

In a study done in coronary heart disease we discribed the existence of increased VLDL (IV+Lat.IV) in 45,7% of patients 3 months after myocardial and in 52,3% six months after (table V).

In cerebro-vascular disease and Peripheral Artherial Disease we observed similar results (23). (table VI and VII).

We calculate the risk factor for these dyslipidemia (24) (table VIII).

The existence of increase skin cholesterol (25) in these situations (table IX) let us the question if VLDL are atherogenic as itself or associated to an abnormal cholesterol transport.

Table VIII. Dyslipoproteinemia as a risk factor (%)

	Normal	Type IV	Lat. T.IV	IIb
PAD	o,85	0,85	1,25	1,70
CVD	0,63	1,10	1,70	0,85
CHD	-	1,20	2,50	0,10

Table IX. Skin cholesterol as tool to atherosclerosis study (in mg/g of skin) Reference values > 1,4 mg/g)

	Normal	Type IV	Lat.T.IV	IIa	IIb
CHD	-	2,8 ± 0,4	2,0 ± 0,1	1,9	1,9
PAD	2,0 ± 0,1	2,7 ± 0,2	1,9 ± 0,1	-	-
CVD	1,6 ± 0,1	2,0 ± 0,0	2,0 ± 0,0	-	2,5

REFERENCES

1. Miguel M.J.P.; Halpern M.J. - Am. Biol. Clin. 3:145 (1983)
2. Serra Silva P. et all. - Halpern M.J. (ed) Lipid Metabolism and its Pathology (Excerpta Medica International Congress Series 670) 175 (1986)
3. Martins F.M. et all - Brit Med. J. II:544 (1973)
4. Amaral JMV et all - Halpern MJ (ed). Lipid Metabolism and its Pathology (Excerpta Medica International Congress Series 670:143 (1986)
5. Mesquita M.F. et all - These Proceedings
6. Halpern M.J. - Am. J. Cardiol 95:540 (1973)
7. Halpern M.J. - Acta Med. Scand. 202:335 (1977)
8. Halpern M.J. and Mesquita M.F. - Beynen A (eds) - Effects on cholesterol Metabolism (transmundial, Voorthinzen) 81 (1986)
9. Halpern M.J. (eds) Lipid Metabolism and its Pathology (Plenum Press) 153 (1985)
10. Halpern M.J. - Nouvelle Press Med. - 2:3124 (1973)
11. De Gennes J.L. Turpim G. and Truffert J. - Nouvelle Press Med. - 1:1627 (1972)
12. Halpern M.J. in De Gennes et all (eds)- Latent Dyslipoproteinemia and atherosclerosis (Raven Press NY 1 41 (1984)
13. Vallek J et all - Nutr. Metab. 16:193 (1974)
14. Patterson D. and Slack J - Lancet I:393 (1972)
15. Nikkila EA and Aro A - Lancet I 954-958 (1973)
16. Castelli B, Doyle JI, Gordon I et all - Circul. 55:767-772 (1977)
17. Carlson LA and Bottiger LE - Ather. 38:287 (1981)
18. Aberg H et all - Ath. 54:84-97 (1985)
19. Carlson LA and Aberg H - NEJM 312:12271 (1985)
20. Henefeld M et all - Ather. 53:47-58 (1984)
21. Weldius G - in Fidge NH and Nestel PJ (ed)- Atherosclerosis (Excerpta Medica) 205-208 (1986)
22. França, A et all - in Halpern M.J. (eds) Lipid Metabolism and its Pathology (Excerpta Medica International Congress series) 670 -119 (1986)
23. Manso Ribeiro et all - in Halpern M.J. (eds) Lipid Metabolism and its Pathology (Excerpta Medica International Congress Series 670-127 (1986)
24. Mesquita et all - in Halpern M.J. (eds) Lipid Metabolism and its Pathology (Excerpta Medica International Congress series 670-133 (1986)
25. Halpern et all - in Malmendier C.L. and Alaupovic P. (eds) Lipoproteins abd atherosclerosis (Advances in experimental Medicine and Biology) Plenum Press vol. 210-213 (1987)

EXCHANGE AND TRANSFER OF APOLIPOPROTEINS AND LIPIDS:

IMPACT ON LIPOPROTEIN METABOLISM

Gert M. Kostner[§], Karin Schaupp[#)] and Gerhard Stvarnik[§)]

§) Institute of Medical Biochemistry, University of Graz
A-8010 Graz, Austria
#) Research Laboratories of Leopold & Co Graz, Austria

INTRODUCTION

At present time a minimum of 20 different apolipoproteins have been characterized in human plasma (for a review see Ref.1-2). Most if not all of them play a key role in the metabolism of plasma lipids and Lp: Among other functions they activate enzymes or modulate their activity; they mediate specific binding of Lp to cell surface receptors ant they promote exchange or unidirectional transfer of lipids from one Lp-class to another.

Lp are biosynthesized in the liver and the intestine and secreted in the form of "nascent particles" into the blood stream. Although detailed knowledge on the actual nature of nascent lipoproteins is still lacking, there seems to be little doubt that certain classes of apolipoproteins are taken up by nascent Lp upon their entry in the extracellular compartment, or during circulation in blood and lymph and the action of enzymes and transfer proteins .These latter processes also lead to a redistribution and exchange of apo-Lp during the whole intravascular catabolism.

The apo-Lp which are known to behave in such a manner are the apoC's apoE apoAIV and apoH. There is no doubt that the physiological rationale of this apo-Lp movement is the reutilization of these factors on one hand and the targeting of certain Lp classes to specific organs. With that respect the B/E-receptors and the E-receptors play a role of utmost importance.

Lipoproteins which are turning over very fast and which have been connected to atherogenesis are chylomicrons (CYM) and large VLDL. They consist to a considerable degree of TG and are secreted in their nascent form probably only together with apoB and possibly few other apo-Lp. Since it is known that artificial TG-rich emulsions under optimal conditions behave in vivo similar to CYM, we addressed in this study the question what factors may mediate specific apo-Lp binding to large TG-rich particles, and as a consequence influence the Lp metabolism.

MATERIAL AND METHODS

Five different artificial TG-rich emulsions have been used. Three of them were commercial products and 2 were prepared in the research laboratories of one of us (K.S.). All emulsions had a starting TG-concentration of 10 % (w/v). Some of their other features are listed in Table I.

TABLE 1. Composition of the 5 Emulsions (E) Used in this Study.

Nr.	Code	Source of TG	Emulsifier	SIZE(μ)[1]	Stabilizer
1	LF	Soy bean oil 10% w/v	Soya lecithin 0.75 % w/v	0.35	Glycerol 2.5% w/v
2	IL	Soy bean oil 10% w/v	Egg lecithin 1.2% w/v	0.40	Glycerol 2.5% w/v
3	VL	Soy bean oil 10% w/v	Soya lecithin*) 1.2% w/v	0.15	Glycerol 2.5% w/v
4	L1	Soy bean oil 10% w/v	Soya lecithin*) 1.2% w/v	0.20	Glycerol 2.5% w/v
5)	L2	Soy bean oil 10/ w/v	Soya lecithin*) 1.2% w/v	0.25	Glycerol 2.5% w/v

*) The lecithin used was partially hydrogenated.
1) Size distribution was determined by photon correlation spectroscopy.

Only E-2 contained lecithin from egg yolk, and thus was contaminated with measurable amounts of cholesterol. All other emulsions were prepared with plant lecithin. E- 3-5 were essentially identical with respect to their composition but differed significantly in size.

Measurement of the substrate properties against PHLA

PHL-active plasma samples were prepared from healthy normolipemic volunteers 10 - 15 min after the administration of 100 USP heparin per kg of body weight. We considered it as essential to measure LPL activity by acidimetry; in this way all FFA hydrolyzed from PL and glycerides were accounted for. Radioactive or fluorometric methods were avoided because they strongly reflect the distribution of the tracer within the emulsion.

The emulsions were incubated at 37°C with PHLA-plasma at increasing concentrations ranging from 120 to 6000 mg/dl of TG. The FFA generated during this process were extracted by standard procedures and titrated with 1 mM NaOH.

The stability of the emulsions was tested in the following way: Emulsions were mixed with normal fasting plasma at a concentration of 1200 mg/dl, incubated for 30 min in a shaking water bath at 37°C , and centrifuged for 10 min at 20 000 g. The floating creamy layer was washed severeal times with saline and the breakage of the emulsion was tested by visual inspection.

The binding of apo-Lp was studied by incubating the emulsions at 500 mg/dl TG concentrations with normal fasting plasma for 1h at 37°C. The mixture was centrifuged in a swinging bucket rotor whereby tubes were

filled approx. one third of their height, and overlayered with phosphate buffered saline. The floating particles were lyophilized and delipidated with ethanol:ether 3:1 as described earlier (3). The remaining apo-Lp were separated by SDS and urea- polyacrylamide gel electrophoresis. After staining with Coomassie blue, gels were evaluated by densitometry. Apo-E isoforms were determined by isoelectric focussing in a pH gradient of 5-7 following standard procedures.
In vivo metabolism: The in vivo metabolism of the emulsions was studied by infusing them intravenously into 5 fasting (14h) normolipemic volunteers at a rate of 0.1g/kg/h for 8h. Plasma lipids and lipoproteins were measured before and at the time points 4, 8, 11 and 24 hours. The probands did not eat any food during this 24h study period. Only E-1 E-2 and E-3 were used for the *in vivo* experiments. Each volunteer received all 3 emulsions with intermissions of 1 wee in a random sequence.

RESULTS AND DISCUSSION

I) Stability of the emulsions

The 5 emulsions were mixed with human plasma, incubated and washed several times with saline. During this process, E-1 and E-5 started to disintegrate after 3 washes, whereas the other emulsions remained stabile for at least 5 washes. E-1 tended to exhibit the lowest stability. The reduced stabilits of E-1 may be caused by the relatively low content of emulsifier.

II) Binding of apo-Lp and plasma proteins to the emulsions

After incubation with plasma, the TG-rich particles were floated by centrifugation and investigated by gel electrophoresis. The relative distribution of proteins associated with the particles is shown in Tab.2.

TABLE 2A. Relative Distribution of Various Apo-Lp Associated with the Different TG-emulsions: Results from PAGE in 10% urea containing gels. Incubations were performed at 500 mg/dl of TG.

EMULSION	SCANNER UNITS IN % OF TOTAL				
	Alb.	ApoCII	$ApoCIII_1$	$ApoCIII_2$	ApoCII/tot.ApoC
1	56	14	18	12	0.331
2	49	14	24	13	0.269
3	56	12	21	11	0.269
4	60	12	16	12	0.309
5	23	19	32	20	0.271

TABLE 2B. The same as Tab.2A but the associated proteins were separated by SDS-PAGE using 12.5% gels.

EMULSION	SCANNER UNITS IN % OF TOTAL				
	Alb.	ApoH	ApoE	ApoAI	ApoH/ApoE
1	68	16	9	7	1.613
2	80	4	7	9	0.550
3	93	<1	<1	7	---
4	94	<1	<1	6	---
5	57	8	26	9	0.302

From these experiments we conclude that E 1-4 differed little with respect to apoC binding. The total relative amount of apoCII and apoCIII bound by E-5 however was significantly higher whereas the CII/CIII ratio, the factor which is believed to regulate the LPL activity in *in vitro* assays was comparable with the other 4 emulsions.
Of interest was also the relative amount of the other apo-Lp bound by the emulsions: There was no difference in apoAI binding. E-5 exhibited the largest relative apoE binding in the order of 25%. The greatest fraction of apoH binding was observed with E-1. E-3 and E-4 bound only trace amounts of apoE and apoH.
From these results we may conclude that apoC binding is little affected by size , composition and stabilizer with the exception of E-5. E-5 apparently bound relatively little albumin reflecting the high fractional amount of apoC found by scanning the gels.
There were, however striking differences in apoE and apoH binding. As the 2 smallest fractions E-3 and -4 were almost devoid of apoE and apoH we assume that binding afinity of these apo-Lp increases with particle size.

III) Substrate properties for PHLA

In these experiments the emulsions were incubated with post heparin plasma and the hydrolysis of PL + glycerides was determined by titration. We do believe that this method reflects closely the in vivo situation. The results are shown in Tab.3.

TABLE 3. Substrate Properties of the 5 Different Emulsions: E 1-5 were incubated with post heparin plasma at various TG concentrations and the hydrolysis was followed by titrating the liberated FFA. The activities are expressed in: µM FFA per ml per min.

EMULSION	T G - CONCENTRATIONS IN THE INCUBATE 120 mg/dl	400 mg/dl	1200 mg/dl	6000 mg/dl - PS*)	+PS
1	0.08	0.16	0.25	0.39	0.22
2	0.16	0.31	0.40	0.94	0.21
3	0.03	0.08	0.20	0.21	0.17
4	0.10	0.10	0.12	0.23	0.20
5	0.12	0.10	0.22	0.23	0.18

*) In one experiment, the mixture was incubated with protamin sulfate (final concentration 10 mg/ml)

The results of the experiments with PHLA may be interpreted as follows:

1) The kinetics of hydrolysis using PHLA are strikingly different with the various E. E-2 exhibited the greatest hydrolysis at all concentrations tested.
2) The maximal activity against E-1 and E- 3-5 is in the order of 0.2 µM/ml/h FFA. With E-2 a 5-fold higher reactivity ws observed at substrate saturation.
3) Increasing the size of the particles leads to a higher reactivity at concentrations below saturation.
4) Inhibition of the LPL by protamin sulfate had little influence on the emulsions 3-5 but reduced the hydrolysis of E-1 markedly and that of E-2 drastically.

From this latter observation one may speculate, that the hepatic lipase comprises most of the hydrolysis of E 3-5, and the major part of E-1; E-2 in contrast may be hydrolyzed at least at high substrate concentrations primarily by tissue lipase, which can be inactivated by the addition of protamin sulfate.

5) The emulsions 3-5 were hydrolyzed especially at low substrate concentrations to a markedly lower degree as compared to E-1 and E-2. Since E 3-5 contained partially hydrogenated lecithin (in order to reduce peroxydation during storage), one may speculate that the lower fluidity of the surface layer may be responsible for the reduced hydrolysis by PHLA. Similar findings have been published in earlier work (4).

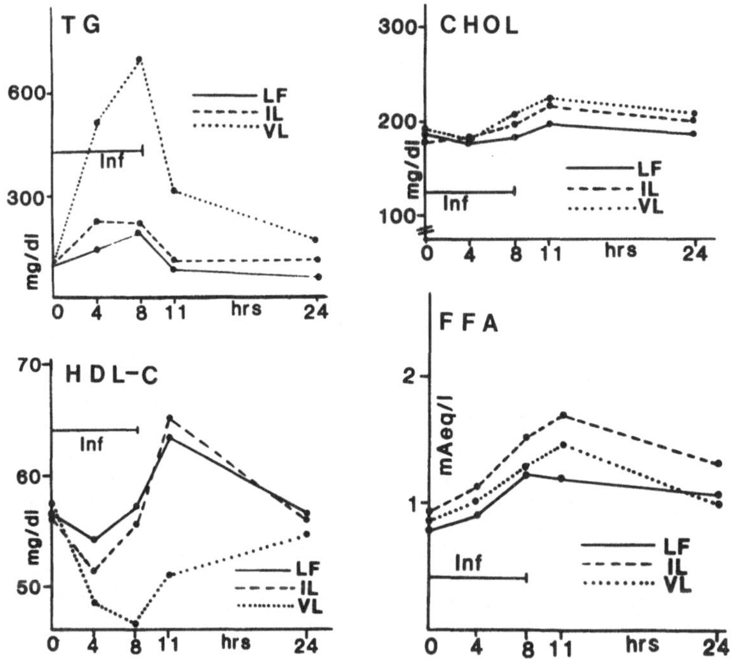

Fig 1: Time course of several plasma lipids during the infusion of 3 different artificial TG-emulsions into humans. The emulsions were intravenously infused for 8 h at a rate of 0.1 g /kg /h and plasma samples were drawn at the indicated time points. The volunteers were fasting for 14 h before the experiment and did not take any food orally during the experiment. Each individual received all 3 emulsions at weekly intervals. Values represent means obtained with 6 different volunteers.

IV. In vivo experiments

The *in vivo* experiments can be treated here only very shortly and are subject to a separate report. Only E1-3 were studied. Fig. 1 shows the time course of plasma TG, TC, FFA and HDL-C during the 24 h period

With E-2 we observed the highest increase of plasma FFA with a peak 3h after the infusion had been stopped. E-1 and E3 behaved similarly. There was however a very striking difference with respect to increments of plasma TG levels. The lowest increase was observed with E-1. With E-2 we observed a somewhat higher increase of TG. With E-3 plasma TG values reached very high levels peaking at the end of the infusion (> 700 mg/dl).
The cholesterol values of E-2 and E-3 also raised markedly. The changes of HDL-C in the course of infusion correlated negatively with the changes of TG. E-1 had no apparent influence on HDL-2 whereas a striking reduction was observed with E-3.
Apolioprotein-E isoforms:
One of the study subjects exhibited an $apoE^{2/3}$ pattern, three were homocygous for $apoE^{3}$ and one was heterozygous $E^{3/4}$.
The proband with the $apoE^{2/3}$ pattern exhibited a significantly higer increase of plasma TG as compared to the others. This was particularly true for E-1 and E-2 and not significant for E-3. From these experiments one may conclude that apoE binding to the emulsions has important implications for the in vivo catabolism of artificial TG emulsions or the biotransformation products thereoff.

CONCLUSION

Artificial TG-rich emulsions may serve as excellent models for nutritional lipid and chylomicron metabolism. If mixed with fasting plasma, emulsions bind specifically apo-lipoproteins (5,6) . These are especially apoC's, apoE, apoH and some apoAI. Probably not so much the apoC binding but rather size and the surface PL layer mediate the accessibility by post heparin lipases. There seems to be also a difference in interaction between hepatic- and tissue lipases.
There was a most marked difference of apoE binding which strongly correlated with particle size. In *in vivo* experiments we observed that emulsions which do not bind apoE lead to a striking increase of plasma TG and an markedly impaired catabolism. Since nutritional lipids strongly reflect the composition and size of lymph and plasma chylomicrons, we believe that our model might be a valuable tool for studying nutritional factors influencing plasma lipid concentrations.

REFERENCES

1. Kostner GM .Apolipoproteins and lipoproteins of human plasma:Significance for health and diseases. Adv.Lipid Res. 20 (1983)1-44.
2. Kostner GM. Isolation and characterization of high densitylipoproteins. In: High Density Lipoproteins, C.E.Day Ed. DEKKER press (1981) 133-143.
3. Kostner G, & Holasek A. Characterization and quantitation of apo-lipoproteins from human chyle chylomicrons. Biochemistry 11 (1972) 1217 - 1223.
4. Bennet SC & Derksen A. Biochim Biophys Acta 920 (1987) 37 - 46.
5. Erkelens DW & Mocking JAJ. The effect of apoA on in vitro apoC binding and in vivo removal of artificial TG-rich particles. Metabolism 34 (1985) 222 - 226.
6. Polz E, Kostner GM. The binding of ß-glycoprotein-I (apoH) to Intralipid: Determination of the dissociation constant. Biochem. Biophys. Res. Commun. 90 (1979) 1305 - 1312.

HYPERTRIGLYCERIDEMIA AND OMEGA-3 FATTY ACIDS

H.U. Kloer* and C. Luley**

*3rd Medical Department
University of Gießen
Gießen, FRG

**Clin. Chemistry Department
University of Mainz
Mainz, FRG

Recent studies by Bang and Dyerberg (1) have stimulated interest in the metabolism of omega-3 fatty acids and have suggested a link between the ingestion of these fatty acids in a diet and the low death rate from athersclerotic disease in Eskimos. In Japan, where fish consumption has traditionally been high, a concommitant shift in tissue lipid consumption favouring omega-3 polyunsaturated fatty acids has been interpreted as causative of a relatively low incidence of cardiovascular disease (1). When analyzing Eskimo food consumption, it became clear that the consumption of omega-3 PUFA is much higher besides a 50% reduction of saturated fat and a relatively high content of monounsaturated fat of chain lengths of more than 18 carbon atoms. The total fat intake of the Eskimos is in the order of 40% of total calories per day, which is as high in Western diets. In addition, cholesterol intake averages almost 800 mg, which is even higher than that usually found in Western diets.

When comparing lipid and lipoprotein levels of Greenland Eskimos and Danes, Dyerberg and Bang (2) found significantly lower LDL and VLDL concentrations leading to drastically lower triglyceride and total plasma cholesterol concentrations in Eskimos. The HDL levels were about the same.

Between 1956 and 1963 8 studies of fish oil feeding were carried out in normal volunteers (3-9). Although they varied in the kinds of subjects studied, in the degree of dietary control and in the source of fish oil, there has been good agreement that fish oils were at least as hypocholesterolemic as polyunsaturated vegetable oils. Two interesting features of these studies were not appreciated at the time. Since natural, unrefined fish oils contain large amounts of cholesterol (300-500 mg/dl) and vegetable oils contain none, the majority of investigators fed higher levels of cholesterol during the fish oil phase than during the vegetable oil phase. Also, daily intake of omega-6 fatty acids, even with the fish-oil-enriched diets, was greater than the intake of omega-3 fatty acids, and yet, similar or greater reductions in plasma cholesterol levels occured from omega-3 fatty acids. These reductions occured when omega-3 fatty acids constituted from 1 to 8% of total calories. In light of the fact that linoleic acid intakes of 15-20% of calories were needed to achieve similar depressions in plasma cholesterol levels, the dietary omega-3 fatty acids were roughly 2-5 times more potent than the omega-6 PUFAs.

Dramatic reductions in the concentrations of plasma triglycerides and VLDL have been observed in both normolipidemic (10-14) and hyperlipidemic subjects (3,15) fed diets supplemented with fish oils. Harris et al. (14) compared the effect of salmon oil and vegetable oil and found that plasma cholesterol levels were reduced similarly with both salmon and vegetable oil-rich diets. In contrast, plasma triglyceride levels fell 33% in the salmon oil diet, but were unchanged after the vegetable oil diet. VLDL concentrations fell 50% and LDL concentrations 60% without any change in HDL concentration. In a number of recent studies in hyperlipoproteinemic patients, the significant fall of VLDL and, in some groups, also of LDL was confirmed. The most dramatic effects occured in hyperlipoproteinemia Type V, which is characterized by elevated VLDL and by fasting chylomicronemia. Phillipson and colleagues (15) reported triglyceride reductions of 79% and cholesterol decreases of 45% in this group of patients. Giving a vegetable oil to these patients produced a rapid and significant rise in plasma triglyceride levels. Similar results in Type V hyperlipoproteinemia have been reported by Sanders et al (16) and Simmons et al. (17).

In contrast to extensive studies on the mechanism of action of saturated and omega-6 polyunsaturated fatty acids, there are only a few studies on the mechanism of action of omega-3 PUFA. Illingworth et al.(18) showed in a group of 7 healthy volunteers that omega-3 fatty acid administration caused a reduction of total cholesterol from 160 to 124 mg/dl and of LDL from 103 to 82 mg/dl. Triglyceride levels fell almost 50% from 91 to 52 mg/dl. In these normal subjects, kinetic studies of iodinated LDL metabolism disclosed a significantly lower rate of synthesis of LDL ApoB on the omega-3 diet. In contrast, the fractional catabolic rate of LDL was similar in both diets. These authors concluded that dietary omega-3 fatty acids lowered plasma LDL in normal human subjects by reducing the rate of synthesis of ApoB.

Nestel and coworkers (19) recently showed that the reason for the reduced VLDL concentrations was that VLDL synthesis was markedly reduced after 3 weeks of fish oil as compared to a safflower oil diet. There was also a significant reduction of daily production of VLDL triglyceride, as calculated from radiolabelled glycerolinjection studies. The fractional removal rate as a measure for VLDL catabolism was not consistantly affected, which would mean that the plasma lipolytic system was probably not activated. In a recent experiment, Harris and coworkers (20) were able to show that the hypertriglyceridemia induced by high carbohydrate intake could be prevented by dietary omega-3 fatty acids. When a high carbohydrate diet was given to normal volunteers, the plasma triglyceride level rose from 105 to 194 mg/dl. When fish oil was given, the plasma triglyceride levels fell from 194 to 75 mg/dl, and the VLDL triglyceride and cholesterol levels were reduced from 156 to 34 and from 34 to 12 mg/dl, respectively. These effects were noted by 2 or 3 days of fish oil diet indicating an extremely rapid onset of the fish oil effect.

As mentioned above, evaluation of the Eskimo diet showed a fairly high intake of dietary cholesterol in the presence of very low average plasma cholesterol and LDL cholesterol level (21). This raises the question whether omega-3 fatty acids might be able to counteract the hypercholesterolemic effect of cholesterol. In a recent study by Nestel (22)this question was addressed in 6 normolipidemic subjects, who received first a Western diet, then a fish oil-enriched diet providing approximately 13 g of omega-3 fatty acids per day besides a lowered cholesterol and lowered/saturated fat content. Switching to that diet resulted in a 20% decrease of plasma cholesterol, VLDL and LDL cholesterol as well as a highly significant decrease in HDL cholesterol. Plasmatriglycerides decreased by more than 60%. Maintaining that same diet but adding 800 mg of cholesterol per day as egg yolk did not lead to a substantial rise in the plama lipid or lipoprotein parameters measured. Although the dietary periods were perhaps not long enough (3-week periods) and too many variables were changed at the same time (saturated fat, cholesterol content, fish oil), this study still raises the interesting question whether pure supplementation with omega-3 fatty acids in relatively small amounts might be able to counteract the hypercholesterolemic effect of saturated fat and cholesterol-containing food items.

We recently studied the effect of supplementation of an ordinary Western diet with 8 g of omega-3 fatty acid as purified fish oil capsules in 20 healthy volunteers. In particular, the rise of plasma triglycerides induced by an oral fat load was studied before and after a 4-week period of omega-3 fatty acid supplementation. As expected, a fat load consisting of 100 g of butter led to a significant postprandial increase of triglyceride over a period of 5 to 7 hours. However, only half of the participants showed a very significant rise in triglyceride of close to 100% postprandially. After 4 weeks of fish oil capsule treatment, the fat load with butter was repeated. This time, none of the participants showed any significant increase in postprandial triglycerides (Fig.1)

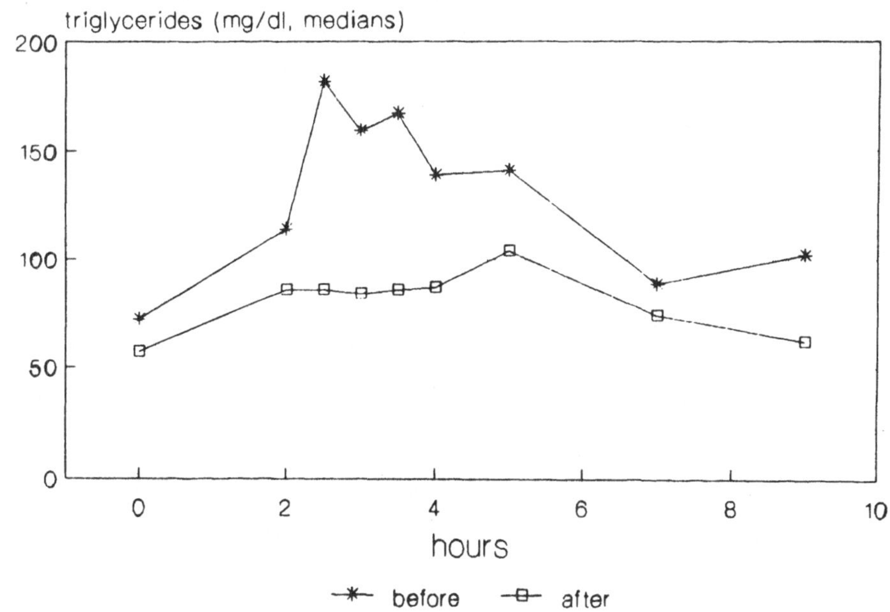

Fig. 1. Butter Load before and after Fish Oil Treatment (Responders only, n = 8).

When a fat load consisting of 100 g of fish oil (60 g of omega-3 fatty acids) was given, there was no increase in postprandial TG values, indicating an apparent lack of response to the large load of omega-3 fatty acids. In contrast to the large fat load with butter there was significant diarrhea developing in all 20 participants with up to 10 bowl movements over the next 24 hours. The first bowl movement occurred already 5-7 hours after fat ingestion indicating a much shortened intestinal transit time due to maldigestion and/or malabsorption. Neither the butter load nor an equivalent fat load with 100 g of safflower oil rich in omega-6 fatty acids produced such an effect.

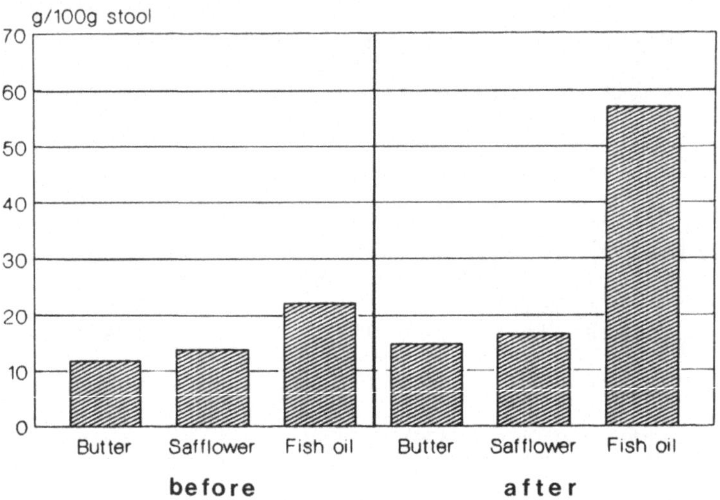

Fig. 2. Fecal Fat Excretion after Fat Loads (n = 20).

When fecal fat excretion (Fig.2) was determined over the 24 hour period following the ingestion of either butter or safflower oil or fish oil before and after the 4 weeks period of fish oil capsule treatment, there were clear indications that the ingestion of large quantities of fish oil in form of a fat load caused severe steatorrhea. This was particularly evident after the treatment period with fish oil capsules.

We had hypothesized that a possible mechanism leading to an attenuated response to an acute fat load of saturated fat might be a temporary retention of TG in the intestinal mucosa during the process of chylomicron formation and release. In order to evaluate that possibility we took small intestinal biopsies by an upper endoscopy 5 hours after the beginning of the fat load.

Fig.3 shows the mucosal concentrations of triglyceride, cholesterol ester and free cholesterol as analysed by neutral gas chromatography. The mucosal concentrations of these lipids were compared in the fasting state, after the butter load and before the fish oil capsule treatment, after the butter load following a 4 week omega-3 fatty acid treatment and after the large fish oil load. The large fish oil load caused a significant increase - as might be expected - of mucosal triglyceride. However, the mucosal cholesterol ester content did not increase at all as compared to the fasting state. The free cholesterol remained the same as well. When comparing the mucosal lipid content after the butter load before and following the omega-3 fatty acid capsule treatment, there was no change

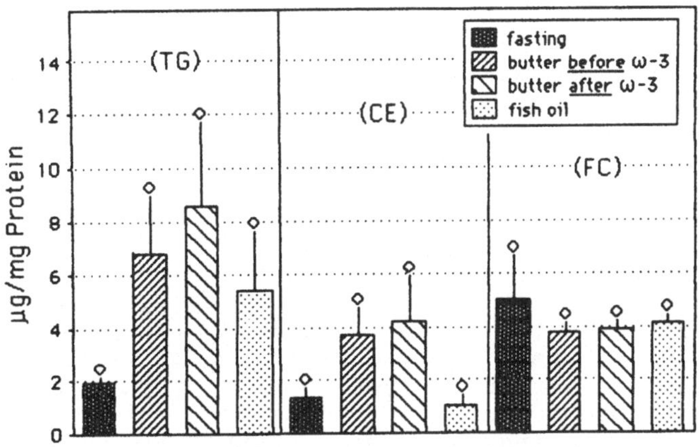

Fig. 3. Intestinal lipids: TG, CE and FC in the fasting state as compared to butter loads before and after ω-3 treatment and fish oil load (n = 15).

in free cholesterol concentration. It was interesting to note that the mucosal triglyceride content was somewhat higher when the butter load was given after the fish oil capsule treatment. Such a tendency could also be observed in the mucosal cholesterol esters. Although these data are not conclusive, they seem to indicate that some retention of intestinal triglyceride and cholesterol ester might occur after pretreatment with low dose omega-3 fatty acids.

When comparing the plasma triglyceride results shown in Fig. 1 and the mucosal triglyceride concentrations shown in Fig. 3 at 5 hours after the fat load, it is interesting to note that the plasma triglyceride is highest at that time point. This could be interpreted as a reflection of a delayed release of stored mucosal triglyceride in the form of chylomicrons. Indeed, chylomicron triglyceride concentration at that time point was higher than at earlier time points after fish oil capsule treatment. Without fish oil capsules given before the acute butter load the chylomicron triglyceride concentration reached its peak at 3 hours after the fat load and was significantly lower at 5 hours.

In conclusion, our preliminary results indicate that ingestion of large quantities of omega-3 fatty acid (60 g at a time) do not induce postprandial hypertriglyceridemia at all. One of the apparent reasons for this total lack of response seems to be the induction of either maldigestion or malabsorption and subsequent steatorrhea. The mechanism of this malassimilation is unclear. The analysis of mucosal lipid content at a relatively late time point in the process of lipid absorption and processing in the mucosa seems to indicate that triglyceride synthesized from the fish oil accumulates to some extent in the mucosa and is not released in the form of chylomicron triglyceride. The total lack of mucosal cholesterol ester may be interpreted as the consequence of malabsorption of luminal cholesterol during the large fish oil load.

The chronic administration of small quantities of omega-3 fatty acids attenuates the postprandial response to saturated fat. Our mucosal lipid studies seem to indicate that some retention of saturated triglyceride might occur in the intestinal mucosa before the triglyceride is released in the form of chylomicrons. This delayed release might improve the substrate situation for the intravascular lipolytic system and thereby help to maintain a relatively low postprandial triglyceride level. This effect of chronic omega-3 fatty acid administration may be responsible for the triglyceride lowering effect seen in Type V hyperlipoproteinemia, where chylomicron concentrations are very high even in fasting plasma. The observation that chronic ingestion of omega-3 fatty acids can attenuate the postprandial lipemia induced by saturated fat merits confirmation in a larger group of individuals over a prolonged period of time, because postprandial lipemia may be one of the key events in atherogenesis in Western populations living predominantly on fat-rich diets.

References

1) Bang, H,O., Dyerberg, J. In: Draper, H.H. ed.: Advanced nutrition research, vol. 3 New York: Plenum Press, 1980 1-22
2) Dyerberg, J., Bang, H.O. Scand. J. Clin. Lab. Invest 42 Suppl. 161: 7, 1982
3) Kingsbury, K.J., Morgan, D.M., Aylott, C., Emmerson, R. Lancet 1: 739, 1961
4) Bronte-Stewart, B., Antonis, A., Easles, L., Brock, J.F. Lancet 1: 521, 1956
5) Keys, A., Anderson, J.T., Grande, F. Lancet 1: 66, 1957
6) Worne, H.E., Smith, L.W. Am.J.Med.Sci. 237: 710, 1959
7) Kinsell, L.W., Michaels, G.D., Walker, G., Visintine, R.E. Diabetes 10: 316, 1961
8) Kingsbury, K.J., Morgan, D.M., Aylott, C., Emmerson, R. Lancet 1: 739, 1961
9) Imaichi, K., Michaels, G.D., Gunning, B. et al. Am.J.Clin. Nutr. 13: 158, 1963
10) Rhoads, G.G., Gulbrandsen, C.L., Kagan A. N.Engl.J.Med. 294: 293, 1976

11) von Lossonczy, T.O., Ruiter, A., Bronsegeest-Schoute,H.C. et al. Am.J.Clin.Nutr. 31: 1340, 1978
12) Bronsegeest-Schoute, H.C., van Gent, C.M., Luten, J.B. et al. Am.J.Clin.Nutr. 34: 1752, 1981
13) Sanders, T.B., Vickers, M., Haines, A.P. Clin.Sci.61: 317, 1981
14) Harris, W.S., Connor, W.E. Trans. Assoc. Am. Physicians 43: 14, 1980
15) Phillipson, B.E., Rothrock, D.W., Connor, W.E. et al. N.Engl.J.Med. 312: 1210, 1965
16) Sanders, T., Sullivan., D.R., Reeve, J., Thompson, G. Arteriosclerosis 5: 459, 1985
17) Simons, L.A., Hickie, J.B., Balasubramaniam, S. Atherosclerosis 54: 75, 1985
18) Illingworth, R., Harris, W.S., Connor, W.E. Arteriosclerosis 4: 270, 1984
19) Nestel, P., Connor, W.E., Reardon, M.F. et al. J.Clin.Invest. 74: 82, 1984
20) Harris, W.S., Connor, W.E., Inteles, S.B., Illingworth,R. Metabolism 33: 1016, 1984
21) Dyerberg, J. Nutr. Reviews 44: 125, 1986
22) Nestel, P.J. Am.J.Clin.Nutr. 43: 752, 1986

CONTRIBUTORS

Numbers in parentheses indicate the pages on which the author's contributions begin.

L.L. ADAMS-CAMPBELL (81)	Department of Epidemiology, University of Pittsburgh, Pittsburgh, PA 15261, U.S.A.
G. AILHAUD (271)	Centre de Biochimie CNRS, Université de Nice, Parc Valrose, F-06034 Nice, France
P. ALAUPOVIC (289)	Lipoprotein and Atherosclerosis Research Program, Oklahoma Medical Research Foundation, Oklahoma City, OK 73104, U.S.A.
J.J. ALBERS (213)	Departments of Medicine and Pathology, University of Washington School of Medicine and Northwest Lipid Research Center, Harborview Medical Center, Seattle, WA, U.S.A.
A. ATHIAS (263)	Laboratoire de Biochimie des Lipoprotéines, Faculté de Médecine, Hôpital du Bocage, CHRU Dijon, France
D. ATKINSON (123)	Biophysics Institute, Housman Medical Research Center, Department of Biochemistry & Medicine, Boston University School of Medicine, Boston, MA 02118, U.S.A.
P.O. ATTMAN (289)	Lipoprotein and Atherosclerosis Research Program, Oklahoma Medical Research Foundation, Oklahoma City, OK 73104, U.S.A.
M. AYRAULT-JARRIER (149)	CNRS UA 524, UFR Saint-Antoine, Laboratoire de Chimie Biologique, F-75571 Paris, France
R. BARBARAS (271)	Centre de Biochimie CNRS, Parc Valrose, F-06034 Nice and Laboratoires Fournier, F-21121 Fontaine-les-Dijon, France
J.M. BARD (289)	Lipoprotein and Atherosclerosis Research Program, Oklahoma Medical Research Foundation, Oklahoma City, OK 73104, U.S.A.
A. BARKIA (271)	Institut Pasteur, SERLIA, F-59019 Lille, France

S. BASTIRAS (263) Laboratoire de Biochimie des Lipoprotéines, Faculté de Médecine, Hôpital du Bocage, CHRU Dijon, France

E. BEKAERT (149) CNRS UA 524, UFR Saint-Antoine, Laboratoire de Chimie Biologique, F-75571 Paris, France

F. BERTHEZENE (165) Laboratoire de Métabolisme des Lipides, Service d'Endocrinologie et des Maladies de la Nutrition, INSERM U 197, Hôpital de l'Antiquaille, F-69321 Lyon, France

A. BIRCHBAUER (239) Institute of Medical Biochemistry, University of Graz, A-8010 Graz, Austria

J.M. BOEYNAEMS (13) Institute of Interdisciplinary Research, School of Medicine, Campus Hôpital Erasme, Université Libre de Bruxelles, Brussels, Belgium

W.C. BRECKENRIDGE (75) Department of Biochemistry, Faculty of Medicine, Dalhousie University, Halifax, Nova Scotia, Canada B3H 4H7

M.J. BROEKMAN (31) Division of Hematology-Oncology, Department of Medicine, New York Veterans Administration and Cornell University Medical College, New York, NY 10010, U.S.A.

C. CACHERA (283) SERLIA and INSERM U 279, Institut Pasteur, Lille, and Laboratoire de Recherches Néphrologiques, Hôpital Calmette, Lille, France

C. CALVO (165) Laboratoire de Métabolisme des Lipides, Service d'Endocrinologie et des Maladies de la Nutrition, INSERM U 197, Hôpital de l'Antiquaille, F-69321 Lyon, France

A.D. CARDIN (157) Merrell Dow Research Institute, Merrell Dow Pharmaceuticals Inc., Cincinnati, OH 45215, U.S.A.

H. CHAP (255) INSERM U 101, Biochimie des Lipides, Hôpital Purpan, F-31059 Toulouse, France

C. CLADARAS (107) Section of Molecular Genetics, Cardiovascular Institute, Department of Medicine & Biochemistry, Boston University Medical Center, Boston, MA 02118, U.S.A.

Mi. CLERC (67) Laboratoire de Biochimie Médicale A, Faculté de Médecine, Université de Bordeaux II, F-33076 Bordeaux, France

Mo. CLERC (67) Laboratoire de Biochimie Médicale A, Faculté de Médecine, Université de Bordeaux II, F-33076 Bordeaux, France

C. CORDER (289) Lipoprotein and Atherosclerosis Research Program, Oklahoma Medical Research Foundation, Oklahoma City, OK 73104, U.S.A.

H. CZARNECKA (225)	Clinical Research Institute of Montreal, Montréal, Québec H2W 1R7, Canada
C. DACHET (179)	INSERM U 32, Hôpital Henri-Mondor, F-94010 Créteil, France
J-L. DE COEN (133)	Laboratoire de Chimie Générale I, Université Libre de Bruxelles, B-1050 Brussels, Belgium
P. de KNIJFF (87)	Gaubius Institute TNO, Leiden, The Netherlands
C. DELCROIX (133,299)	Fondation de Recherche sur l'Athérosclérose and Université libre de Bruxelles, B-1000 Brussels, Belgium
A. DEMBINSKA-KIEC (21)	Department of Pharmacology, Copernicus Academy of Medicine, PL-31-531 Cracow, Poland
D. DEMOLLE (13)	Institute of Interdisciplinary Research, School of Medicine, Université Libre de Bruxelles, Campus Erasme, B-1070 Brussels, Belgium
L. DE ROY (299)	Hôpital Militaire, Brussels, Belgium
N. DOUSSET (255)	INSERM U 101, Biochimie des Lipides, Hôpital Purpan, F-31059 Toulouse, France
L. DOUSTE-BLAZY (255)	INSERM U 101, Biochimie des Lipides, Hôpital Purpan, F-31059 Toulouse, France
D. DOWNS (289)	Lipoprotein and Atherosclerosis Research Program, Oklahoma Medical Research Foundation, Oklahoma City, OK 73104, U.S.A.
M. DRACON (283)	Laboratoire de Recherches Néphrologiques, Hôpital Calmette, Lille, France
P. DROUIN (279)	Hôpital Jeanne d'Arc, Dammartin-les-Toul, France
D. DUBOIS (299)	Fondation de Recherche sur l'Athérosclérose, B-1000 Brussels, Belgium
R.P.F. DULLAART (247)	Department of Internal Medicine, University Hospital, NL-3511 GV Utrecht, The Netherlands
M-F. DUMON (67)	Laboratoire de Biochimie Médicale A, Université de Bordeaux II, F-33076 Bordeaux, France
K. EQUAGOO (283)	Laboratoire de Recherches Néphrologiques, Hôpital Calmette, Lille, France
D.W. ERKELENS (247)	Department of Internal Medicine, University Hospital, NL-3511 GV Utrecht, The Netherlands
J.R. FALCK (31)	Department of Molecular Genetics, University of Texas Health Science Center, Dallas, TX, U.S.A.
R.A. FAUST (213)	Department of Pathology, University of Washington School of Medicine and Northwest Lipid Research Center, Harborview Medical Center, Seattle, WA, U.S.A.

R.E. FERRELL (81)	Department of Biostatistics, Human Genetics Division, University of Pittsburgh, Pittsburgh, PA 15261, U.S.A.
C.J. FIELDING (219)	Cardiovascular Research Institute, School of Medicine, University of California Medical Center, San Francisco, CA 94143-0130, U.S.A.
C. FIEVET (279)	Service de Recherche sur les Lipoprotéines et l'Athérosclérose (SERLIA), Institut Pasteur, F-59019 Lille, France
S. FISCHER (31)	Medizinische Klinik Innenstadt der Universität München, Munich, F.R.G.
R.R. FRANTS (87)	Department of Human Genetics, Sylvius Laboratories, University of Leiden, Leiden, The Netherlands
J-C. FRUCHART (271,279,283)	Service de Recherche sur les Lipoprotéines et l'Athérosclérose (SERLIA), Institut Pasteur, F-59019 Lille, France
D.J. GALTON (95)	Medical Professorial Unit, Diabetes and Lipid Research Laboratory, St Bartholomew's Hospital, London EC1A 7BE, United Kingdom
P. GAMBERT (263)	Laboratoire de Biochimie Médicale, CHRU de Dijon, Hôpital du Bocage, F-21034 Dijon, France
P. GHEZZI (193)	Mario Negri Institute, Milan, Italy
M. GOERIG (55)	University of Heidelberg, Medical School, D-6900 Heidelberg, F.R.G.
P. GRIMALDI (271)	Centre de Biochimie (CNRS), Parc Valrose, F-06034 Nice, France
J.E.M. GROENER (231)	Department of Biochemistry I, Erasmus University Rotterdam, NL-3000 DR Rotterdam, The Netherlands
S.M. GRUNDY (311)	University of Texas Southwestern Medical Center at Dallas, TX 75235, U.S.A.
R.J. GRYGLEWSKI (21)	Department of Pharmacology, Copernicus Academy of Medicine, 31-531, Cracow, Poland
A. HABENICHT (55)	Medizinische Klinik, Abteilung Innere Medizin I, Ruprecht-Karls-Universität Heidelberg Klinikum, D-6900 Heidelberg, F.R.G.
M. HADZOPOULOU-CLADARAS(107)	Section of Molecular Genetics, Cardiovascular Institute, Department of Medicine & Biochemistry, Boston University Medical Center, Boston, MA 02118, U.S.A.
D.P. HAJJAR (37)	Department of Pathology, Cornell University Medical College, New York, NY 10021, U.S.A.

K.A. HAJJAR (37) Department of Medicine and Pediatrics, Cornell University Medical College, New York, NY 10021, U.S.A.

M.J. HALPERN (327) Centro de Bioquimica dos Lipidos e Departamento de Bioquimica, Faculdade de Ciencias Medicas, Palacio Burnay, P-1100 Lisbon, Portugal

J.A. HAMILTON (123) Biophysics Institute, Housman Medical Research Center, Department of Biochemistry & Medicine, Boston University School of Medicine, Boston, MA 02118, U.S.A.

L.M. HAVEKES (87) TNO Gaubius Institute for Cardiovascular Research, NL-2313 AD Leiden, The Netherlands

E. HERVAUD (149) CNRS UA 524, UFR Saint-Antoine, F-75571 Paris, France

C.B. HESLER (225) Department of Medicine, Columbia University College of Physicians & Surgeons, New York, NY 10032, U.S.A.

G. HUSBY (185) Institute of Clinical Medicine, University Hospital of Tromsø, N-9012 Tromsø, Norway

A. HUSEBEKK (185) University Hospital of Tromsø, N-9012 Tromsø, and University of Oslo 3, Oslo, Norway

M.M. HUSSAIN (107) Section of Molecular Genetics, Cardiovascular Institute, Department of Medicine & Biochemistry, Boston University Medical Center, Boston, MA 02118, U.S.A.

N. ISLAM (31) Division of Hematology-Oncology, Department of Medicine, New York Veterans Administration and Cornell Unversity Medical College, New York, NY 10010, U.S.A.

R.L. JACKSON (157) Merrell Dow Research Institute and University of Cincinnati Medical Center, Cincinnati, OH 45215, U.S.A.

B. JACOTOT (179) Unité de Recherche sur les Dyslipidémies et l' Athérosclérose, INSERM U 32, Hôpital Henri-Mondor F-94010 Créteil, France

M.A. JOHNS (193) Boston University School of Medicine, Boston, MA 02118, U.S.A.

A.M. JULIA (255) INSERM U 101, Biochimie des Lipides, Hôpital Purpan, F-31059 Toulouse, France

M.I. KAMBOH (81) Human Genetics Division, University of Pittsburgh, Pittsburgh, PA 15261, U.S.A.

D. KARDASSIS (107) Section of Molecular Genetics, Cardiovascular Institute, Department of Medicine & Biochemistry, Boston University Medical Center, Boston, MA 02118, U.S.A.

D.C. KING (61) — Lipid Metabolism Laboratory, USDA Human Nutrition Research Center on Aging at Tufts University, Boston, MA 02111, U.S.A.

H. KLOER (339) — Medizinische Poliklinik, 3rd Medical Department, Universität Giessen, D-6300 Giessen, F.R.G.

G. KNAPSCHAEFER (193) — Boston University School of Medicine, Boston, MA 02118, U.S.A.

C. KNIGHT-GIBSON (289) — Lipoprotein and Atherosclerosis Research Program, Oklahoma Medical Research Foundation, Oklahoma City, OK 73104, U.S.A.

G. KNIPPING (239) — Institute of Medical Biochemistry, University of Graz, A-8010 Graz, Austria

R. KORBUT (21) — Department of Pharmacology, Copernicus Academy of Medicine, PL-31-531 Cracow, Poland

E. KOREN (289) — Lipoprotein and Atherosclerosis Research Program, Oklahoma Medical Research Foundation, Oklahoma City, OK 73104, U.S.A.

E. KOSTKA-TRABKA (21) — Department of Pharmacology, Copernicus Academy of Medicine, PL-31-531 Cracow, Poland

G.M. KOSTNER (239,333) — Institute of Medical Biochemistry, University of Graz, A-8010 Graz, Austria

A. KOUVATSI (107) — Section of Molecular Genetics, Cardiovascular Institute, Department of Medicine and Biochemistry, Boston University Medical Center, Boston, MA 02118, U.S.A.

L. LAGROST (263) — Laboratoire de Biochimie des Lipoprotéines, Faculté de Médecine, Hôpital du Bocage, CHRU Dijon, France

C. LALLEMANT (263) — Laboratoire de Biochimie des Lipoprotéines, Faculté de Médecine, Hôpital du Bocage, CHRU Dijon, France

M. LAPRADE (149) — CNRS UA 524, UFR Saint-Antoine, F-75571 Paris, France

J. LARRUE (47) — Unité de Recherche de Cardiologie INSERM U 8, F-33600 Pessac, France

J-F. LONTIE (133,203,299) — Fondation de Recherche sur l'Athérosclérose, B-1000 Brussels, Belgium

C. LULEY (339) — Clinical Chemistry Department, University Mainz, Mainz, F.R.G.

B-Y. LUO (165) — Laboratoire de Métabolisme des Lipides, Service d'Endocrinologie et des Maladies de la Nutrition, INSERM U 197, Hôpital de l' Antiquaille, F-69321 Lyon, France

J. MAGNUS (185) University Hospital of Tromsø, N-9012 Tromsø and University of Oslo 3, Oslo, Norway

T. MAGOT (299) Laboratoire de Physiologie de la Nutrition, Université de Paris-Sud, Orsay, France

C.L. MALMENDIER(133,203,299) Fondation de Recherche sur l'Athérosclérose, et Faculté de Médecine, Université Libre de Bruxelles, B-1000 Brussels, Belgium

Y.L. MARCEL (225) Laboratoire du Métabolisme des Lipoprotéines, Institut de Recherches Cliniques de Montréal, Montréal, Québec H2W 1R7, Canada

A.J. MARCUS (31,37) Divisions of Hemato-Oncology, Department of Medicine, New York Veterans Administration and Cornell University Medical College, New York, NY 10010, U.S.A.

G. MARHAUG (185) University Hospital of Tromsø, N-9012 Tromsø and University of Oslo 3, Oslo, Norway

L. MEJEAN (279) INSERM U 308, Nancy, France

M.F. MESQUITA (327) Center for Lipid Research and Department of Biochemistry, Faculty of Medical Sciences, Campo Santana, P-1100 Lisbon, Portugal

R.W. MILNE (225) Clinical Research Institute of Montreal, Montréal, Québec H2W 1R7, Canada

S. MONCADA (1) The Wellcome Research Laboratories, Langley Court, Beckenham, Kent BR3 3BS, United Kingdom

C. MOTTA (179) Laboratoire de Biochimie, Hôtel-Dieu, F-63000 Clermont-Ferrand, France

D. NEUFCOUR (179) INSERM U 32, Hôpital Henri-Mondor, F-94010 Créteil, France

T. NISHIDE (213) Department of Medicine, University od Washington School of Medicine and Northwest Lipid Research Center, Harborview Medical Center, Seattle, WA, U.S.A.

J.M. ORDOVAS (61) Lipid Metabolism Laboratory, USDA Human Nutrition Research Center on Aging at Tufts University, Boston, MA 02111, U.S.A.

F. PAOLUCCI (149) CNRS UA 524, UFR Saint-Antoine, Laboratoire de Chimie Biologique, F-75571 Paris, France

H.J. PARRA (283) SERLIA and INSERM U 279, Institut Pasteur, Lille, France

D. PASTIER (149) CNRS UA 524, UFR SAINT-ANTOINE, Laboratoire de Chimie Biologique, F-75571 Paris, France

B. PAU (149) Centre de Recherches Clin-Midy (Sanophi), F-34000 Montpellier, France

E. PETIT (149)	CNRS UA 524, UFR Saint-Antoine, Laboratoire de Chimie Biologique, F-75571 Paris, France
S. PIROTTON (13)	Institute of Interdisciplinary Research, School of Medicine, Université Libre de Bruxelles, Campus Erasme, B-1070 Brussels, Belgium
J. POLONOVSKI (149)	CNRS UA 524, UFR Saint-Antoine, Laboratoire de Chimie Biologique, F-75571 Paris, France
K.B. POMERANTZ (37)	Department of Medicine, Cornell University Medical College, New York, NY 10021, U.S.A.
G. PONSIN (139,165)	INSERM U 197, Laboratoire de Métabolisme des Lipides, Hôpital de l'Antiquaille, Université de Lyon, F-69321 Lyon, France
H.J. POWNALL (173)	Division of Atherosclerosis and Lipoprotein Research, Baylor College of Medicine, The Methodist Hospital, Department of Medicine, Houston, TX 77030, U.S.A.
P. PUCHOIS (271)	SERLIA, Institut Pasteur, F-59019 Lille, France
F. PUYGRANIER (165)	Laboratoire de Métabolisme des Lipides, Service d'Endocrinologie et des Maladies de la Nutrition, INSERM U 197, Hôpital de l' Antiquaille, F-69321 Lyon, France
P. SALBACH (55)	University of Heidelberg, Medical School, D-6900 Heidelberg, F.R.G.
L.B. SAFIER (31)	Divisions of Hematology-Oncology, Department of Medicine, New York Veterans Administration and Cornell University Medical College, New York, NY 10010, U.S.A.
E.J. SCHAEFER (61)	Lipid Metabolism Laboratory, USDA Human Nutrition Research Center on Aging at Tufts University, Boston, MA 02111, U.S.A.
K. SCHAUPP (333)	Research Laboratories of Leopold & C°, Graz, Austria
L.M. SCHEEK (231)	Department of Biochemistry I, Erasmus University Rotterdam, NL-3000 DR Rotterdam, The Netherlands
B.S. SEPEHRNIA (81)	Human Genetics Division, University of Pittsburgh, Pittsburgh, PA 15261, U.S.A.
J.D. SIPE (193)	Arthritis Center, Boston University School of Medicine, Silvio O. Conte Medical Research Center, Boston, MA 02118, U.S.A.
B. SKOGEN (185)	University Hospital of Tromsø, N-9012 Tromsø, and University of Oslo 3, Oslo, Norway
K. SLETTEN (185)	University Hospital of Tromsø, N-9012 Tromsø, and University of Oslo 3, Oslo, Norway

D.M. SMALL (123) — Biophysics Institute, Housman Medical Research Center, Department of Biochemistry & Medicine, Boston University School of Medicine, Boston, MA 02118, U.S.A.

M. SMIT (87) — Department of Human Genetics, Sylvius Laboratories, University of Leiden, Leiden, The Netherlands

E. STEYRER (239) — Institute of Medical Biochemistry, University of Graz, A-8010 Graz, Austria

G. STVARNIK (333) — Institute of Medical Biochemistry, University of Graz, A-8010 Graz, Austria

V. SYVERSEN (185) — University Hospital of Tromsø, N-9012 Tromsø, and University of Oslo 3, Oslo, Norway

A. TACQUET (283) — Laboratoire de Recherches Néphrologiques, Hôpital Calmette, Lille, France

A.R. TALL (225) — Department of Medicine, Columbia University College of Physicians & Surgeons, New York, NY 10032, U.S.A.

M. TAVELLA (289) — Lipoprotein and Atherosclerosis Research Program, Oklahoma Medical Research Foundation, Oklahoma City, OK 73104, U.S.A.

J.H. TOLLEFSON (213) — Department of Medicine, University of Washington School of Medicine and Northwest Lipid Research Center, Harborview Medical Center, Seattle, WA, U.S.A.

H.L. ULLMAN (31) — Divisions of Hematology-Oncology, Department of Medicine, New York Veterans Administration and Cornell University Medical College, New York, NY 10010, U.S.A.

A. van TOL (231) — Department of Biochemistry I, Faculty of Medicine, Erasmus University Rotterdam, NL-3000 DR Rotterdam, The Netherlands

G.L. VEGA (311) — University of Texas Southwestern Medical Center at Dallas, Dallas, TX 75235, U.S.A.

C. VON SCHACKY (31) — Medizinische Klinik Innenstadt der Universität München, Munich, F.R.G.

M.T. WALSH (123) — Biophysics Institute, Housman Medical Research Center, Department of Biochemistry & Medicine, Boston University School of Medicine, Boston, MA 02118, U.S.A.

C.S. WANG (289) — Lipoprotein and Atherosclerosis Research Program, Oklahoma Medical Research Foundation, Oklahoma City, OK 73104, U.S.A.

P.K. WEECH (225) — Clinical Research Institute of Montreal, Montréal, Québec H2W 1R7, Canada

K.M. WEISS (81) Department of Anthropology, Pennsylvania State University, University Park, PA 16802, U.S.A.

V.I. ZANNIS (107) Section of Molecular Genetics, Cardiovascular Institute, Boston University School of Medicine, Boston, MA 02118, U.S.A.

INDEX